Personal, Portable, Pedestrian

Personal, Portable, Pedestrian

Mobile Phones in Japanese Life

edited by Mizuko Ito, Daisuke Okabe, and Misa Matsuda

The MIT Press
Cambridge, Massachusetts
London, England

MIT Press books may be purchased at special quantity discounts for business or sales promotional use. For information, please e-mail ⟨special_sales@mitpress.mit.edu⟩ or write to Special Sales Department, The MIT Press, 55 Hayward Street, Cambridge, MA 02142.

This book was set in Stone Serif and Stone Sans on 3B2 by Asco Typesetters, Hong Kong. Printed and bound in the United States of America.

Library of Congress Cataloging-in-Publication Data

Personal, portable, pedestrian : mobile phones in Japanese life / edited by Mizuko Ito, Daisuke Okabe, and Misa Matsuda.
    p.   cm.
Includes bibliographical references and index.
ISBN 0-262-09039-2 (hc : alk. paper)
1. Technology—Social aspects—Japan. 2. Cellular telephones. I. Ito, Mizuko. II. Okabe, Daisuke. III. Matsuda, Misa, 1968–

HN727.P47   2005
303.48′33′0952—dc22   2004065594

10 9 8 7 6 5 4 3 2 1

To our children, Amane Matsuda, Madoka Matsuda, Taisei Okabe, Luna Ito-Fisher, and Eamon Ito-Fisher, among the first generation of kids born into a *keitai*-saturated society.

Contents

Acknowledgments

This book represents an extended collaborative effort by a great number of individuals and institutions in Japan and the United States. We would like to acknowledge here the people who most directly supported the writing and production of this book.

Most of the papers collected in this volume were initially presented at a workshop sponsored by the DoCoMo House design cottage at Keio University Shonan Fujisawa Campus. DoCoMo House also provided financial support for translation and editing. The initial translation into English of the papers written in Japanese was in the capable hands of Yoko Takahashi. At the other end of the editing process, Justin Hall contributed a keen eye as well as mobile media expertise in polishing the papers into final form. Kunikazu Amagasa reformatted and prepared most of the figures and tables and provided invaluable help in the final production details. We would also like to thank our editors at MIT Press, Douglas Sery and Deborah Cantor-Adams, for their ongoing help and support in guiding us through the editorial process.

Editors' Note on Translation

With the exception of the chapter by Miyata, Boase, Wellman, and Ikeda and the chapters where Ito is the first author, all were written originally in Japanese. These were translated by Yoko Takahashi or Mizuko Ito, including quotes. Although the authors have worked closely with Ito in editing the English versions of the chapters, Ito bears responsibility for any faults in the translation.

Japanese names are written in the Western format, first name first, to avoid confusion in the frequent juxtaposition of Japanese and Western names.

Certain key terms with no direct counterpart in English have been written in romanized Japanese, including *keitai* (mobile phone), *deai* and *deai-kei* (encounters, encounter/dating type), and *kogyaru* (street-savvy teenage girls). Japanese nouns have no plural form, thus *keitai* and *kogyaru* (like samurai) are treated as both singular and plural. See Matsuda's chapter 1 for further discussion of the term *keitai*.

Personal, Portable, Pedestrian

Introduction: Personal, Portable, Pedestrian

Mizuko Ito

The three terms *personal, portable, pedestrian* point to a technological imaginary[1] embedded in the social and cultural specificities of Japanese mobile phone use, interpreted on a transnational stage. In contrast to the *cellular phone* of the United States (defined by technical infrastructure), and the *mobile* of the United Kingdom (defined by the untethering from fixed location) (Kotamraju and Wakeford 2002), the Japanese term *keitai* (roughly translated, "something you carry with you") references a somewhat different set of dimensions. A *keitai* is not so much about a new technical capability or freedom of motion but about a snug and intimate technosocial tethering, a personal device supporting communications that are a constant, lightweight, and mundane presence in everyday life.

This introduction serves to locate the *keitai* as a particular sociocultural object in relation to the international state of mobile communications adoption and sociocultural research. Beginning with an overview of Japanese mobile society and culture in the transnational arena, this chapter introduces the theoretical themes and papers represented in this volume. Roughly corresponding to different methodological and disciplinary frameworks, these themes, corresponding to the book's parts, are the social and cultural construction of technological systems (social history), cultures and imaginaries (cultural studies), social networks and relationships (sociological surveys), practice and place (ethnography), and reports on emergent developments. The conclusion identifies emergent dimensions of *keitai*-enabled social life that cross-cut these disciplinary divisions and theoretical debates.

Technological and Intellectual Geopolitics

Ever since NTT DoCoMo launched its i-mode mobile Internet service in 1999, which was received with surprisingly high adoption rates, international attention has been focused on Japan as defining the future of "the mobile revolution." Although the United States and Scandinavia initially held the lead in the deployment and adoption of mobile phones (Agar 2003), the rapid spread of Japan's mobile Internet services, the popular uptake of mobile devices, and innovative handset design by Japanese companies stole the wireless limelight in the twenty-first century. In *Smart Mobs* (2002), a book that catapulted mobile cultures into heightened visibility in the West, Howard Rheingold opens with a scene of texters eyeing their mobile phones as they navigate Shibuya crossing in Tokyo, allegedly the site of the highest mobile phone density in the world. A BBC reporter writes in a piece titled "Japan Signals Mobile Future," "If you want to gaze into the crystal ball for mobile technology, Tokyo is most definitely the place to come to" (Richard Taylor 2003). The heavy use of the *keitai* Internet and text messaging as well as a particular variant of gadget fetishism has made Japan distinctive in the transnational arena.

The view of Japan as a curiously urbanized incubator for the future of mobile technology is based on an international appreciation of how Japan has pushed the envelope on mobile technology design, business practice, and use—an appraisal that seems well-placed given Japan's unparalleled levels of adoption of the *keitai* Internet (see chapter 1) and its steady march into new areas such as camera and video phones, location-based services, broadband *keitai* Internet, and m-commerce. Portable gadgets and wireless business models are a cornerstone of Japan's emergent "gross national cool" (McGray 2002), helping define a hip Japanese popular culture embodied in animation, video games, comics, food, and other Japanese cultural exports. Tamagotchi, Game Boys, Pokémon cards, and *keitai* are intimate, personal, and often cute media technologies scoring high on both Japanese cultural distinctiveness and global appeal (Iwabuchi 2003; Kinsella 1995; Tobin 2004). While recognizing the persuasively globalizing image of Japanese technoculture, however, I would like to insert some cautionary notes about using Japan as a template for a mobile future in other countries. We argue collectively in this book for the international importance and even centrality of the Japan case without losing sight of the specificities of social, cultural, and historical contexts in structuring the development and deployment of mobile phones.

The current Euro-American fascination with Japanese technoculture has deeper roots than the recent turn to the *keitai* Internet. Invoking Japan as an alternative technologized modernity (or postmodernity) is nothing new. At least since the late 1970s, with rapid industrialization and emergence as an economic and electronics powerhouse, Japan has confounded Western models of modernization and technologization (Miyoshi and Harootunian 1989). Harking back to the international attention focused on Japanese management and electronics in the 1980s and 1990s, current Western interest in Japanese mobile phones and technoculture echoes a familiar mix of fascination and unease. On the one hand, i-mode is held up as a technological and business model to be emulated; on the other hand, discourse abounds on the cultural strangeness of Japanese technofetishism that casts it as irreducibly foreign. William Gibson's inspired cyberpunk Tokyo landscapes, *Wired*'s steady stream of oddities in its "Japanese schoolgirl watch" column, ongoing coverage of the Japanese video game industry in

Euro-American gaming magazines—Japan provides fertile fodder for a wide range of techno-imaginings that are valued at least in part because of their cultural distinctiveness. As Tim Larimer (2000) writes in his cover article for a special issue of *Time Asia* on "Gizmo Nation," "More than any other country on earth, Japan has put its faith—and future—in the hands of technology."

Despite the high-tech and postmodern trappings, transnational cultural politics retain many of the same contours of fascination and unease as in the 1980s: emulation of a "Japan as Number One" (Vogel 1979) economic success, coupled with the popularization of the image of the inscrutable Japanese salaryman. Coining the term *techno-orientalism*, David Morley and Kevin Robins (1995) have argued, "Japan has come to exist within the Western political and cultural unconscious as a figure of danger, and it has done so because it has destabilised the neat correlation between West/East and modern/premodern" (160). And just as international attention fed back into a revitalized Japanese nationalism in the 1980s (Yoshino 1999), mobile businesses have become a source of national pride even in the face of the post-bubble recession and failing political system since the 1990s (see chapter 1). Marilyn Ivy (1995, 8) describes a "coincident modernity" of Japan and the West that has led to an image of Japan as composed of a core of tradition surrounded by protean outer trappings of modernity.

Despite the ascendancy of the vision (and fears) of a Japonesque mobile future for the world, there are reasons to question an eventual global technology upgrade to the latest and greatest Japanese version. The United States, the supposed vanguard of the information society, has been stubbornly resistant to the allures of mobile messaging, and NTT DoCoMo's exported i-mode model has been facing challenges in Europe. Perhaps most significant are countries outside the high-tech Euro–American–East Asian axis, particularly those using wireless to leapfrog from a struggling land line infrastructure into the information age. Famously armed with cheap prepaid phones, the mobilization of the Filipino "Generation Text" against Estrada demonstrates the explosive alchemies of a newly informatted generation mixed with repressed political tensions (Rheingold 2002; Agar 2003). In her ethnographic survey of international mobile phone use, Sadie Plant (2002, 75) writes about how wireless telephony is being introduced to villages previously lacking land lines in places such as Swaziland, Somalia, and Bangladesh.

Hsain Ilahiane (2004) has found that among Moroccan urban poor who make their living with freelance service work, mobile telephony has become a means to organize a newly networked work life, often resulting in income increases of over 200 percent. In this context, mobile phones are as indispensable as they are for Japanese teens, but with striking differences in how and to what effect the phones are used. The features and services valued by these Moroccan users also differ from those used most frequently by Japanese *keitai* aficionados. There, voice telephony is the dominant modality rather than text, which is the central modality for heavy mobile phone users in

Japan and many other settings. These stories provide an important counterbalance to the weighting of international attention toward Japanese mobile culture. While this book contributes to this weighting, it also aims toward a certain parochialization and grounding of the Japanese case. The development of *keitai* uses and cultures is a complex alchemy of technological, social, cultural, economic, and historical factors that make wholesale transplantation difficult.

Following the lead of prior international studies of mobile phone usage (Agar 2003; Katz and Aakhus 2002; Plant 2002; Rheingold 2002), this book seeks to examine the social and cultural diversity in mobile phone use. Our strategy, however, is to approach this issue not through a comparative or global survey of mobile phone use but rather through a multifaceted and sustained engagement with one national context. An important dimension to our approach is to draw primarily from Japanese intellectuals rather than surveying the scene from a more sweeping viewpoint. All the first authors in this book can stake some claim to being "native" intellectuals. Despite a vibrant business and policy literature and prolific coverage of the topic in the popular media, research literature on the social and cultural dimensions of *keitai* use in Japan has been curiously absent in the English-speaking world. Three edited volumes on social studies of mobile phones surveyed use across a wide range of national contexts, including various European countries, the United States, Korea, China, Philippines, Russia, Israel, and Bulgaria (Brown, Green, and Harper 2002; Katz 2003; Katz and Aakhus 2002), but these volumes included no research on Japan. There are only a few articles in English on Japanese mobile phone use from a social or cultural perspective (Hjorth 2003; Ono and Zavodny 2004; Holden and Tsuruki 2003). The absences are not accidental (though still striking) given barriers of language and academic practice. Although a steady stream of English social science texts are translated into Japanese, the reverse flow is relatively rare. We seize the opportunity presented by the current fascination with Japanese mobile phones to showcase native intellectual production as well as the intricacies and range of Japanese mobile cultures and social life.

Emergent Technologies, Emergent Theoretical Conversations

Current social and cultural study of mobile phone use is reminiscent of the state of the study of the Internet ten years ago. As new technical capabilities have entered popular consciousness and use, a small entrepreneurial community of researchers has been galvanized into developing methods and frameworks for studying practices supported by the new technologies. Many of the same writers and researchers have entered this new field. Much as Rheingold's book *The Virtual Community* (1993) heralded a groundswell of popular and academic interest in online social life, his *Smart Mobs* (2002) propeled popular and academic interest in mobile communications. Academic researchers have been drawing the connections between Internet and mobile communications, juxtaposing and integrating articles in both camps through edited volumes such as *Virtual Society?* (Woolgar 2002), *Machines That Become Us* (Katz 2003), *The Internet in Everyday Life* (Wellman and Haythornthwaite 2002), and *Japanese Cybercultures* (Gottlieb and McLelland 2003). Many researchers have moved from Internet studies to mobile communications studies.

At the same time, mobile communications demand a set of engagements at various methodological and theoretical points that differ substantially from Internet study. This has meant building interdisciplinary theoretical and methodological bridges between technology, media, and communication studies as well as forging alliances across regional and national boundaries. Although we are still just beginning to define the conceptual frameworks for understanding the role of *keitai* in Japanese life, already a series of theoretical problematics are emerging that challenge frameworks from Internet and communication studies. What follows is a discussion of these theoretical problematics in relation to the chapters that form the body of this book. The framing is inflected by my own background in Internet research and positioning between Japan and the United States.

The Social and Cultural Construction of Technological Systems

As described, the intellectual and technological geopolitics of mobile media have foregrounded sociocultural diversity in a way that was not, at least initially, evident in Internet studies. Unlike the Internet, where the United States has dominated both development and adoption trends, contemporary mobile communications have been driven forward most prominently by Asian and European countries, upsetting the geopolitics of information technology advancement. This disruption of the status quo, combined with the diversity in implementation of mobile communications infrastructures, has meant that wireless technology has been seen from the start as located in specific social, cultural, and historical contexts rather than as a cross-culturally universal solution (as Internet protocols are often portrayed). Perhaps the most frequent question that I have received in relation to my work on Japanese *keitai* use has been, "Yes, but to what degree is all this specific to Japanese culture?" It is difficult to imagine a similar question being asked with such frequency about the Internet as an artifact of U.S. culture.

A growing cadre of researchers has been insisting that PC hardware and Internet protocols and infrastructures also rest on a set of social and cultural predispositions. For example, Jason Nolan (2004) has argued that the "hegemony of ASCII" systematically discriminates against certain languages. Related critiques have been mounted from the viewpoint of gender difference (Cherny and Weise 1996) and generational identity (Ito et al. 2001). Studies of Internet use outside of North America got off to a slow start, but now constitute a growing body of literature (e.g., Miller and Slater 2000; Gottlieb and McLelland 2003; Wellman and Haythornthwaite 2002; Woolgar 2002). In their contribution to this volume, Kakuko Miyata, Jeffrey Boase, Barry Wellman, and Ken'ichi Ikeda describe the continued international variability in forms of Internet uptake despite original expectations that other countries would follow the U.S. model. In contrast to Internet research, the first collections of mobile phone studies represent a wide range of national contexts (Brown, Green, and Harper 2002; Katz 2003; Katz and Aakhus 2002). Our book exploits this international perspective. Rather than having to argue for it, we can build on the recognition that technology is not independent of social and cultural setting. Further, the Japan mobile Internet case represents a counterweight to the notion that PC-based broadband is the current apex of Internet access models; ubiquity, portability, and lightweight engagement form an alternative constellation of "advanced" Internet access characteristics that stand in marked contrast to complex functionality and stationary immersive engagement.

In the case of the PC Internet, differences in adoption were most often couched in terms of a digital divide, of haves and have-nots in relation to a universally desirable technological resource. By contrast, mobile media are frequently characterized as having different attractions depending on local contexts and cultures. The discourse of the digital divide has been mobilized in relation to Japanese *keitai* Internet access (see chapter 1) and is implicit in the discourse suggesting that the United States needs to catch up to Japanese *keitai* cultures. At the same time, uptake of mobile communications has tended to be viewed less as a single trajectory toward a universal good than as a heterogeneous set of pathways through diverse sociotechnical ecologies. For example, among post-industrial countries, the United States has been characterized by slow uptake of texting and mobile Internet use. This could be attributed as much to the greater presence of PCs and broadband access as to inadequate business models and technological standards. In the United States mobile phones are not universally heralded as an advance but have been questioned as a problematic technology that erodes personal space. Meanwhile, mobile messaging and Web access have been questioned as second-rate versions of their PC counterparts. The metaphor of a digital divide does not fully describe an arena that was from the start characterized by multiple deployment trajectories.

Rather than seeking to explain or transcend national differences in uptake of a technology, we take cultural, social, and technological specificity as a starting point. We critique a pervasive assumption that society and culture are irreducibly variable but technologies are universal. In this, we join ranks with various approaches to the social construction of technological systems (e.g., Bijker, Hughes, and Pinch 1993; Bijker and Law 1992; Callon 1986; Dourish 2001; Hine 2000; Suchman 1987; Wellman 1999). These approaches posit that technologies are both constructive of and constructed by historical, social, and cultural contexts, and they argue against the analytic separation of the social and technical. Although not necessarily intentionally, research agendas that survey a wide range of national contexts can have the effect of producing a perception that we are dealing with a single technology deployed across multiple settings. James Katz and Mark Aakhus (2002, 310) state this stance in positing an "*Apparatageist*" of "perpetual contact": "*Apparatageist* can be broadly vocalized because universal features exist among all cultures regarding PCT [personal communication technologies].... Regardless of culture, when people interact with their PCTs they tend to standardize infrastructure and gravitate towards consistent tastes and universal features." By contrast, our approach is that technological universality, rather than being a structural given, is a contingent production of a wide range of actors, including governments, technologists, and scholars.

The current variability in wireless deployment is not necessarily on its way to becoming standardized toward universal access but is a symptom of fundamentally heterogeneous and resilient sociotechnical formations that vary across lines such as gender, nation, class, institutional location, and age. Our narrative is not of a single technology disseminated to multiple contexts but of the heterogeneous co-constitution of technology across a transnational stage. Although we may see a transnational alliance push for the emergence of technological standards that integrate the current international patchwork of protocols and infrastructures, this does not mean a homogenizing of imagining, use, and design. Unlike the Internet, created by a relatively narrow and privileged social band (predominantly educated, white, male, North American), mobile technology owes not only its uptake but its actual form to people more on the social and cultural peripheries: Scandinavian texting teens, pager cultures of Japanese teenage girls, multitasking housewives, Filipino youth activists, mobile service workers. Clearly cross-national similiarities in the form and use of these new technologies abound. I would argue, however, that they result from similarities in structural conditions and transnational articulation rather than from a context-independent technological form. For example, the international boom in youth texting cultures stems from the similar position that youth occupy in postindustrial societies (Ito forthcoming).

Chapters 1, 2, and 3 describe the development and deployment of *keitai* in Japan as structured by particular social and cultural contexts and historical junctures. Misa Matsuda describes the cultural and intellectual history of discourse on *keitai*, beginning with the etymology of the term, and going on to trace how Japanese researchers and journalists have engaged with the topic of *keitai* and society. She follows how popular and research discourse began with a focus on business-related uses of *keitai*, shifting to a focus on youth relationships in the late nineties, to the current nationalistic celebration of Japan's lead in *keitai* technology. Tomoyuki Okada describes how *keitai* cultures developed out of the fertile ground of youth street practices and visual cultures and a history of text messaging that extended back to youth pager use in the early 1990s. Drawing from his own interviews with youth as well as historical materials, he illustrates how Japanese youth cultures pushed mobile media in the direction of personalization and multimedia functionality, presenting an alternative design paradigm that differed from the original thrust toward networking business institutions. From the perspective of an engineer and executive at NTT DoCoMo, Kenji Kohiyama focuses on a pivotal decade in the development of Japanese mobile communication technologies between 1993 and 2002. He describes the historically contingent details of certain key junctures, particularly the competition between the pager, Personal Handyphone (PHS), and *keitai* in the mid-1990s. The details describing the emergence of cellular-based DoCoMo and the i-mode model as a national standard in the late 1990s demonstrate the local vagaries of technological implementation and concretize the contention of constructivist orientations to technology; it could have been otherwise.

Cultures and Imaginaries

Returning to Internet studies as a useful point of theoretical contrast, we again find an intriguingly different set of issues in the field of cultural analysis. Much early Internet research grew out of theoretical interests in virtual reality and cyberspace. In their review of Internet ethnography, Daniel Miller and Don Slater (2000, 4) write, "[An] earlier generation of Internet writing ... was concerned with the Internet primarily through concepts of 'cyberspace' or 'virtuality.' These terms focused on the ways in which the new media seemed able to constitute spaces or places *apart from* the rest of social life ('real life' or offline life)." Christine Hine (2000, 27) has a similar view of "virtual ethnography": "A focus on community formation and identity play has exacerbated the tendency to see Internet spaces as self-contained cultures, as has the reliance on observable features of social organization." Miller and Slater (2000, 5) suggest that we start from an assumption, now well-established, "that we need to treat Internet media as continuous with and embedded in other social spaces, that they happen within mundane social structures and relations that may transform, but that they cannot escape into a self-enclosed cyberian apartness." By integrating ongoing work on social networks with studies of new communication technologies, Wellman (1999) has made complementary arguments about how the Internet articulates with existing personal and communal networks.

The extroverted, out-of-doors nature of mobile communication, as well as its low-profile origins in the pedestrian technology of telephony, has meant that the online component of mobile communications has not been experienced as cut off from everyday reality, places, and social identities. Internet studies have been tracing the increasing colonization by real-life identity and politics of the hitherto "free" domain of the Net; *keitai* represent the opposite motion of the virtual colonizing more and more settings of everyday life. Haruhiro Kato analyzes latent themes in video productions by Japanese college students on the subject of *keitai* (see chapter 5). Narratives center on imagining of a life without *keitai*. In their comments and their productions, students portray *keitai* as a thoroughly mundane and indispensable aspect of their everyday lives. Narrative tension and drama are created simply through the imaginary force of extracting *keitai* from their lives; very few narratives explored the possibility of alternative online identities and worlds through the medium of *keitai*.

Rather than something inherently disjunctive, *keitai* suggests a vision of the virtual seamlessly integrated with everyday settings and identities. With a broader brush, Kenichi Fujimoto surveys what he describes as a "new cultural paradigm" in mobile communications defined by youth street cultures (see chapter 4). He frames *keitai* as a business-oriented technology that was hijacked by popular youth consumer cultures in the late 1990s. For Fujimoto, *keitai* supports a *nagara* (while-doing-something-else) culture of kids simultaneously cohabiting online and physical worlds, and is an "anti-ubiquitous territory machine" (Fujimoto 2003a) that carves out spheres of personal space within the urban environment. Rather than being conceived by an elite and noncommercial technological priesthood and gradually disseminated to the masses, as was the Internet, *keitai* came of age as a mass consumer technology framed by cultures of gadget fetishism and technofashion. Unlike the immersive and often escapist idioms of Internet social life, Fujimoto suggests, *keitai* functions more as a medium of lightweight "refreshment" analogous to sipping a cup of coffee or taking a cigarette break. It is a street-level device packaged and mobilized in the ongoing status displays of everyday life.

Social Networks and Relationships

Studies of mobile communications lie between computer-mediated-communication studies and personal communication studies. In contrast to Internet studies, which initially focused on community online (Jones 1995; 1998; Rheingold 1993; Smith and Kollock 1998), mobile communication studies have focused on private communications and connections between intimates. Several chapters in this book suggest how, for most heavy users, *keitai* reinforces ties between close friends and families rather than communal or weaker and more dispersed social ties. In line with research findings in other countries (Grinter and Eldridge 2001; Kasesniemi 2003; Kasesniemi and Rautiainen 2002; Ling and Yttri 2002), Japanese youth send the majority of their mobile text messages to a group of three to five intimates (see chapters 6 and 12; Ito forthcoming). These findings contradict moral panics over fast and footloose *keitai* street cultures (see chapter 1). Far from *keitai* being a tool for producing indiscriminate social contact in an undisciplined public urban space, most youth use *keitai* to reinforce existing social relations fostered in the traditional institutions of school and home (Matsuda 2000a). As Richard Harper (2003, 194) found in a survey of use in the United Kingdom and Germany, "people who knew each other before the onset of GSM [a mobile phone standard] now use the technology to call each other more often" in a process of "invigorating" social relationships.

Miyata, Boase, Wellman, and Ikeda found that *keitai* use correlated with a greater volume of e-mails to people geographically and socially closer than those sent by PCs (see chapter 7). These findings tie into conclusions that build on Wellman's theory of "networked individualism": the trend towards individualized over more traditionally communal and spatially defined social ties. "The person has become the portal." Matsuda's description of "selective sociality" (see chapter 6) also notes that *keitai* users participate in a similar trend towards contact with chosen intimates at the expense of both given and serendipitous relationality. While new communication technologies offer the possibility of an expanded range of partners and means of communicating, most communication gets channeled into a narrow and highly selective set of relationships. Matsuda also locates *keitai* in an ecology of personal communication in Japan that has increasingly valued the discursive production of intimacy, particularly between family members and couples. Telephone calls and *keitai* messages become one way to create a "full-time intimate community" (Nakajima, Himeno, and Yoshii 1999). Further, *keitai* enables the maintenance of close friendships fostered in community institutional contexts like neighborhood play groups and middle schools even after people may have dispersed to different schools or workplaces.

These intimate circles of contact are what Ichiyo Habuchi describes as "telecocooning," the production of social identities through small, insular social groups. Habuchi and Tomita contrast this more prevalent mode of relating with new forms of meeting and relationship building through *deai-kei* (encounter/dating) sites; *bell-tomo*, or *beru-tomo* (relationships fostered through pagers); and *mail-tomo*, or *meru-tomo* (relationships fostered through mobile e-mail). Habuchi provides an overview of *deai* practices among Japanese youth, analyzing the current role of *keitai* communications (see chapter 8). She focuses in particular on the new practices that emerged with pager cultures, particularly relationship building between people who never meet face-to-face. While only 7.9 percent of young people report they engage in such relationships through pagers or *keitai*, they represent a significant subcultural trend among youth who seek relationships outside traditional peer structures. Habuchi analyzes the emergence of this minority as a side effect of the growing mainstream reliance on telecocooning to define social identity. The narrowness and intimacy of peer groups produces a claustrophobic reaction in some. Tomita focuses more specifically on *deai* cultures through telecommunications and the emergence of the "intimate stranger" as a new and compelling kind of social relationship (see chapter 9). Reviewing the development of anonymous encounter sites in voice mail and telephone clubs of the 1980s, Tomita provides historical context to contemporary *deai* practices that flourish on both the PC and *keitai* Internet. Although consistently marginalized and denigrated by the mainstream, anonymous *deai* sites represent a variant form of *keitai*-mediated relationality that is an inseparable shadow of the more prevalent forms of *keitai* use.

Taken together, the chapters in this book suggest that *keitai* are implicated in a heterogenous set of shifts keyed to social and cultural differentiation and growing out of prior forms of practice. In other words, we see reason to be skeptical of sweeping claims describing a shift to a new mobile society characterized by dispersed and fragmented networks rather than localized and integrated ones; or most classically, the shift from *Gemeinschaft* to *Gesellschaft* (Tonnies 1957). While we do see the strengthening of discourses and bonds of intimacy and selective relationality, the forms that these take in everyday practice are so varied that they cannot be reduced to a single model of sociability. In fact, the chapters in this book point to the resilient salience of structuring institutions such as life stage, school, workplace, and home in contextualizing and differentiating social relationships. Even among youth there are significant variants in the way social networks are created and maintained. In the ethnographic cases that compose the next part, studies of different social groups such as workers (chapter 12) and housewives (chapter 11) provide even more evidence of how stratification of cultures and practices demand different forms of communication and relationships. As Harper (2003, 187) has argued, "The mobile age is not rendering our society into some new form; it is, rather, enabling the same social patterns that have been in existence for quite some time to evolve in small but socially significant ways."

Practice and Place

While community studies drove early Internet ethnography, studies of mobile use in public spaces constituted the initial focus of mobile phone ethnography (Ling 2002; Murtagh 2002; Plant 2002; Weilenmann and Larsson 2001). Consistent across both of these is the tendency for ethnography to gravitate toward sites of publicly observable communal action, whether online or physically located. Studies of mobile phone use in settings such as public transportation (chapter 10) and restaurants (Ling 2002; Taylor and Harper 2003) continue to provide important insights on topics such as identity display, manners in public space, and the situational constraints on mobile phone use. At the same time, it is becoming clear that mobile communication studies need to engage with a reconfigured methodological toolkit that takes into account both public and private communications and their layering within any given setting. This means designing new observational methods to document private communication and trace practices that span physically demarcated localities. In contrast to the questionnaire and interview methods that are basic to studies of social networks and relationships, practice-based studies of mobile sociality demand a substantial revision of traditional ethnographic method and theory. The ethnographic papers in this part of the book represent a range of approaches, including diary-based study (chapter 13), shadowing of users (chapter 12), visits and interviews in domestic space (chapter 11), and observations in public places (chapter 10).

Chapter 10 takes on the issue of manners in public places, the entry point for much ethnographic work on mobile phone use. The ethnographic core of the chapter is a series of observations on public transportation, framed by ethnographic interviews regarding *keitai* manners and a survey of public discourse on the topic. The chapter describes the evolution of the current social consensus that silent uses of *keitai* (e-mail, Web) are permissible on public transport but that voice communication is not. It also documents how this social order is instantiated at an interactional level, describing the subtle strategies for how users handle gaze and voice in disciplining and managing social transgressions. In contrast to the public social setting of the train, Dobashi (chapter 11) examines the domestic setting of the home, focusing on the practices and identity construction of the Japanese housewife in relation to *keitai*. The public transportation work is a study of *keitai* regulation and discipline; Dobashi provides a parallel case in the domestication of technology in the private sphere. He describes how *keitai* becomes integrated into the existing political and social configuration of the home and the resilient identity of the housewife. In contrast to the PC, which demands a certain amount of focused engagement, *keitai* Internet fits into housewives' need for constant "microcoordination" (Ling and Yttri 2002) in managing family relationships and the fragmented temporal demands of domestic work. Dobashi argues that housewives' *keitai* use provides an important counterpoint to the focus on social change and youth street cultures. Domestication demonstrates variable uptake depending on social location as well as the conservative dimensions of the new technology. Taken together, chapters 10 and 11 represent case studies in the maintenance of social orders through the regulation and domestication of new technologies.

Tamaru and Ueno's study of copier service technicians represents another counterpoint to the field's focus on youth cultures (see chapter 12). Their fieldwork involved interviews and shadowing technicians as they traveled about their service areas. Tamaru and Ueno describe how a simple mobile Internet bulletin board system transformed the service technician's experience of place and the work of social coordination by providing constant and lightweight access to information about colleagues' location and dispatching. Again, a new technology has been domesticated by a highly structured social order, but it has also enabled new forms of communication and uses of place. This study joins the ranks of a small but growing corpus of work that documents how mobile phones are part of an assemblage of technologies that constitute a distributed workplace (Brown and O'Hara 2003; Churchill and Wakeford 2002; Laurier 2002; Schwarz, Nardi, and Whittaker 1999; Sherry and Salvador 2001).

Chapter 13 rests on a notion of "technosocial situation" as the frame for practices that hybridize technological, social, and place-based infrastructures. For example, mobile texters have developed practices for conducting online chats that are keyed to their motion through different physical locales. Ito and Okabe relied on a diary-based method of data collection adapted from prior mobile communication studies. Chapters 12 and 13 both represent an attempt to conduct fieldwork and theorize practice and place as constructed through an interaction between physical and geographically based structures and technologically mediated remote connections. These two studies both build on frameworks from face-to-face interaction, which have been foundational to practice-based study; we continue to attend to the details of local social orders but do not take for granted grounding in physically co-present encounters.

In contrast to Internet communication and community studies (where most work continues to focus on the online setting without taking into account the physical locale of the users), mobile communication studies are tied to a revitalized attention to locality and place. As referenced in part II, Culture and Imaginaries, the *keitai* Internet has never been imagined as a domain of "cyberian apartness" from everyday physical reality but has always been a site of tension and integration between the demands of face-to-face encounters and footwork and the demands of the remotely present encounter and visual attention to the handheld screen. *Keitai* users are characterized by their attention to and immersion in the physical environment and social order, even as they increasingly maintain contact with distant personal relations through an intimate portable device. The *keitai* both colonizes and adapts to the structures of existing practices and places. A crucial emergent area of inquiry is the need to theorize the layering of different forms of social and physical presence and to study interactional practices for managing simultaneous presence in multiple social situations.

The social life of *keitai* resonates with research traditions in computer science of "pervasive" or "ubiquitous" computing, which have argued for a model of computing more seamlessly integrated with a range of physical objects, locations, and architectures (Dourish 2001; Grudin 1990; McCullough 2004; Weiser 1991; Weiser and Brown 1996). In many ways, contemporary *keitai* use is an instantiation of these visions of computation as it has migrated away from the desktop and into settings of everyday life. Yet contemporary *keitai* use differs substantially from many of the visions of sensors, smart appliances, and tangible interfaces that characterize the field of ubiquitous computing. What the work in this book demonstrates is that ubiquitous computing might best be conceptualized not as a constellation of technical features but as *sociotechnical practices* of using and engaging with information technologies in an ongoing, lightweight, and pervasive way. Paul Dourish's (2001, 3) phenomenological stance is "more concerned with interaction than with interfaces, and more concerned with computation than with computers." In this formulation, the features of portable, personal, and pedestrian refer not to technologies but to action and experience that can be altered and enhanced by new media technologies.

Emergent Developments

Part V of this book contains two chapters that report on new developments in the area of *keitai* use. Chapter 14 reports on the growing adoption of mobile phones by elementary and middle school children, the reasons cited for adoption, and use patterns. Chapter 15, on camera phones, is a preliminary foray into how these new devices are beginning to be used and might be used in the future. The camera phone is tied to new visual literacies that are being articulated in relation to the pervasive presence of social connectivity.

Conclusion

This book has been organized by theoretical and disciplinary categories at the expense of highlighting interdisciplinary themes. In conclusion, I would like to use some broader strokes to invoke a more speculative picture of the patterns emerging from the interdisciplinary linkages this book represents and to bring the research discussion back to the issues surrounding Japan in the transnational arena.

One cross-cutting theme is the salience of "the personal" and discourses of intimacy in *keitai* communications. Decisive was the shift in the late 1990s from *keitai* primarily identified as a business tool to its identification as a tool for personal communication and play. Now, even when being used for serious work purposes, *keitai* in the work-place and in public places generally (and often negatively) invoke "personal business." Even before the *keitai* Internet, voice communications created a juxtaposition between private affairs and public places, tagging the *keitai* as a narcissistic device that invaded the communal with the demands of the personal. Now, widespread mobile e-mail and other online communication tools lead to these intimate spheres' being even more pervasively present; mobile text and visual communication can colonize even communal places where telephony would be frowned upon (public transportation, classrooms, restaurants). The microcoordination between family members and the ubiquitous spaces of intimacy between young couples and peers are the most evocative of these new dimensions of always-on intimate connection. Even workplace studies have documented *keitai*'s now indispensable role in coordinating small and tightly coordinated work groups. These telecocoons and full-time intimate communities represent an expansion of the long-standing sphere of intimate relations. The papers in this book have only begun to explore the profound implications for the production of social identity, the experience of public and urban spaces, and the structuring of institutions such as households, couples, and peer groups.

This dimension of the pervasively personal is tied to an out-of-doors and low-profile vision of informational and communication networks that goes against the metaphors of indoor, immersive experience that have dominated our imaginings of virtual reality, cyberspace, and Internet social life. *Keitai*'s social value is tied to its colonization of the small and seemingly inconsequential in-between temporalities and spaces of everyday life. Whether it is the quick text reminder sent by a multitasking housewife, the service technician who wants to keep track of which team members are out to lunch, or young couples texting sweet nothings as they take the bus to school, *keitai* connectivity is a membrane between the real and virtual, here and elsewhere, rather than a portal of high fidelity connectivity that demands full and sustained engagement. Metaphors of *keitai* engagement are as often side-by-side as they are face-to-face, as much about ambient and peripheral awareness as they are about demanding attention in the here and now.

The mostly young natives of the *keitai*-pervaded world experience social presence through pulsating movement between foreground and background awareness rather than through clearly demarcated acts of "logging in" or "showing up" to a sociotechnical space. This is a view of the mobile universe that sees remote and networked relations as a pervasive and persistent fixture of everyday life rather than something that is specifically invoked through intentional acts like making a phone call or powering up a networked PC. This is about the seamless and unremarkable integration of this "virtual domain" into more and more settings of everyday life, simultaneously residing both here and elsewhere as a comfortable and unremarkable social subjectivity.

The chapters in this book document *keitai*'s incorporation and domestication into a wide range of social practices and institutions. Even within Japanese society, we see *keitai* use stratified along lines of age, gender, and profession. Mobile phones are characterized by malleability in uptake, while they also serve as an articulation of a distinctive new model of communication. As we work to identify stratification factors in *keitai* use, we may find that certain social categories trump national identity as a predictor of use. Already, cross-cultural comparison of youth *keitai* use indicates that social structural location can determine uptake more than the specifics of technology deployment or business models. Youth texting cultures have caught on in a wide variety of postindustrial social contexts despite very different technological infrastructures and deployment trajectories. These types of resonances suggest how we can frame and define the study of technological systems in ways that differ from the nation-based frame that we have developed for this particular book.

As I discussed earlier, our aim with this book is not to hold up Japan as a nation that defines the mobile future for other countries, nor to suggest that Japan is irreducibly culturally other in its approach to technology. Rather, by locating Japanese *keitai* use and discourse in historical, social, and cultural contexts, our hope is, somewhat paradoxically, to move beyond national identity as the primary tag for social and cultural distinctiveness. Stressing heterogeneities *within* Japanese culture and society, the papers collected describe current *keitai* use as contingent on a wide range of social, technical, and cultural factors, some of which might be shared with certain social groups elsewhere, others of which may not be. In other words, we argue against the idea that variable technology use is an outcome of a universal technology (the mobile phone) encountering a particular national culture (Japan); both technology and culture are internally variable and distinctive. Japanese *keitai* use is not a transparent outcome of Japanese culture but emerges from a historically specific series of negotiations and contestations within and outside of Japanese society.

Notes

This chapter has benefited from comments by Barry Wellman, Justin Hall, and members of the Southern California Digital Culture Group faculty seminar.

1. I use the term *imaginary* in the sense that George Marcus (1995, 4) describes in the introduction to *Technoscientific Imaginaries*, "a socially and culturally embedded sense of the imaginary that indeed looks to the future and future possibility through technoscientific innovation but is equally constrained by the very present conditions of scientific work." In other words, the term references shared imaginative projection of technological futures as grounded in everyday practices and the cultural present.

I   The Social and Cultural Construction of Technological Systems

1 Discourses of *Keitai* in Japan

Misa Matsuda

The mobile phone, or more appropriately *keitai*, is indispensable in Japanese society today. Take a quick stroll and you will see a youth in front of a convenience store, a mother watching her child play in the park, even somebody riding a bike—all are staring intently at a small terminal in the palm of the hand. Although these are completely commonplace scenes now, just a decade ago they did not exist.

Today, cellular phone subscription numbers in Japan are at over 82 million. In addition, Japan had over 5 million PHS (personal handyphone system) subscriptions as of the end of May 2004. If we assume that each subscription is held by one individual, then adoption rates are at over 70 percent of the entire population. By comparison, in 1994 cellular phone subscriptions were at 2.13 million. Although other countries have similar adoption rates, *keitai* and Japanese society have unique characteristics that deserve study, such as the permeation to a wide age demographic and the high rates of mobile Internet use.

The chapters in this book analyze and theorize on popular discourse about *keitai* as well as the concrete details of how people relate to *keitai*. This informs not only understandings of Japanese society but also the relationship between technology and society more generally. This opening chapter provides some background context for the chapters that follow. I describe evolving discourses surrounding *keitai* in Japanese society and trends in related sociological research. Since the 1990s, phenomena related to *keitai* came to be described as "youth problems." Of course, the first users of the cellular phone were executives and then businessmen (not women), who were required to carry them. These groups did not stop using *keitai*; however, *keitai* use became explosively popular among young people, drawing the attention of the general public as well as the interest of researchers. In the following sections, I trace how popular and research discourse has shifted from a focus on business uses to youth relationships to a technonationalistic celebration of Japan's leadership in the *keitai* arena.

Keitai, Not Mobile Phone

First, I would like to explain briefly why we use the term *keitai* in our work rather than "cellular phone" (*keitai denwa*) or "mobile phone" (*idou denwa*).

Four contributors to this volume, Hidenori Tomita, Kenichi Fujimoto, Tomoyuki Okada, and I began research in *keitai* studies in 1995. We published the results of our research in two edited volumes, *Poke-beru Keitai Shugi!* (*Pager and Keitai Manifesto!*) (Tomita et al. 1997), and *Keitai-Gaku Nyumon* (*Understanding Mobile Media*) (Okada and Matsuda 2002). In the introduction to the latter, Okada cites Ivan Illich's (1981) concept of the vernacular in explaining our use of the term *keitai*.

From the start, we referred to the objects of our research—cellular phones and personal handyphones (PHS)—by the shared moniker of *keitai*. By then, *keitai denwa* (cellular phone) had been abbreviated to *keitai* in everyday speech. We chose to use this colloquial term to make clear our position on cellular phones and PHS: they are not "new technologies/media introduced from the outside" but rather "technologies/media that come to be embedded in society."

To study *keitai* rather than the cellular phone, mobile phone, or mobile communications media means examining these devices as they are embedded within a particular society we call Japan, and by extension to examine Japan as a society with *keitai*. Our position, however, is not that the phenomena surrounding *keitai* are exclusive to Japanese culture. As Ito writes in her introduction, we argue against theorizing technology and society, or technology and culture, as separate entities; instead we stress their indivisibility. By examining the wide range of phenomena surrounding *keitai*, we gain insight into modern global society in which mobile phones are unevenly distributed.

Considering how the word *keitai* is constituted raises some interesting issues. In Japanese, *keitai denwa* (portable phone) is a combination of two different two-character *jukugo* (compounds of Chinese characters): *denwa* (telephone) and *keitai* (portable), creating a new four-character *jukugo*. Masao Aizawa (2000) points out that, in Japanese, when abbreviating composite *jukugo*, generally the first character of each word is used. In the case of *keitai* and *denwa*, following this norm would result in *kei-den*. The term *kei-den* would retain both the meaning of "portable" and "telephone." However, *kei-den* was never taken up, and *keitai* became the established term. "Telephone" was eliminated.

It almost seems as though the popularization of the name *keitai* foretold the subsequent development of the mobile phone. As many have noted, today's mobile phone, with functions such as e-mail, Internet, and digital camera, is still a phone but is not merely a phone because phones solely support one-on-one voice communication. Particularly among young people, the *keitai* is not so much a phone as an e-mail machine. Further, in practice, cellular phones and PHS are not distinguished from one another and are most commonly both called *keitai*. Only at times when it is necessary to distinguish the two is the PHS noted as "PHS" or *picchi* (in the case of the younger set).[1]

We feel that *keitai* is the most appropriate term to use in our inquiry into mobile phones in contemporary Japan. In this book we use the term *keitai* or "mobile phone" to refer to both cellular phones and PHS. When there is a need to distinguish them, we use the terms "cellular phone" and "PHS" (see chapter 3).

The *Keitai* Research Network

Here I briefly introduce the research background of our contributors to situate this book within the broader discourses of *keitai*. I have already described how Tomita, Fujimoto, Okada, and I have been conducting *keitai*-related collaborative research since 1995. Tomita's interests focus on intimacy and youth cultures, and he has written a book on the 900 number service Dial Q[2] (Tomita 1994). Okada has a background in media studies, and he has conducted participant observation research on the voice mail service, *dengon dial* (Okada 1993). Fujimoto studies popular culture, broadly conceived, and at the time that we began our research together, he was researching watches and pagers as gadgets. With the addition of my own background in communication studies, we began collaborative research based on a shared background in sociology but with different sets of interests. Ichiyo Habuchi, with her interests in intimacy, later joined this group.

In contrast to the explosive spread of *keitai* and the high degree of public interest, in the latter half of the 1990s *keitai* research was not exactly flourishing. Most scholars of media and communications had turned their attention to the Internet, which was also expanding in Japan during the same period. But the irrepressible spread of the mobile Internet finally drew researchers to study *keitai*.

In the spring of 2002, Mizuko Ito and I discussed putting together an edited collection as an English language publication. As a first step, we organized a workshop and invited contributions from my existing research team as well as Yukiko Miyaki, a researcher at a private think tank who had been conducting surveys related to *keitai*; Haruhiro Kato, who had researched computer-mediated communications since the early days of Japanese computer networking; and Shingo Dobashi, with a background in media and technology studies specializing in computer-mediated communications. We also received contributions from Daisuke Okabe, who had been collaborating with Ito based on a shared interest in situated learning theory; Eriko Tamaru and Naoki Ueno, who had been conducting workplace research from an ethnomethodological perspective; Kenji Kohiyama, an engineer who had been engaged in the research and development of various radio communication systems, including the development of the PHS at NTT; Fumitoshi Kato, with an interest in communication theory and the sociocultural context of digital media technologies; and Kakuko Miyata, Jeffrey Boase,

Barry Wellman, and Ken'ichi Ikeda, researchers who had been collaborating on a study of Internet adoption in Japan from a social networks perspective.

The varying backgrounds of the authors are reflected in their contributions. This diversity was not only effective in advancing the discussions held during the workshop but also indispensable in considering the multiple dimensions of *keitai* as a media form embedded in everyday life.

Chapters 2 and 3 discuss the details of the *keitai* diffusion process in the 1990s, with Okada discussing the relation with youth culture and Kohiyama approaching the issue as a telecommunications industry insider (see also Matsuda 2003). Here I describe the evolution of the more popular discourses surrounding *keitai* and review related theories of media, communication, and sociology that emerged in parallel to these developments.

The first section examines the process by which *keitai* was transformed from its initial form as a businessmen's tool to a youth media technology. It describes the initial image of *keitai* as "uncool" as well as issues about the use of *keitai* in public spaces. In the second section, I discuss the moral panic that accompanied the construction of *keitai* as a "youth problem," introducing the main actors: *kogyaru* (street-savvy high school girls) and *jibetarian* (young people who sit and congregate on the street). Next, I summarize how researchers began to study the effects of *keitai* on the interpersonal relationships of young people, once *keitai* became their medium. Finally, I outline the complete transformation of the image of *keitai* after the diffusion of the *keitai* Internet, touching on technonationalism and the growth in *keitai* research.

From a Business Tool to a Youth Medium

The car phone was introduced in Japan in 1979, followed in 1985 by the shoulder phone, which could be taken out of the car, and in 1987 by handheld cellular phone service. However, the user population did not grow steadily, and serious growth did not happen until 1993 (see chapter 2). Although there are numerous reasons why adoption was slow, I believe the most salient was cost. For example, in 1991 new subscriptions cost just under ¥50,000 (approximately US$450), reduced to ¥36,000 in 1994. On top of this, subscribers were charged a ¥100,000 deposit (done away with at the end of September 1993). At the time, subscribers were primarily men who needed cellular phones for their work (see Nakamura 1996a; Matsuda 1996a).

"Uncool" *Keitai*

What was the image of *keitai* and its users during this period? Yasuyuki Kawaura (1992) conducted one of the first sociological inquiries focused on *keitai*. Drawing from newspaper articles and letters from readers regarding *keitai*-related problems, he discusses *keitai* as a media technology that enables people to become free from place. The cases presented describe the dangers of using *keitai* while driving and the issue of manners in public places. Among the articles he cites, a number are interesting when viewed from the vantage point of the present:

In a feature article titled, "These are the kinds of men we hate!" a women's magazine puts "men with *keitai*" on the opening page.

"Since he is not really busy at all, the phone never seems to ring. The only calls are ones that he makes himself, and even then, it is not very important communication, something like, 'Oh, thanks for dealing with that.'" The next illustration goes even further: "When he has a free minute he makes a call on his *keitai* asking, 'Were there any calls for me?'" (Kawaura 1992, 307)[2]

A few years later, in an essay entitled *"Keitai-girai no Dokuhaku"* ("Monologue of a *Keitai* Hater"), Junko Sakai (1995) describes how *"keitai* are tasteless." She writes, not only do *keitai* "seem to be used mostly by garishly dressed young women, people who had happened on some quick cash, and people who work in fields that society tends to view as frivolous," just having a *keitai* is a display to others of "a kind of over-eager mentality of 'I want to be reached and be able to reach people at all times and places.'"

In light of the subsequent explosive growth of *keitai* use, we might view this negative image of *keitai* users, as Kawaura (1992) does, as simply "envy on the part of those who don't have one." In the early 1990s, the period of early adoption, *keitai* were still expensive and functioned as a status symbol subject to being cast as "uncool" and "tasteless." In 1991, after the economic bubble burst, "people who happened on some quick cash" may have stood out from the crowd a bit more than usual. *Keitai* were disliked as a device that conspicuously broadcasted "I am wealthy," "I am busy," "I am needed by others."

During this period, even *keitai* users did not have a very positive image of *keitai*. Most use centered on work, and users experienced *keitai* as a new "shackling medium" that succeeded the pager. Ritomo Tsunashima (1992), in his column for a weekly magazine, describes seeing somebody using a *keitai* in a public toilet stall. He writes, "Do you really have to take a call on the toilet?!" "In this busy Japan, there will always be some people who have to take the stance that they will take a call at any time or place," he writes with some sympathy.

This negative image quickly changes after the sudden growth in popularity among young people in the last half of the 1990s. The image continues to be negative, but the nature of the characterization shifts. I discuss this evolution in subsequent sections that deal with young people and *keitai*, but first I discuss the public debate surrounding *keitai* in public spaces from the early adoption years to the present.

Keitai in Public Spaces

It is often said that mobile phones blur the boundaries between public and private. From the early 1990s in Japan, use of *keitai* in public spaces was considered a social problem, specifically "inconsiderate *keitai* use on trains" and "poor manners in speaking loudly on a *keitai*." In March 1990, Central Japan Railway started playing an announcement on their bullet trains asking passengers to refrain from using *keitai* while at their seats. Other public transportation facilities soon followed suit and began playing similar announcements despite complaints that the announcements themselves were a nuisance.

In his review of letters sent by readers to newspapers, Kawaura (1992) identifies four sources of discord surrounding *keitai* use in public spaces: (1) physical noise (voice, ringing phones), (2) violations of privacy in having to listen in on conversations one doesn't want to hear, (3) the general creepiness of conversations with people who are not sharing the same space, and (4) the formation of a new kind of hybrid space—the privatization of public space and the impression that personal conversations are out of place there. Drawing from Stanley Milgram's (1970) concept of "norms of noninvolvement," Tomita (1997, 69) makes the following analysis:

Although others in the vicinity are "pretending not to hear," the person talking on the *keitai* seems totally oblivious to the consideration of others around them. Because others are "pretending not to hear," the speaker should also be "pretending they are not being heard." ... But *keitai* users ignore this rule and appear to those around them as if they really do not care. In this way, the norms of noninvolvement in trains have been thrown into disarray.

The physical noise is not the problem. Rather, *keitai* conversations disrupt the order of urban space. *Keitai* lay waste the unspoken agreements determining behavior in public spaces.

At first, those transgressing these norms were businessmen making work-related calls; they constituted the majority of early adopters. Hence the settings for social discord were indicative of this initial user base: mostly the first-class cars in bullet trains, hotels, golf courses, and the like. In the last half of the 1990s, however, as young people begin to adopt *keitai*, bad *keitai* manners become reframed as a problem specific to youth. In this transitional period, an article discussed the PHS the month before it was introduced: "Because PHS are relatively inexpensive, it is likely that young people with bad manners will be using them. I think people will come forward to insist on listeners' right not to have to listen to conversations" (*Nihon Keizai Shinbun*, evening edition, June 17, 1995, 7).

As youth emerged as the predominant *keitai* user base, "young people's inconsiderate use of *keitai* in public spaces" probably did increase. However, even before it became a reality, public discourse was already constructed around the "youth problem" of bad *keitai* manners. I discuss some of the reasons behind this in the following sections, but here I would like to trace the course of the issue of youth and *keitai* manners as it was taken up in the last half of the 1990s.

At the same time, the issue of electromagnetic waves from *keitai* rose to public consciousness,[3] and there was a growing concern that *keitai* use in crowded trains could affect the functioning of pacemakers (see chapter 10; Mori and Ishida 2001). As a result, public transit facilities have been moving from the stance of self-regulation ("please do not use your *keitai*") to prohibition ("please turn off your *keitai*") during rush hour and in train cars flagged as no-*keitai* zones. This quarantining approach is the latest iteration of trial-and-error efforts to regulate *keitai* on public transit.

One result of these ongoing efforts is the perception in recent years that "Japanese mobile phone manners are good." Ikuo Nishioka, the president of Mobile Internet Capital Inc. and a world business traveler, writes in his Web column, "Around the world, people are very tolerant. Only in Japan are people exceptionally strict in regulating use. If you imagine what it would be like if everyone in a packed train car in over-populated Japan used their *keitai*, it is understandable that it would be considered poor manners. That is the reality in Japan" (Nishioka 2003). Others have commented on the oppressiveness of the social regulation against *keitai* on public transit, where the issue of pacemakers gets trotted out more as a rationalization for regulating youth behavior on trains (Hoshino 2001; Takeda 2002).

This concern with manners has been affirmed by a number of empirical studies. For example, based on research with college students, Masato Ishikawa (2000) found that the sense of discord that comes with *keitai* use is a result not so much of violating the communal accord of the train but more a matter of sheer noise level. Toru Suematsu and Hitoshi Joh (2000), writing about personal space issues, suggest that one source of the displeasure is related to the fact that *keitai* users have a lowered awareness of the physical space they are occupying. Based on a questionnaire study, Shunji Mikami (2001) found a relation between perceptions about *keitai* manners and public manners in general; according to Mikami, men and younger people tended to have a more tolerant attitude toward public manners than women and older people. Okabe and Ito's observations of *keitai* use in trains are also discussed in relation to public manners (see chapter 10).

Moral Panic

Okada describes in detail how first the pager and then the PHS and *keitai* became widely adopted by youth in the 1990s as entry-level prices for the devices dropped (see chapter 2). A kind of moral panic (Cohen 1972) emerged as a reaction to the spread of these technologies among young people. This reaction was based on the perception that young people transgressed social norms through *keitai* use. In particular, their *keitai* use did not conform to an earlier norm that "phone calls should be made when there are specific tasks to be dealt with." From this perspective, youth use of *keitai* represented trivial and useless chatter, and it was wrong to be using expensive voice calls for such conversations. A number of broader social issues emerged from this initial reaction, particularly in relation to how young people relate to others. Here I organize issues surrounding the interpersonal relationships of young people into two types of "social problems." One is concern about youth's tendency to avoid relationships with others. The other is the discomfort with young people's forming "relationships with anonymous others." At the root of both of these concerns is the anxiety that traditional human relationships are being lost as a result of new media.

Kogyaru and Jibetarian

At the peak of the pager years, around 1996, *kogyaru* appeared widely in the mass media. A few years later, as *keitai* became widely adopted in 1997 and 1998, it was *jibetarian* who became the focus. Both refer to young people who gather in city centers, but with one major difference. *Kogyaru* refers only to young women, but *jibetarian* is used for both genders. The distinguishing features of *kogyaru* are their bleached brown or blond hair, heavy makeup, very short miniskirt high school uniforms worn with loose socks, and their practices of taking lots of pictures with *puri-kura* (sticker photos from a photo booth) and disposable cameras, and keeping in ongoing touch with their friends with their pagers and *keitai* (see chapter 4). These girls were behind the relabeling of the act of dating men for money from *shoujo baishun* (young girl prostitution) to *enjo kousai* (literally, compensation for companionship). *Jibetarian*—a contraction of *jibeta* (the ground) and the English suffix "-arian"—refers to these people's signature practice of squatting or sitting on curbs or in a corner of a shopping street.[4] The term was adapted to cover both *kogyaru* and young men of a similar age who were inhabiting the city, seemingly without a particular task, actively maintaining their social life with communications media.

For both these groups, "bad manners" became the target of criticism. "*Kogyaru* who put on makeup and change clothes in the train" and "*jibetarian* who not only sit on the street but eat and drink there"—young people nonchalantly doing things that one should be too embarrassed to do based on traditional sensibilities. In order to "understand" these *kogyaru* and *jibetarian*, characterizations were put forth such as "young people who can't distinguish public and private," "young people who are only concerned with their friends," or "young people who transform public places into private space."

Their *keitai* use was described in similar terms. The reason they "engage in idle chatter on the train" is because they don't have any interest in "others" who are not their friends. At the same time, they are criticized for "denying the importance of others who are right in front of them by connecting with a partner in a remote location." Does the *keitai* detract from the relationship with the "others who are right in front of them," or do they use *keitai* because they have no interest in these strangers? Without really querying the causal links, public discourse stresses the affinity between *kogyaru*, *jibetarian*, and *keitai* and extends these characterizations to young people as a whole.

I am not suggesting that "young people's manners really aren't so bad," or that *"keitai* use does not have an impact on young people's interpersonal relationships." I am not working to call attention to the reality of the matter. Rather, I wish to examine the *perception* that young people have bad manners, how their interpersonal relationships and *keitai* use were intuitively linked and taken up as "social problems." Based on 2001 data from the Mobile Communication Research Group, Habuchi (2002b) found that there are no particularly distinguishing features of *keitai* users who make voice calls in public places. In other words, young people do not have a stronger tendency to make these calls, and we could draw the conclusion that the "problem with young people's *keitai* manners" is more a reflection of the older generation's "wanting to characterize youth as having no shame" (Habuchi 2002b, 50).

In every era, young people's attitudes, actions, and values are viewed as inappropriate or incomprehensible by older people. In Japan, particularly since the 1980s, new media have played a role in this framing. For example, when devices such as Walkmans, Famicon game machines, and videocassette recorders appeared in rapid succession, young people in Japan were labeled and discussed as *shinjinrui* (the new breed), and one of their distinguishing features was their mastery of these new media devices. Then, near the end of the 1980s, *otaku* (media geeks) make their appearance. Although the meaning of the label diversified in later years, at the time it was a negative term that referred to youth who "grew up immersed in media and so are unskilled at human communication." Saeko Ishita (1998) describes how both youth and media represent something new to society and get discussed in the same context, where each becomes a basis for explaining the other. She calls this relationship "the unhappy marriage of youth and media theory." This unhappy marriage, of course, was applied to pagers and *keitai* through the 1990s.[5] In contrast to this view of the older generation, Fujimoto (chapter 4) and Kato (chapter 5) analyze how young people themselves understand pagers and *keitai* in their lives.

Anonymous Relationships and Media: From *Beru-Tomo* to *Deai-Kei*

With the popularity of pagers in the mid-1990s, mass media attention focused on *bell-tomo*, or *beru-tomo*, friendships built through the exchange of pager (*poke-beru*) messages. After meeting once face-to-face, friends, strictly speaking, cease to be *beru-tomo*. Some young people would exchange several tens of messages a day. Among these youth, some felt they could share troubles that, with *beru-tomo*, they would not share with their friends, and this was reported with astonishment by the mass media.

Tomita (chapter 9) describes how relationships with anonymous others preexisted the *beru-tomo*, enduring in the social lives of youth as they transitioned from call-in and voice mail services accessed by home telephones to pay phones and pagers, and now to the Internet and *keitai* and the *deai-kei* (encounter/dating) sites that are the current subject of public concern. I would like to point out here that anonymous relationships supported by these different media forms have always been discussed in negative terms.

In a value system that sees face-to-face relationships as most vital, baring one's soul to a *beru-tomo* whom one has never met in person is characterized as strange. Often these young people are described as getting into mediated communication because "they are not skilled at face-to-face communication." Alternatively, they might be characterized as choosing anonymous relationships because they don't want to get hurt by real ones. Kato (chapter 5) analyzes how this "mythology of the face-to-face" has deep roots not only for the older generation but for young people as well. So what distinguishes the youth who get into *deai-kei* and those who avoid them? Habuchi (chapter 8) takes up this question.

Crimes associated with call-in services and *deai-kei* were eagerly reported in the media, stressing danger and risk and reinforcing the negative images associated with these anonymous relationships. These services have also had a strong association with the sex industry. Trade magazines for educators and police have recently taken up the theme of how to deal with *deai-kei* sites as problematic and deviant behavior.

Keitai and the Transformation of Interpersonal Relations

What research has taken up the issues surrounding youth, pagers, and *keitai*?[6] As public interest in the topic grew, researchers also became interested in the unique uses of *keitai* developed by young people. For example, young people transformed the pager from a technology designed to get somebody to initiate a call to a medium for one-on-one communication (see chapter 2). Youth use changed the technology. Technology does not unilaterally change society, nor does society simply take up a technology unchanged as a useful tool. Researchers' interests were drawn to the case of the pager first and then to *keitai* as a visible and concrete example of the interaction between technology and society. In addition, a number of practices such as *wan-giri* or *wan-ko* (making a call and hanging up after the first ring)[7] or *ban-tsuu sentaku* (call-screening using caller ID) (see chapter 6) are examples of *keitai* use unique to or primarily utilized by young people.

Most of the chapters in this book follow the trend set in the mid-1990s of focusing on youth use. These studies analyze *keitai* use by young people because they foreshadow subsequent developments and highlight what is distinctive about *keitai*. A smaller number of the contributions represent emergent research attention toward other groups of *keitai* users (chapters 7, 11, and 12).

Pagers and a New *Yasashisa* among Youth

Let me begin by introducing the psychiatrist Ken Ohira's (1995) theory of *yasashisa* (kindness, sweetness, gentleness). Ohira posits that prior meanings of *yasashisa* changed in the nineties. In contrast to the earlier meaning of *yasashisa* based on sensing and responding to another's feelings and creating a congenial relationship, for 1990s young people, *yasashisa* means not interfering with another's feelings by "carelessly doing something that might seem considerate." And Ohira relates this "new *yasashisa*" to pagers. Pagers are "a tool for taking a passive role, placing the first step of mutual communication in another's hands" (88). Unlike the phone, which burdens the receiver with the demand of an immediate and direct response, the pager conforms to a "new *yasashisa*" in allowing the receiver to read the message when it is convenient. Ohira argues that receivers also have the option of cutting the power or making an excuse that they did not see the message, alleviating stress for the pager owner.

A popular theory goes, "Young people today have many friends that they always spend time with, but because they don't argue or share their problems, because they associate without getting too intimate, young people today are isolated." As did Ohira, the mass media made similar points about young people's relationships, tracing continuities between pagers, PHS, and cellular phones: "The switch can be turned off for a pager or *ketai* at any time; this fits with young people who want to avoid conflicts in interpersonal relationships." "Young people's relationships are maintained completely by media communication and are superficial." "Because they feel isolated, young people look for constant connection with the *keitai*."

In the pager heyday of 1995 and 1996, the media "discovered" a youth with several hundred *beru-tomo* who was taken up as a "real case" of shallowness in interpersonal relationships, combined with a sense of isolation. On top of this, the exchanges were not based on any necessity, being merely exchanges of simple feelings or to confirm a sense of connection, like "good morning," "how are you?" "good luck." This case was mobilized as "evidence" of young people's social disconnection.

In response to this situation, Isao Nakamura (1996b) draws from a questionnaire survey of college students to challenge the view that young people have superficial relationships and that pagers foster this tendency. Nakamura (1997, 27) explores the influence of pagers and PHS on young people's relationships through a panel survey and refutes the view that pagers lead to superficial relationships: "[T]he users of pager and the users of PHS have tendency [sic] to have active face to face interaction.... The using of mobile telecommunication media doesn't encourage the mobility in daily life. On the other hand, there is a tendency for the mobile telecommunication media to enhance the face to face interactions."

From a media theory perspective, Okada (1997) argues that depending on the nature of the relationship, the pager can be a "violent medium," such as when a company demands pager use for an employee. He argues against Ohira's stance that the pager is inherently a medium of "new *yasashisa*" and that research needs to attend to the settings in which media are used. Further discussion of this issue can be found in Okada's contribution in chapter 2, particularly his review of Takahiro's work (1997a; 1997b).[8]

Selective Sociality and Full-Time Intimate Community

By 1996 subscriptions to pagers had started to fall off, and young people were transitioning to *keitai*. In 1995 I began working with Tomita, Fujimoto, Okada, and Habuchi, conducting fieldwork and interview studies on mobile communications using pagers and *keitai*. Based on our research from 1996 to 1998 in urban centers in Tokyo and Osaka (Matsuda et al. 1998), we found that young people's use of caller ID services reflected a highly selective and discriminating approach to interpersonal relationships. Hashimoto's (1998) analysis of public statistics from 1970 on also refutes the popularly held notion that young people's relationships are becoming superficial; he finds indications that heavier *keitai* and pager use correlates with a preference for deeper and broader friendships (see also Hashimoto et al. 2000; Hashimoto et al. 2001).

Building on this prior research, I have argued that young people's relationships are tending not toward superficiality but toward *selectivity*, where friends are chosen based on context rather than on some other broad or shallow footing (Matsuda 1999b; 2000a; also see chapter 6). In a similar vein, Daisuke Tsuji (1999) has described a "flipper orientation," where youths favor flexible switching of communication to engage and disengage from relationships. He argues that young people's contemporary conception of self can only be understood using a multifaceted model of self with multiple centers rather than a more conventional model of a single self embedded in a set of concentric relational circles. Debate continues on the issue of superficiality and selectivity of young people's relationships, as other researchers have stepped in to challenge and expand on both views (e.g., Iwata 2001; I. Tsuji 2003; Mikami et al. 2001; Miyata 2001; Kotera 2002).

Habuchi's (2002a) work on *keitai* and self also represents an important dimension to this debate (see chapter 8). She describes how the continuous connectivity of *keitai* feeds into the reflexive construction of self that is dependent on feedback from peers. Akihiro Kitada (2002) describes how *keitai* actualizes social connections through text messages that are not oriented so much toward actual communicative content but toward the metamessage "I am trying to communicate with you." The exchange of these messages, in turn, transforms urban environments into a superfluous ritual space where young people need to constantly ward off the risk of "I who may not be connected."

I would like to introduce one final concept on the topic of *keitai* and young people's interpersonal relationships. Based on four surveys of young *keitai* users, Ichiro Nakajima, Keiichi Himeno, and Hiroaki Yoshii (1999) developed the concept of "full-time intimate community," which describes how young people use *keitai* to maintain communication with a select group of no more than ten close friends. These close friends are also people that they see on an everyday basis. The addition of *keitai* communication means that the select peer group has a sense of being in psychological contact twenty-four hours a day, thus forming a full-time intimate community. Research and debate is ongoing on this issue (Hashimoto et al. 2000; Mikami et al. 2001; H. Ishii 2003).

Effects of *Keitai* and Characteristics of Users

Keitai influences more than communication. Yoshii (2001) writes that nearly 50 percent of college students who take up a *keitai* report having less money to use freely, 20 percent report that their purchases of clothing and accessories, comics, and CDs have gone down, and 33 percent report that their time doing part-time work has increased. *Keitai* has become a social necessity, and this is having repercussions on youth consumption patterns and time rhythms.[9] Although the popular media have widely linked the recent decline in restaurant business, karaoke parlors, and CD sales to *keitai* costs (e.g., Oda 2000), the picture is actually more complex.

A study of female college and junior college students finds that heavy *keitai* users tend to do more part-time work, have more disposable income, and consume more fashion items such as clothing and cosmetics (Watanabe 2000). Although these results appear contradictory, they can be read as indicating that people who are more proactive and outgoing tend toward heavy *keitai* use. Other studies find that *keitai* users are more sociable than nonusers (Nakajima, Himeno, and Yoshii 1999; Okada, Matsuda, and Habuchi 2000; Hashimoto et al. 2000) and that heavy users are even more sociable (Miyaki 1999; Okada, Matsuda, and Habuchi 2000; Matsuda 2001b). *Keitai* is used by sociable people and augments their sociability. In Yoshii's (2001) survey of young people, 63 percent reported meeting with friends more often after starting to use *keitai*; only 1 percent reported doing so less.[10]

Keitai, Family, and Gender

The public's interest in the effects of *keitai* on the family has grown as the age of *keitai* users has dropped from those in their twenties to teens, from high school to middle school students. The central question has been whether *keitai* pushes family members apart or draws them together.

Those arguing for the fragmenting effects of *keitai* point out that family members are each making connections to external relations through the personal medium of *keitai*. In particular, it becomes difficult for parents to monitor their children's relationships, so there may be cases where children become involved in crimes or delinquency through their mobile phones.[11] The opposing view recognizes that *keitai* can push family members apart but also suggests that it can strengthen their existing connections. In families with an existing tendency to communicate, *keitai* promotes more communication, but it does not promote communication for families that tend not to communicate to begin with (see chapter 6). Daisuke Tsuji's (2003a; 2003b) research with 16–17-year-olds suggests that the children's peer communications, strengthened by *keitai*, are increasingly happening outside of parental oversight but that this has not had a significant impact on at least the qualities of trust and satisfaction in family relationships. In chapter 14, Miyaki describes a trend toward purchasing *keitai* for children for safety and other reasons, despite reservations about its effects on the family (see also Cabinet Office 2002b).

Other research examines how social structural positions tied to institutions of family and gender relate to differences in *keitai* use. Chapter 6 analyzes gender differences in patterns of use, and chapter 11 describes how *keitai* has been incorporated into the sociotechnical entity "housewife." Rakow and Navarro (1993) argue that just like other media, *keitai* have "the potential to disrupt old social and political conventions, to rearrange hierarchies, and to reconfigure the boundaries of the public and the private" (144), but "It is gender ideology, operating within a particular political and economic context, that leads to women and men living different lives and using technology differently" (155).

The Success of i-mode and Technonationalism

Around the year 2000 the negative image of *keitai* as a medium uniquely tied to youth began to shift. Starting with i-mode in February 1999, *keitai*-based Internet services started to spread across the broader population: "*Keitai* are saviors of the Japanese economy." "*Keitai* represent a Japanese-style IT revolution." Suddenly, Japan began to pin high hopes on *keitai*.

By the end of 2001, 72.3 percent of *keitai* users were subscribed to the *keitai* Internet, and by the end of 2002, this percentage was 79.2. Comparable percentages in other countries are much lower: South Korea 59.1, (74.9 in 2002), Finland 16.5 (29.1 in 2002), Canada 13.8 (20.0 in 2002), and the United States 9.4 (8.9 in 2002). The White Paper on Information and Communications in Japan (2002) which reported these statistics, labeled this topic the "World-Leading *Keitai* Internet."

Moreover, NTT DoCoMo reported that the average monthly revenue per unit (ARPU) for i-mode alone was ¥1,970 (voice ARPU, ¥5,640) in January through March 2004, and Web access was the driving force for packet traffic, at 88 percent in comparison to 12 percent for e-mail.[12] The *keitai* Internet attracted attention not only because it boosted revenue for DoCoMo but also because it represented a new model for e-business, which had stagnated in Japan because of security concerns.

The IT Revolution and the Digital Divide
The *keitai* Internet has been a vessel for hope that goes beyond its potential as a new business model. Japan had been suffering with a prolonged recession after the economic bubble of the 1980s burst. In contrast, the U.S. economy flourished in the 1990s after emerging from its recession in the previous decade, and information technology (IT) was identified as one key element in this recovery. Although the U.S. "new econ-

omy" is now subject to suspicion after the dot-com bust, in the late 1990s Japan looked toward IT as a key component of economic recovery.

But the primary IT industry driver, the desktop PC Internet, did not spread as rapidly as hoped in Japan. This stood in marked contrast to the unexpected popularity of *keitai*. Some factors for the low rates of growth were slow connections, high costs, and low penetration of PCs. Improving Internet infrastructure continued to be a political goal. For example, the November 2000 IT Law stipulates the "formation of a superior information communications network with the highest global standard that can be widely used at low cost."

This was the backdrop to the introduction and spread of the *keitai* Internet. Mari Matsunaga (2001), a developer of DoCoMo's i-mode service, describes how the service was based on a concept of a "regular telephone," targeting general users instead of business users. The *keitai* Internet became an extremely popular service and gave a substantial boost to Japan's Internet participation; at the end of 1999, Internet penetration stood at only 21.4 percent; by the end of 2001, the rate had doubled to 44.0 percent.

The *keitai* Internet has also been characterized as an antidote to the central social concern of the IT era: the digital divide between people who can and cannot use IT. Compared to the costly and difficult process of gaining PC Internet access, the *keitai* Internet is accessed through inexpensive terminals; it is simple to subscribe to and navigate. The barriers of technical knowledge and pricing that inhibited PC Internet adoption do not apply to the *keitai* Internet, and many Japanese first connected to the Internet through *keitai*. For example, Dentsu Soken (2000) has categorized survey respondents into those with high technical literacy and affinities to the PC Internet, and those with low technical literacy and affinities to *keitai*. This latter group is nonetheless characterized by high rates of access to the *keitai* Internet, and the report concludes that this group will "lead the evolution of the Web phone world."[13]

Technonationalism

The slogan *"keitai* IT revolution" emerges from this context. The White Paper on Information and Communications in Japan (2000) reports, "The NTT DoCoMo group has more subscribers than the other major ISPs in Japan." This celebration could be characterized as an effort to gain leadership in the IT revolution through *keitai* after lagging behind the United States in PC Internet use. Dentsu Soken (2000, 18) describes this sentiment:

In the world of the "PC + Internet," the US, the birthplace of the Internet, has always taken the lead over other countries. Today's economic prosperity in the US is also tied to this ongoing IT leadership. With the Web phone, however, Japan has taken the initiative, and if a new IT social model takes hold, Japan could claim a substantial international contribution, and most importantly, could contribute to a reactivation of the Japanese economy.

In his book *Keitai ga Nihon wo Sukuu* (Keitai *Will Save Japan*), Kiyoshi Tsukamoto (2000, 47) relates Japanese *keitai* achievements to a kind of national character reminiscent of earlier *Nihonjinron* (Japanese theory) texts, which claimed the uniqueness of Japanese culture and people:

Americans are strong in the world of large memory and fast connections, as for the PC, but the Japanese are adept in the world of small capacity and miniature displays, as for the *keitai*. This may be a reflection of the difference between a country with rich resources and one with few, but somehow, along the way, *keitai* have become a master craft of the Japanese.

This book does not mention Motorola's 1989 MicroTAC despite the dramatic breakthrough it represented in defining the trend toward miniaturization of terminals. Even before *keitai*, Japanese transistor radios and cars were characterized as beating international competition with "compact high performance" grounded in Japanese "top-grade technology." The current celebration of *keitai* arises from a technonationalist sentiment typical of the self-characterization of postwar Japan.

Of course, Japan was not always a high-tech nation. Akio Morita, the founder of Sony, writes in *Made in Japan* that Japanese products were of poor quality until the 1950s and that it was not until the late 1970s that Japanese companies started to dominate high-tech markets (Morita, Reingold, and Shimomura 1986). Further, Shunya Yoshimi (1998) describes how U.S.-driven global standards in the 1970s started to stress light and compact technologies and precision instruments rather than heavy and massive technology. This shift was as much a factor in Japanese success as any high standard that Japanese technology came to represent in the international arena.

As soon as Japan became associated with high-tech in the international arena, some Japanese began interpreting characteristics intrinsic to Japan as a source of these technologies. For example, in *Made in Japan*, Morita describes technology as a survival mechanism and locates Japan's technological aptitude in the characteristics of an island country with harsh weather and few natural resources, inhabited by deeply spiritual people "who tend to believe that God resided [sic] in everything" (226). He suggests that the Japanese believe that frugality is a virtue and that they "also seem naturally more concerned with precision. It may have something to do with the meticulousness with which we must learn to write the complicated characters of our language" (223). In short, the popular characterization of the *keitai* Internet as the product of superior technology crafted by Japan's "intrinsic cultural qualities" expressed much more than an economic hope for overcoming the recession; it represented an effort to salvage a national pride damaged by the economic downturn in the early 1990s.[14]

Keitai E-mail and Internet Use
In contrast to business success stories, from the user perspective *keitai* Internet use has been neither frequent nor diverse (see chapter 6; Mikami et al. 2001; Hashimoto et al.

2001). Use centers primarily on *keitai* e-mail, while the Web is used primarily for wall-paper and ring-tone downloads.

Keitai text messaging first became possible in Japan in April 1996 through a service launched by the DDI Cellular Group (now called au). The other carriers soon followed suit in launching this type of service, which has come to be called, collectively, short message. At the time, it was not possible to send messages between *keitai* subscribed to different service providers, and each provider had a unique service name, such as DDI Cellular's Cellular Moji Service, DoCoMo's Short Mail, and J-Phone's SkyWalker. In other words, these services were not the same as short message services (SMS) because they did not rely on a shared Global System for Mobile Communications (GSM) standard.

J-Phone (now Vodafone) introduced Internet e-mail for *keitai* in November 1997; Japanese users gradually chose this type of e-mail over short messages because it is cross-platform and allows for longer messages. From a user perspective, however, there is little difference between text messages sent as short messages and those sent as Internet e-mail. Users refer to both as *meiru* (mail), and in this book we call both *keitai* e-mail.

Just as with pagers, *keitai* e-mail has been adopted by more women than men, and use is heaviest among those in their teens and twenties. How is *keitai* e-mail being used by young people and with what kinds of social effects? Survey data show that students who are heavy users of *keitai* e-mail tend to have more friends and to be more sociable (Matsuda 2001b).[15] Further, they tend toward self-disclosure, are not insecure about relating to others, and are not lonely (Tsuji and Mikami 2001). While reporting similar survey results, Nakamura (2003) also states that heavy *keitai* users exhibit a fear of loneliness and a lack of strength in enduring loneliness. As a result, he fears that acclimation to "convenient" human relationships through *keitai* e-mail—where one can contact people 24 hours a day without worrying about their availability—can lead to an inability to tolerate being alone.

Nakamura (2001a) considered the messages sent over *keitai* e-mail and concluded that they are characterized by "casual and colloquial expression," including onomatopoeic words, childlike expressions, and regional dialect. As well, young people often use *emoji* (emoticons) that are programmed into the handsets, and send long messages (enabled by the 250–3,000 character length of *keitai* e-mail) despite the difficulties of one-thumb input. They also mix graphics, video, sound, and Web links into their messages (Miyake 2000; *Nihongo-Gaku* 2001).

In chapter 13, Ito and Okabe look in detail at what kind of messages are exchanged and with what frequency. In general, users see *keitai* e-mail as a medium that allows for "an appropriate sense of distance" based on the immediacy of transmission coupled with a format that is less intrusive than voice (Kurihara 2003). But young people still feel that "it is inconsiderate not to respond immediately." One outcome of this is the mobile text chat that Ito and Okabe describe, where users may send a great number of messages back and forth over a short span of time. The use of *emoji* (pictorial icons provided by operators) in these messages reflects not only "wanting to effectively convey feelings that can't be conveyed by text only" and "wanting to make messages cute with *emoji*" but also a sense of consideration toward others, because "they might mistakenly think I am angry if I don't include *emoji*" (Tsuzurahara 2004). Just as with pagers, the *keitai* is not always a medium of *yasashisa*.

Turning to the *keitai* Internet, Kenichi Ishii (2004, 57) offers the following summary of his analysis of the World Internet Project Japan survey (2002):

The mobile Internet has positive effects on sociability with friends, while the PC Internet does not have such effects. Email via a mobile phone is exchanged mainly with close friends or family, whereas email via a PC is exchanged with business colleagues. These results suggest that PC diverge in terms of social functions; in other words, mobile Internet use has more in common with time-enhancing home appliances such as the telephone, while PC Internet use has more in common with the time-displacing technology of TV. . . .

The experiences in Japan show that neither technological advantages nor telecommunication policy promote a new type of telecommunication service. Japanese experience after 1995 demonstrates that user needs have brought about the high penetration rate and unique usage patterns (e.g., *beru-tomo* and picture mail) of the mobile Internet in Japan. The Japanese government has placed political importance more on broadband than on mobile phones.

See chapter 7 for a comparison of *keitai* Internet and PC Internet use.

From Rejection to Utilization of *Keitai*

I have described how, in the late 1990s, *keitai* were criticized by the general public as devices used by young people for trivial personal communication, closely linked to poor manners and superficial relationships.[16] High schools generally prohibited *keitai* because of its association with problematic and deviant social behavior. As *keitai* continued to spread among young people, however, it has come to be tacitly tolerated, and schools are now taking the stance that "as long as it doesn't ring during class, it is okay." In recent years, there have even been efforts to educate students on how to relate to *keitai* from a media literacy perspective. Since practically all college students have *keitai* now, there is a project under way to use *keitai* for real-time teaching evaluation as part of a faculty development effort (Takeyama and Inomata 2002; Hara and Takahashi 2003).

Surveys indicate that *keitai* use has not been and is not currently exclusive to young people or to private uses. Although those in their twenties have the highest ownership rates, those in their thirties rank next. In the Mobile Communication Research Group (2002) survey, 10.2 percent of respondents indicated they had multiple *keitai* handsets; these respondents were primarily male managers and officers of corporations and other organizations in their thirties and forties. These users generally have separate *keitai* handsets for personal and business use. Heavy users of voice calls tend to be men and those working full time; 26.0 percent responded that 100 percent of their voice calls were personal, and 15.5 percent responded that 90 percent of their voice calls were, but 12.8 percent responded that 10 percent of their calls were personal, and 12.0 percent responded that only half of their calls were. As it was in the early adoption years, *keitai* continue to be used as media for work.

In the late 1990s, public discourse surrounding *keitai* deviated from the realities of use reflected in survey data: it viewed *keitai* as "a youth problem" and "emblematic of the private." This discourse changed after the peak of rapid *keitai* adoption in the 2000s; *keitai* are now firmly established and the *keitai* Internet widely adopted. Now that *keitai* have become unremarkable, its use by those other than youth and for business purposes is being "rediscovered." For example, a year and a half after the introduction of i-mode, the *Asahi Shinbun* (2000) newspaper carried an article titled "i-mode has made its appearance in the workplace—easy to use and carry around":

Just as I thought that there are fewer young people talking loudly on their *keitai* in the train, now I can't help but notice people intently moving their thumbs [over their keypads]. Services such as "i-mode" have appeared and greatly expanded the pleasures of e-mail and information access. But, if you think about it, these "any time, any place" features that can be accessed with just one thumb have got to appeal to business uses as well. After initial tentative usage, the technology has demonstrated a surprising power for managers who are in demand for speedy decision-making and for the intensely competitive front lines of sales and marketing. As the youth culture origin technology of "i-mode" starts to spread among the older population as a business tool, there are hints that business styles are changing.

The article goes on to describe e-mail exchanges between the president of a company and the head of personnel as the president is in transit on a train. In this way, *keitai* e-mail has been spreading from the base of business e-mail use that started with the PC Internet. Currently, there is virtually no research on such business uses of the *keitai* Internet; Tamaru and Ueno's contribution in chapter 12 is a rare inquiry into this space.

Conclusion

I have explored various developments in the public discourse concerning *keitai* from the 1990s. To borrow from Ito's introductory framing, I have described the state of domestic Japanese mobile communications not as "the cellular phone," defined by technical infrastructure, nor as "the mobile phone," defined as an untethering from fixed location, but rather as *keitai*, an artifact located in a specific national context. Through this process, I have contextualized the contributions to this book as well as reviewed additional pertinent popular and research literature from Japan.

Our collective effort is to describe *keitai* not as something exclusive to Japanese society but as something deeply embedded in the contexts of Japan. In contemporary societies characterized by the presence of mobile media, the studies in this book will speak to the specific contexts that readers bring to bear, and will no doubt provide insight into contemporary human relations and social systems.

Notes

1. In the early years, PHS users were mostly youth, and PHS were used in ways similar to the cellular phone. As young users transitioned to the cellular phone, however, the low-cost PHS came to be used differently from the cellular. Research shows that 44.6 percent of PHS users have more than one *keitai* (cellular and/or PHS) (Mobile Communication Research Group 2002).

2. In this article, from the magazine *Kuriiku*, May 20, 1991, as well as the others cited by Kawaura, the word *keitai* is not yet being used.

3. In other work (Matsuda 1996b; 1997), I analyze statements that *keitai* have personal health effects from the perspective of rumor and urban legend, describing the process through which *keitai* is received by society.

4. *Jibetarian* have a deep relationship with *keitai*. In an article about *jibetarian*, a reporter describes reasons why they sit on the street: "(1) They have no endurance and get tired. (2) The sidewalks have gotten clean. (3) They want to make voice calls in peace. (4) They want to save money for their *keitai* communication costs" (*Asahi Shinbun*, evening edition, October 24, 1997).

5. Sociological inquiry into Japanese youth in the 1990s include Shinji Miyadai (1994; 1997) and Tomita and Fujimura (1999). In addition, Satoshi Kotani (1993) provides an outline of the evolution of youth studies since the 1970s.

6. Hashimoto (2002) has produced an English-language paper on the influence of *keitai* on youth.

7. By cutting the call after one ring, the caller does not incur a toll but is able to leave a record that they called through caller ID, thus sending a free message that "I am thinking about you."

8. Other works on pagers and youth include those by Fujimoto (1997) and Tomita et al. (1997). Matsui (1998; 1999a; 1999b) has analyzed the changing image of pagers.

9. The NHK Broadcasting Culture Institute has conducted a series of surveys on the influence of *keitai* on life rhythms. For example, *keitai* do not reduce sleep or television viewing time as much as the PC Internet does (Kamimura and Ida 2002). Further, young people generally use *keitai* at the same time as they are engaged in other activities, such as watching TV, eating, or conversing with friends (Nakano 2002).

10. More information on user demographics, adoption, and use patterns can be found in studies by Nakamura (2001b), Mikami et al. (2001), Mobile Communication Research Group's (2002) national survey, and Hashimoto et al. (2000), Hashimoto et al. (2001). The World Internet Project, ⟨http://media.asaka.toyo.ac.jp/wip/⟩, has survey results available in English.

11. In the past few years, the term *puchi iede* (*petite* running away from home) has become popular. This refers to teens who stay at places like their friends' homes and don't return to their own homes for several days or weeks. Family members can "reach their child through *keitai*," so they don't file a missing person report, and after a while the child generally returns home without incident, so this phenomenon differs from the prior practices of running away from home.

12. NTT DoCoMo Web site, ⟨http://www.nttdocomo.co.jp/info/new/release.html⟩.

13. This "low literacy stratum" is composed mainly of young people, however, so it is unlikely that it will become a window for Internet access for the older age groups that use PCs and the Internet even less. Further, Kimura (2001a) writes that (1) the *keitai* Internet business model is the same as Dial Q^2 and is not particularly innovative; (2) there is a large gap between "possession" and "use" of devices; (3) the *keitai* Internet is typical "conspicuous consumption" that started with the pager and moved onto the PHS and *keitai*. Thus he concludes, "i-mode is not the savior." See also Kimura 2001b; Matsuda 2002.

14. "Compact high-performance" design is also prevalent in other countries besides Japan. Several years ago, U.S. and European manufacturers and mobile entrepreneurs were concerned that "super-compact terminals are too 'Japanese' and might not be well accepted by domestic consumers." (I have also heard similar statements in personal communication.) However, considering that such products actually did penetrate the respective markets, these statements could also could be considered a version of "techno-orientalism" (Morley and Robins 1995).

15. See note 10. Studies of college students, who are heavy *keitai* e-mail users, include Okada, Matsuda, and Habuchi (2000); Matsuda (2001b); and Tsuji and Mikami (2001).

16. At the same time, many continue to describe *keitai* in relation to "youth insecurity in communication" (e.g., Okonogi 2000; Masataka 2003). Masataka cleverly titled his book *Keitai wo Motta Saru: "Ningenrashira" no Houkai* (*Monkeys with Keitai: The Destruction of "Humanity"*); this drew attention from a wide range of newspapers and magazines, which resulted in a bestseller.

2 Youth Culture and the Shaping of Japanese Mobile Media: Personalization and the *Keitai* Internet as Multimedia

Tomoyuki Okada

The social reception and transformation of *keitai* communication and the *keitai* Internet in Japan are closely linked to the use of these media by youth and to youth popular cultures. This chapter examines the history of *keitai* in terms of its social shaping as a medium and a consumer item.

A starting point for this analysis is theories of the social construction of technological systems (Bijker, Hughes, and Pinch 1993). As Bijker and Law (1992, 13) have suggested, "Knowledge is a social construction rather than a (more or less flawed) mirror held up to nature," and further, "Technologies and technological practices are built in a process of social construction and negotiation, a process often seen as driven by the social interests of participants."

Claude Fischer (1992) has extended this approach more radically in examining the domain of telecommunications. In his study of the spread of the telephone and its establishment as a new media form in the United States, Fischer successfully mobilized a social constructivist approach but had the following critique: "Most social constructivism has concentrated on the producers, marketers, or experts of a technological system." He describes how in his own work his intent was "to go further, to emphasize the mass users of technology." In Bijker's study (1992) of the development of the fluorescent lamp, Bijker dismisses the influence of consumers, stating, "The social group of customers does not have its own direct voice in this story." He writes, "The result of market research and an analysis of the popular technical press may be considered to reflect the views of this social group" (1992, 81). By contrast, Fischer (1992) stresses the role of consumption; he believes that in order to understand the social shaping of technology it is crucial to include consumers in the analysis.

My study of Japanese mobile media demonstrates that the role of the consumer is absolutely critical. Taking my cue from Fischer's approach, I build my argument by layering a wide range of materials regarding the particular technological context and the reception by users.

Materials

I draw on a wide range of user surveys and statistical data as well as on interview studies conducted on the streets by our research group. We have been conducting studies since 1995 on mobile media use in Japan (see chapters 6, 8, 9; Okada and Matsuda 2000; Matsuda et al. 1998; Tomita et al. 1997). In this work, we first arrived at an understanding of the users' perspective through interviews. The latter half of the 1990s saw the rapid adoption of mobile phones, and youth were identified as the leaders of this trend. It is extremely difficult to grasp the views of these users through the random abstractions presented in a standard survey form. At that point in time, adoption was still below the 50 percent mark, and survey return rates from the core *keitai* adoption population of young people were relatively low.

Because of these difficulties, our research group approached these new trends by conducting several interview studies on the streets of the two major Japanese cities, Tokyo and Osaka, targeting areas where young people gather in large numbers. In Tokyo these were areas such as Shibuya and Harajuku, and in Osaka, America Village in the Minami district. These areas are popular gathering places for young people in their teens and twenties; they emerged as youth culture centers during the blossoming of consumer culture in the 1980s. We conducted our spot interviews twice at each of the districts: in Shibuya in the summer of 1996 and 1998, and in Minami in the winter of 1997 and the summer of 1998. Our research has continued as an interactive process since then. We conducted a study among college students based on survey questionnaires in 1999 and again in 2001. When referring to interview data, I note the gender, age, location, and interview date for the research subjects.

Personalization

In the history of mobile media in Japan, the first service is generally considered to be the wireless telephone, introduced on merchant ships in 1953. The main purpose of this service was to connect the vessels in port to telephones on land. Three years later, an experimental service that connected trains to land line telephones was implemented on the Kintetsu Express running between Osaka and Nagoya. Many years after that, NTT developed the world's first car phone, in 1979 as the world's first cellular telephone service, and in 1985 as the "Shoulder Phone," a car phone that could be carried outside the vehicle. Subsequently, in 1987, the world's first handheld cellular phone was marketed by NTT. By this time, the concept of *keitai* had become a telephone that could be used away from the location to which it was conventionally fixed, such as an office or an organization.

The *keitai* was originally developed for official uses and for organizational purposes, as manifested by its birth as a shared telephone. Before it became the norm for each individual to have his or her own *keitai* device, a *keitai* was typically provided to one representative within a group of people, for example, to the manager at a construction site where there was no fixed telephone line. Today, unless there is a specific reason, we rarely answer another person's *keitai* when it rings in the owner's absence. However, early *keitai* were typically not owned by an individual but were leased to a group under the name of the representative, and therefore when that person could not answer the phone, it was picked up by somebody else instead.

The same can be said for the pager, which was a popular mobile medium before the *keitai*. Pagers were primarily used by companies and other organizations and were not provided to each individual. Rather, they were shared among a group, and a salesperson would take an available one outside the office as necessary.

The terminals of this period were the tone-only type, which only rang when receiving a call and did not display a callback number. Only co-workers at the individual's own office would call. Consequently, to the person carrying the pager, it was a "binding medium" that seemed to chain him or her to the company (Takahiro 1997a).

Throughout the 1990s, during the adoption of mobile media, uses of the pager were extended into individual and personal purposes. This trend can be described as personalization (Matsuda 1999a). The pager took a major step toward being a personal medium in April 1987, when NTT was struggling to compete against Tokyo Telemessage and other New Common Carriers (NCC) that had entered the market the year before. Before it spun off NTT DoCoMo, NTT introduced the display-type pager, which showed digits and letters on the terminal's liquid crystal display (figure 2.1).

Figure 2.1
NTT's Pocket Bell D-Type (1987), the first display-type pager. Reproduced with permission.

When calling this pager, the caller would input a callback number, and this number would be displayed on the monitor to identify the caller to the receiver. This function changed the pager from a medium limited to receiving calls from one specific individual or location to one that could respond to calls from various sources such as the office, home, and friends. With this change, the pager was extended into private and personal uses outside of the office setting.

As subscription charges dropped significantly, the ages of the users also dropped, and the pager came to be a personal communication tool for female college and high school students. It allowed girls to receive messages from various partners, and a new form of dialogue was constructed through the repeated exchange of pager messages.

In 1987, as business competition began in this area, subscription rates rose over 19 percent in comparison to the same period in the prior year. In the spring of 1992 they rose only 16 percent, and in the spring of 1993, 13 percent, indicating a slow but steady decline. In June 1993, however, the percentage increase started climbing. In September 1993 the cost of a new subscription dropped by a half, to ¥8,000, because of a drop in the rental security deposit. In response, December 1993 saw a growth rate of close to 19 percent (*White Paper on Communications in Japan 1994*). Based on reports aired in industry public hearings in that year, 70 percent of new subscribers were individuals (rather than business subscribers), mostly young users in their teens and twenties.

Isao Nakamura (1996c) writes that until 1990, business hours, particularly around 10 a.m. and between 2 p.m. and 3 p.m., were the peak hours of use for the Tokyo Telemessage service. By 1993, however, 10 p.m. had become the peak hour of use. Nakamura cites this shift as indicating a structural change in pager use between 1991 and 1992, where private uses came to dominate. He also states that in 1993, 80 percent of new subscribers to this service were in their teens and twenties.

In 1995 carriers further lowered costs and introduced a sales model for pager devices, doing away with the prior rental model. Tokyo Telemessage's June 1995 release of the Mola, a new pager that could receive text messages, dramatically expanded youth pager uptake; demand for this device was so strong that new subscriptions had to be suspended. These new pager text-messaging functions are described in more detail in the following section.

In June 1996 pagers subscriptions hit their peak of 10,777,000. Household adoption rates at that time were only 15 percent nationwide and 18.4 percent in urban areas. Among households with children between the ages of 15 and 19, however, adoption rates were the highest, at 35.2 percent. For 20–29-year-olds, this percentage was 25.0 percent. For other age ranges, users 40–49 years old had the highest adoption rates at 13.0 percent, with other age groups consistently under ten percent (*White Paper on Communications in Japan 1997*).

In the same year, a survey of middle and high school students in Tokyo Prefecture indicated that 48.8 percent of female high school students had a pager (Tokyo Metropolitan Government 1997). Further, a 1997 survey by the Ministry of Posts and Telecommunications indicated a rise in the number of pager users responding, "I never use my pager for work." In the prior year, the percentage of responses was only 36.4 percent, but by 1997 it had reached 47.9 percent, indicating a growth in personal uses of the pager (Tokyo Metropolitan Government 1997).

In tandem with the rapid adoption of pagers by youth, *keitai* adoption also grew. Ministry of Posts and Telecommunications surveys show household adoption rates in 1995 and 1996 for cellular phones at 10.6 percent and 24.9 percent, and for personal handyphones (PHS) at 0.7 percent and 7.8 percent (see chapter 3). This growth was fueled by large reductions in subscription and use costs. In April 1994, NTT DoCoMo launched a sales (as opposed to rental) system for mobile handsets, drastically reducing the cost of a new subscription by over half, from ¥24,700 to ¥9,000–¥12,000. In December 1996 all cellular phone providers dropped the rental model. The launch of PHS services in July 1995 also provided users with a device that was cheaper than the cellular phone. PHS providers also pursued a strategy whereby they would provide retailers with free or nearly free handsets as a promotion for attracting new subscribers. Cellular phone providers responded by adopting this strategy as well, creating business competition for new subscribers that further spurred adoption (Matsuba 2002).

Despite these factors, *keitai* adoption rates among young people were still relatively low. In the aforementioned Tokyo study, cellular phone and PHS subscription rates of high school girls were only at 28.3 percent. High school boys were at 26.3 percent (Tokyo Metropolitan Government 1997). For teenagers, pagers were still the mobile devices of choice. Other indicators, however, show personal communications on the rise with *keitai* as well. As with the pager, survey results show an increase in cellular phone subscribers who answered, "I never use my cellular phone for work," growing from 14.0 percent in 1995 to 25.8 percent in 1996 (*White Paper on Communications in Japan 1997*). In this way, mobile media such as the pager and *keitai* were steadily moving toward personal use through the latter half of the 1990s.

Both the pager and *keitai* were at first strongly linked to official places and organizations, but they were gradually transformed into media of direct connection between individual users. This move away from organizational and group use toward directly connecting individuals in the personalization of media had its parallels in the history of land line telephone use. Shunya Yoshimi, who co-authored the first comprehensive sociological research on the social adoption of the telephone in Japan, *Media to shite no Denwa (The Telephone as a Medium)*, observed a change in the location of the fixed telephone in the household. He says that when the telephone first arrived in the average household, it was often placed near the front door of the house. It was later set up in the living room. From the mid-1980s cordless and extension phones were commonly placed in bedrooms as the telephone found its way into the private rooms of each family member (Yoshimi, Wakabayashi, and Mizukoshi 1992). The front door, the physical gateway between the family and the outside world, and the telephone, the communications gateway with the outside world, could both be considered "places" for mediating and regulating social access. The telephone, or the gateway of communications, proceeded deeper into the inner recesses of the household. Yoshimi sees this as indicative of the telephone's turning into a medium of direct connection between the social world and the individual inside his or her private room. Uptake of mobile media carried by individuals moving around in unlimited space could be considered an intensification of this trend.

These dimensions of personalization are also related to the individualization of television, radio, and other forms of mass media since the 1970s. The adoption rate in Japan of the color TV reached 90.3 percent in 1975, and close to 100 percent in 1980. Beginning in 1990 corporations started discussing implementation of individualized measures for TV ratings. In March 1997, Video Research, the primary media ratings firm in Japan, began its "people meter" ratings for Tokyo, tracking individual rather than household viewing. Before the near-universal adoption of the television, radio was the family medium within a household. After it lost its central position to the television, late-night radio programs spearheaded the transformation of radio into a medium constructing a network of youths cloistered in their private rooms (Hirano and Nakano 1975). The popularity of the Walkman also illustrates this shift away from products used by the whole family to ones used by individuals. Takuji Okuno (2000) describes this sequence of events as the shift from household media to individual media.

Another critical factor behind the adoption of mobile media is the advancement of urbanization and consumer lifestyles, related to a decrease in time people spend at home and the increase in time spent outside, especially for youth (table 2.1). Born

Table 2.1
Time Awake Spent at Home, 1975–2000

	1975	1980	1985	1990	1995[a]	2000
16–19-year-old males	7 h 14 m	7 h 05 m	7 h 05 m	6 h 57 m	6 h 35 m	6 h 13 m
20–29-year-old males	5 h 15 m	4 h 55 m	4 h 55 m	4 h 37 m	4 h 53 m	5 h 06 m
16–19-year-old females	7 h 26 m	7 h 45 m	7 h 28 m	7 h 12 m	6 h 59 m	7 h 17 m
20–29-year-old females	9 h 54 m	9 h 09 m	8 h 33 m	7 h 30 m	7 h 17 m	7 h 15 m

Source: NHK Broadcasting Culture Research Institute (2001). *NHK Data Book 2000: National Time Use Survey.* Reproduced with permission.

a. From 1995 on, survey methods changed and results cannot be compared directly.

from the rapid expansion of consumer society between the last half of the 1980s and the early 1990s, the Japanese "bubble economy" is a crucial backdrop to these trends in media consumption. As part of these broader social and economic shifts, youth began to spend more time outside the home, and it became commonplace for leisure hours to be spent in city centers. Returning again to the 1997 Tokyo survey, only 5.1 percent of middle school students responded that they "got home after 9 p.m. after having fun out" "more than once a week," but among high school students the percentage had risen to 18.8. Also, 21.1 percent of middle and high school students responded that they "usually stop somewhere on their way home from school," and 30.3 percent of respondents indicated that "on the street or in town" was the number one "place (other than school) for chatting with friends," followed by 21.4 percent for "fast food restaurants." "My home or a friend's home" was favored by 20.5 percent; "in front of a convenience store" by 6.0 percent (Tokyo Metropolitan Government 1997). These results indicate some of the ways in which young people are seeing the city, broadly conceived, as a place to spend their free time. This is one of the many factors I have described as tied to the expansion of personal uses of mobile media.

Mobile Media as Multimedia

Another trend in the development of mobile media is their process of becoming multimedia. Best known among multimedia is the personal computer, with its relatively long history of multimedia capability. In addition, there is the expanding capability of the multimedia Internet, as well as digital TV broadcasting. The word *multimedia* is defined as the capability of a medium to interact fluently with various modes of communication such as text, sound, and images, both still and moving.

Kouichi Kobayashi (1995) summarized the concept of multimedia in the mid-1990s when interest in multimedia started to increase in Japan. He identifies five trends in the "process of media development or innovation that indicate the transformation into becoming multimedia": (1) having multimodes, (2) interactivity, (3) hypertext properties, (4) a tendency toward digital application, and (5) networking capabilities. *Keitai* with Internet access services like i-mode (NTT DoCoMo), ez-web (au, TU-KA), and J-sky (J-Phone, currently Vodafone's Vodafone Live!) essentially satisfy these conditions and can be considered multimedia. The first three of Kobayashi's factors do not concern the actual content of communication but rather point to the mode of communication. He indicates that transmission of information "requires use of multiple modes, styles, and specifications" and that "they were attainable to a certain level with existing media." In short, Kobayashi suggests that multimedia communication had already been widely established.

The liquid crystal displays for pagers as well as in the implementation of caller ID, and text message functions such as the Short Message Service on *keitai*, are indicative

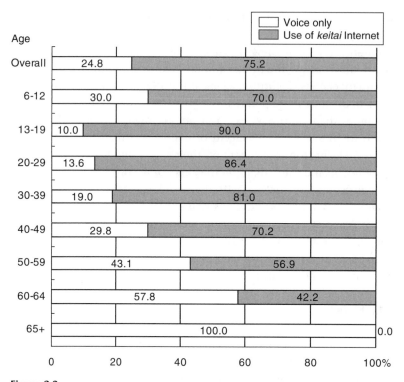

Age

Figure 2.2

Keitai Internet use rate, by age. From Ministry of Public Management, Home Affairs, Posts, and Telecommunications, *Communications Use Trend Survey 2002*. Reproduced with permission.

of early multimedia capability. Now the digital camera has become a standard function of current *keitai*. Vodafone (previously J-Phone) has the largest share of the *keitai* digital camera market. Currently, over half of their subscribers have camera phones. This trend has been influenced by the innovations in technology as well as by use patterns.

The developer of i-mode, Takeshi Natsuno, among others, has noted how the i-mode value chain model helped drive the adoption of the *keitai* Internet in Japan. Natsuno and his colleague Mari Matsunaga have both noted, however, that existing trends and styles in youth communication were also key influences (Natsuno 2003a; 2003b; Matsunaga 2001). Indeed, adoption rates of the *keitai* Internet increase in the younger age ranges (figure 2.2). The most popular uses of the *keitai* Internet are youth-oriented areas such as e-mail and ring tone sites (figure 2.3).

In the following sections I describe the evolution of *keitai* features outside of voice telephony, focusing on the ways in which they were informed by youth cultures.

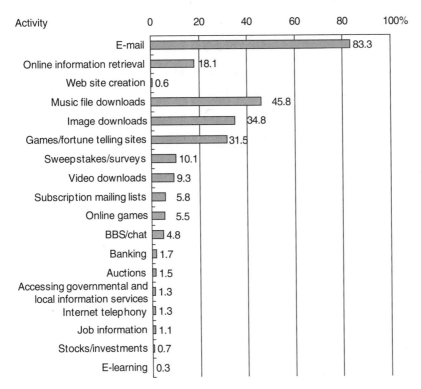

Activity

E-mail		83.3
Online information retrieval		18.1
Web site creation		0.6
Music file downloads		45.8
Image downloads		34.8
Games/fortune telling sites		31.5
Sweepstakes/surveys		10.1
Video downloads		9.3
Subscription mailing lists		5.8
Online games		5.5
BBS/chat		4.8
Banking		1.7
Auctions		1.5
Accessing governmental and local information services		1.3
Internet telephony		1.3
Job information		1.1
Stocks/investments		0.7
E-learning		0.3

Figure 2.3
Activities performed in the past year using the *keitai* Internet. From Ministry of Public Management, Home Affairs, Posts, and Telecommunications, *Communications Use Trend Survey 2002*. Reproduced with permission.

Short Messages

Heavy reliance on mobile e-mail is one of the distinctive features of Japanese youths' *keitai* use. As researchers in other countries have noted, the heavy use of mobile messaging among youth is common in countries with widespread mobile phone adoption (Kasesniemi and Rautiainen 2002; Kasesniemi 2003; Agar 2003). However, the Japanese case is somewhat unique in that text messages far outpace voice calls for young people. Although our survey did not break messaging volume down by age, we consider the category of "students" as indicative of a teenage demographic, since most students in middle school through college range from the teens to early twenties. For the student category, at 27 percent, the most common response was "two to six voice calls a week," and 23 percent responded "one to two voice calls a day." For voice calls, the youth demographic did not differ significantly from the average across all age groups, which had 24 percent responding "two to six voice calls a week" and 27 percent

Purpose

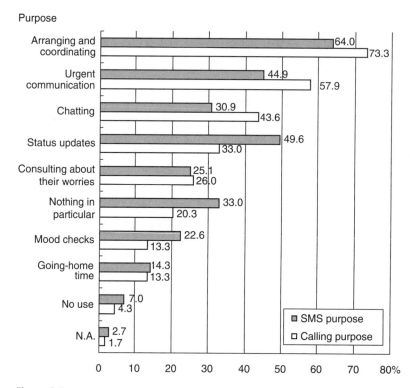

Figure 2.4
Short message users' purposes for making voice calls and sending short messages. From Mobile Communication Research Group (2002).

responding "one to two calls per day." In contrast, the youth demographic deviates significantly from other age groups in *keitai* e-mail volume. The overall average of *keitai* e-mail per week was 28.2 sent and 24.2 received, whereas students report on average sending 66.3 *keitai* e-mail messages per week and receiving 71.8. In other words, students exchange approximately ten *keitai* e-mail messages a day.

What is the content of these communications? Among the general population, our surveys show the content of communication with a friend as follows (figure 2.4): for voice calls, the top categories were arranging and coordinating meetings (73.3%) followed by urgent communication (57.9%), chatting (43.6%), status updates (49.6%), and nothing in particular (33.0%). E-mail had higher proportions of status updates (49.6%), nothing in particular (33.0%), and mood checks (22.6%) but lower proportions of urgent communication (44.9%) and chatting (30.9%). We can conclude that e-mail between friends tends to be used for nonurgent communication (Mobile Communication Research Group 2002).

Pager texting practices were the source of many of these patterns in *keitai* e-mail usage. The technical forerunner of *keitai* text-messaging functions was the numeric display on the pager, a service introduced in 1987 by NTT (now NTT DoCoMo). As described, the concept was to have the caller input the telephone number he or she wanted the pager owner to call back. With the reduction of the subscription deposit in 1993 and the implementation of the terminal purchasing system in 1995, the subscription costs were drastically lowered, boosting the number of young users. These young users began using the pager to exchange short messages in which words were assigned to sequences of numbers and codes.

In light of such trends, Norihiko Takahiro (1997b) applied the three characteristics of new media identified by E. M. Rogers (1986), interactivity, demassification, and asynchronicity, to define pager communications as multimedia/new media. Though the pager can only receive signals, by combining its use with the conventional telephone, the communication process comprised a multimodal exchange of information.

The method of communication popular among high school girls during that period is called *poke-kotoba* (pager lingo), which translates a specific sequence of numbers into specific words, generally using the first syllable of the name of a number as the "reading" of the number. For example, 0840 is *ohayo* (good morning), and 724106 is *nanishiteru* (what are you doing?). The pager, which was designed as a medium to simply request a return call, evolved into a medium of interactive text communication via these girls' using the telephone keypad as a keyboard for sending out messages.

In response to these practices, pager manufacturers added a new function to the pager that converted numbers into phonetic symbols. For instance, 11 became the symbol for *a*, and 21 became the symbol for *ka*. Until then, in order to translate *poke-kotoba*, the users needed a common reference or understanding to decode the digit sequences. With this new function, they were able to send messages that were readable by anybody. This further expanded youth pager use, and in 1996 the number of subscribers topped 10 million. In interviews young people described what these technologies meant for them:

[19-year-old girl, Tokyo, Shibuya district, summer 1996]
When do you usually use your keitai *or pager?*
The pager? Well, it's when it is not something worth calling a *keitai* for, like just "good night" or "good morning" or "how are you?"

[Two 16-year-old girls, Tokyo, Shibuya district, summer 1996]
When do you call [a pager]?
A: When meeting somebody.
B: Yeah, like when meeting somebody.
A: Or just killing time.
B: Yeah, killing time.
A: Or when you just want to say a little something.

So what kind of messages do you send?
B: Hm, well, whatever, anything.
A: Yeah, we send everything.
B: "Good morning," "good night," things like that.
A: Or, like, "I'm tired" or "I'm hungry."
B: That's about it. Probably it is stuff that is easy to send with text.

[19-year-old girl, Tokyo, Shibuya district, summer 1996]
When do you call or send a message?
When? I might send a message like, "Are you free today?" to a friend.

[20-year-old woman, Tokyo, Shibuya district, summer 1996]
When do you call or send a message?
On a Sunday or day off, "good morning" or a greeting like that. Even if you don't call, you can kind of keep a connection. If you send one pager message, you feel connected.

At the time, the cellular phone and the PHS were not very appealing to the youth who drove the pager boom. Basic monthly charges were expensive, and it was felt troublesome to answer the phone each time they received a call. The pager was an irreplaceable medium for sending messages that were not important or urgent enough to be relayed by the telephone.

[Osaka, America Village, Minami district, winter 1997]
Do you want a PHS or a cell phone?
Boy: Yes, if I had the money.
What would you do with your pager if you had a cell phone or a PHS?
Boy: I would keep it.
Girl: I would keep it.
Why?
Boy: I couldn't part with it.
Girl: Yeah, I'd miss it ringing.
The telephones also ring.
Boy: They're not the same. The pager has something special.
Girl: We couldn't just say *ohayo* (good morning) on the phone.
Boy: Yeah, yeah.

As this last example implies, at the time of our street interviews the pager was the preferred medium because it allowed users to communicate through text without bothering people near them (Okada and Habuchi 1999).

Meanwhile, cellular phone and PHS companies were not passively sitting on the sidelines as the pager gained popularity. They were planning their entry into the market by drawing attention to their interactive text transmission services.[1] First, in April 1996, the Cellular Text Service was introduced by the DDI Cellular Group, followed by similar services by IDO and NTT DoCoMo. DDI Pocket started a text-messaging service for the PHS. At the end of 1997, J-Phone (now Vodafone) started its SkyWalker ser-

Figure 2.5
Fujitsu's F501i (1999), the first NTT DoCoMo i-mode handset. Reproduced with permission.

vice using the GSM system, enabling transmission of Internet e-mail (Ohta 2001), and J-Phone temporarily stole some of DoCoMo's young users. Alarmed by the sudden loss of its youth market share, NTT DoCoMo hurried to develop its full-scale mobile Internet i-mode service (figure 2.5), finally introducing it in 1999 to discourage user defection.

The exchange of messages with display pagers, as analyzed by Takahiro (1997a), was merely multimedia-like communication. Consolidating the communication process into a single device, *keitai*, made mobile communication truly multimedia. In this period, youth transitioned from the pager to *keitai* as cell phone and PHS charges were reduced. By the end of the 1990s, those same youth pager messengers were using short messaging to convey peer communications.

[17-year-old boy, Tokyo, summer 1998]
Your keitai can do voice and also send text, right? Do you use the text messaging?
Yes, I do.
Which do you use more, voice or text?
I use text more.
Why is that?
I use text for simple things that are not worth a phone call.
How many messages do you send a day?
Well, only about five or six.

[21-year-old man, Tokyo, Harajuku district, summer 1998]

That one that is compatible with [the keitai *e-mail service], SkyWalker, can do text and also voice, right? Which do you use more?*

I think I must use SkyWalker more.

Why is that?

Well, the first thing is that it's cheap. That is a factor. My crowd tends to have it, and I have a lot of friends, and with new friends it is an easy way to communicate. And I also don't have to worry about the time of day when I send a message. Stuff like that. That it only costs ¥5 per message is probably the big factor.

Is the content of what you say different between text and voice?

Yes, I think it is definitely different, but also because only certain people have *keitai*. The people I am in touch with have to have a PHS or some kind of phone. This is a little different, but I have a friend who has a SkyWalker home page, and he has an e-mail mailing list with over one hundred people. Because of things like this, and wanting to get more information from places like this, I use SkyWalker a lot. For exchanging information and things like that, I use e-mail and only use voice when it is urgent, I need to talk right away, or I need advice, or something like that.

[17-year-old girl, Osaka, Minami district, summer 1998]

What kind of messages do you exchange with [the text message service] P-Mail?

Hm, like, "what's up?" or "where are you now?" When it's just casual, it's, like, "what are you up to?" or something.

Do you talk on the phone more or use P-Mail more?

I think it is must be P-Mail.

As of March 2003 the number of pager subscribers has decreased to approximately 1.13 million. One symbol of youth's shift from pagers to *keitai* is the pop star Ryoko Hirosue's changing roles from being "the face" of NTT DoCoMo's pager in both television and print media to being the poster girl rolling out the new i-mode service (Matsunaga 2001). This latter ad campaign happened after she graduated from high school and was about to start college, and she was represented as also having "graduated" from pagers and moved on to the *keitai* Internet.

Ring Tones

Ring tones (*chaku-mero*) are another key element of the multimedia capabilities of *keitai*. *Chaku-mero* is the term used widely for music that signals the receipt of a transmission to the *keitai*, but it is actually part of a trademark registered by the Astel group in 1997 for the PHS service, "*Chaku-mero yobidashi* service."

The current system requires payment of approximately ¥5 royalty per song downloaded to the Japanese Society for Rights of Authors, Composers and Publishers (JASRAC). In 2002 the revenue from these royalties was over ¥7 billion, more than double the prior year's revenues. Royalty revenues from CDs in the same year were just over

¥30 billion, so ring tone royalties were approximately 20 percent of this amount. Revenue from karaoke sounds downloaded through ISDN lines totaled approximately ¥5.5 billion; ring tones have surpassed this revenue stream (JASRAC 2003). Ring tone distribution has become a key area for Japan's contemporary music industry.

Melodies, however, have been used for ring tones from the days of the pager and the land line phone. Original ring tones became available for *keitai* in September 1996, when IDO (now au) released the D319, a handset that allowed input of original ring tones. Users could use the keypad to input a score note by note and thus register a song. *Chaku-mero* was recognized as a popular trend from approximately December 1997 with the publication of *Keitai Chaku Mero Doremi Book* (*Keitai Ring Tone Do Re Me Book*), considered the first *chaku-mero* composition manual (*Sankei* 1999). Since this first publication, ring tone composition books including hit songs have been popular sellers: *Keitai Chaku Mero Doremi Book* ranked number 9 in 1998's annual bestseller ranking of all books in Japan (reported by Tohan Co.).

In a household or an office the telephone rings in order to notify someone in the area to pick it up. However, since the *keitai* is almost always attached to an individual, it only has to notify the owner. No sound is necessary if the owner can catch the signal by the vibrations in "manner mode" (silent mode). By this logic, the *chaku-mero* function should eventually become obsolete. Instead, the *chaku-mero* function has steadily expanded: devices that could at first only play monophonic sounds can now play three-chord and four-chord melodies.

In August 1999, Astel Tokyo started the first three-chord *chaku-mero* service. Soon competitors entered the field. In September of that year the musical instrument company Yamaha began shipping a ring tone sound chip with up to four chords, including 128 tone types and the capacity for tone combination. The Yamaha chip enabled rich sound similar to that of a PC sound card, a far cry from simple beeping sounds of earlier *keitai*. In June 2000 a chip that could perform up to sixteen chords was released, enabling users to digitally sample and replay live music. This chip was quickly adopted for handsets for various makers, and a sixteen-chord *chaku-mero* service began. In 2001 this was expanded to forty chords, and from December 2002 au rolled out their *chaku-uta* (ring song) service whereby users could use music taken directly from a CD. The latest hit songs can also be downloaded from the Internet. Why has this function made such progress, deviating so far from the features of the conventional telephone?

The concept of "playing one's preferred music outdoors" is similar to audio devices such as headphone stereos and portable stereos. Gary Gumpert (1987) identifies this concept as the creation of an acoustic environment around oneself or the formation of one's "turf" with an "acoustic wall." The music researcher Shuhei Hosokawa (1981) similarly describes these devices as media that "turn urban space into a theater." In other words, *chaku-mero* are media used by *keitai* owners to produce what the music researcher Murray Schafer (1977) calls a soundscape.

With various data compression technologies, operators have experimented with services that distribute music software to *keitai* handsets. In view of a possible fusion of *keitai* and the portable audio devices, *chaku-mero* are being positioned as a characteristic central to *keitai*'s multimedia properties.

Keitai Cameras

According to the *White Paper on Information and Communications in Japan 2003*, as of the end of March 2003 the number of camera phones with active subscriptions in Japan was 22.21 million. This means that 29.3 percent of subscribers had camera phones. (J-Phone/Vodafone has reported that approximately 65 percent of their subscribers use camera phones.) Among those in their teens and twenties, ownership is particularly high, and the percentages drop together with the rise in age of the users. Frequency of *keitai* camera use is also higher among those in their teens and twenties (figure 2.6).

The first *keitai* with a built-in digital camera was the PHS VP-210 (the "Visual Phone"), marketed by Kyocera in July 1999. It was designed as a video phone, capitalizing on the relatively fast transmission speed of the PHS. The press release at the time of the product's introduction featured a scenario of distant grandparents talking to their grandson while viewing his face. However, the terminal weighed 165 grams and was slightly larger than the average handset, attributes not well received by users.

The adoption of the *keitai* digital camera accelerated when J-Phone introduced its first camera handset (figure 2.7) along with a service it called *sha-mail* (photo mail). The Sharp-manufactured handset weighed 74 grams and was the smallest of its kind among the prevailing products. The camera was 110 thousand-pixels, not very high in graphic quality, and it could only capture still images. Developers described the rationale behind the new device: "There is a practice for people to exchange photos, and if there is a camera in your *keitai*, you can make custom wallpaper as well as exchange photos" (Sharp Corporation's Takeo Uematsu, quoted in Fukutomi 2003).

The development of this handset was influenced by the popularity of *Print Club* (*Puri-kura*), a photo booth for making personalized stickers, introduced in July 1995. They were set up in arcades and other entertainment and shopping sites and became a craze among youth. By the end of 1996 over 10,000 units were deployed, and the number increased to 45,000 units by October 1997. The average daily use reached 1,500,000 sheets. In one 1997 survey, 61.3 percent of middle and high school students (81.2 percent of middle school girls and 89.5 percent of high school girls) reported that they had engaged in *puri-kura* collection and exchange (Tokyo Metropolitan Government 1997).

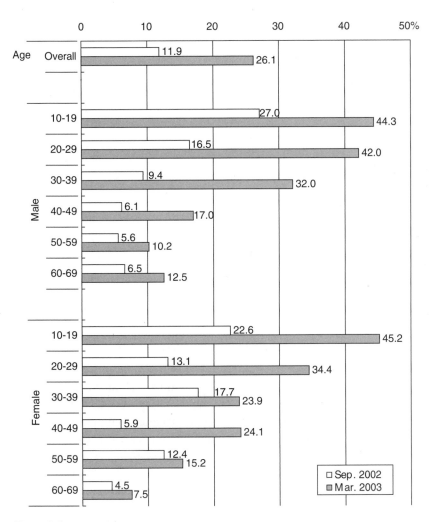

Figure 2.6

Ownership rates of camera-installed *keitai*, 2002 and 2003. From Nomura Research Institute (2003). Reproduced with permission.

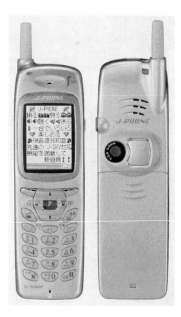

Figure 2.7
Sharp's SH-04 for J-Phone (2000), the first camera cell phone. Reproduced with permission.

The *puri-kura* concept originated in Fuji Film's one-time-use camera *Utsurun-desu* (Quick Snap), which popularized the practice among teenage girls of taking snapshots of friends to keep as mementos. These young women made original photo albums of the snapshots and carried them around as precious lifestyle and friendship records. With the introduction of *puri-kura*, they turned a section of their personal planners into a *puri-kura* album or created an exclusive mini-album for *puri-kura* stickers, which they always carried with them. It also became a very common practice to put a *puri-kura* sticker of a special friend or a boyfriend on their pager or *keitai*.

Nobuyoshi Kurita's 1999 paper is probably the only existing report of a sociological survey of this phenomenon. His research, conducted in 1997, involved collecting *puri-kura* stickers traded by college girls and analyzing them according to the people appearing in the photos, the locations where they were taken and exchanged, and the number of stickers in the collections. According to Kurita, the social function of *puri-kura* photography and exchange can be categorized into four ideal types: (1) everyday confirmation of relationships and fraternity with friends, lovers, family members, and colleagues, (2) confirmation of daily events by memorializing a place visited with a friend, (3) collection of idolized icons in the form of a photo taken with an exceptionally good-looking friend or a celebrity, and (4) collection of rare frames such as special-edition frames keyed to the seasons or tourist sites such as Kyoto or Karuizawa. He sees

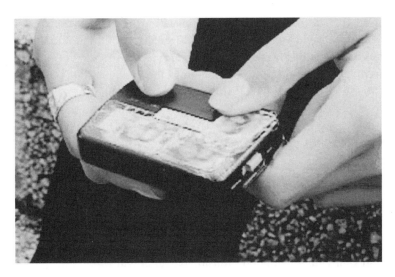

Figure 2.8
Pager with *puri-kura* stickers, September 1996, Harajuku District, Tokyo. Screen capture by author.

mobile media as a key factor that together with age (adolescence) and gender, corre-lates with high volume in *puri-kura* collections. Subjects with pagers, or pagers and *kei-tai,* have exceptionally large numbers of *puri-kura* in their collections, and subjects with only *keitai* have relatively few. At the time of his research in 1997, *keitai* text messaging was rare, and pagers were the primary means for exchanging text messages. In other words, people who were avid users of mobile communications were also avid *puri-kura* collectors. From this, we can deduce a strong affinity between *puri-kura* and mobile media (Kurita 1999).

Through *keitai* adoption, youth tried to express their individuality by putting *puri-kura* stickers on their handsets (figure 2.8) or adorning them with unique accessories and straps. The color display promoted the practice of using favorite illustrations or photographs as *keitai* wallpaper. With the built-in camera, all these functions were incorporated in a single *keitai* terminal. In other words, the *keitai* camera has come to encompass the production of customized wallpaper and some of the functions of *puri-kura* stickers.

Conclusion

I have discussed how the adoption and transformation of mobile media in Japan was influenced by factors such as the uptake by users as well as prevailing communications policy and market conditions. It is in this sense that I see mobile media as shaped by what Fischer has termed social construction.

On the other hand, this case also suggests an application of the concept of domestication, as described by David Morley (1986) and Roger Silverstone and Eric Hirsch (1992), to the process of mobile media reception in modern Japanese society. Domestication involves the integration of media technology into the family and its processes of socialization. As mentioned in chapter 11, this formula helps explicate *keitai* adoption from the viewpoint of the family. However, *keitai* advances have also been driven by young people's distancing themselves from the family order. This reality implies that Japanese *keitai* innovations cannot be explained merely in terms of family and society.

The Japanese *keitai* has taken a unique developmental trajectory by incorporating features and elements widely circulated in youth popular culture. Today such trends are evident not only in Japan but also in other countries. Our next challenge is to analyze adoption and development in these different settings and to develop a new model of media transformation. Further, we must analyze how these changing media forms both grow out of and shape trends in communications. What will third- and fourth-generation mobile systems and *keitai* broadband bring? Will elements of youth culture continue to play a key role in *keitai* adoption and development? We need an even clearer model for the relationship between media, popular cultures, and communication trends.

Note

1. Young women were the core user group for text message exchange with pagers and *keitai*. Setting their sights on the submerged demand among women to "want to do e-mail but not use a PC," NTT DoCoMo developed the Pocket Board based on a concept of "cheap, easy, compact" and released the product in December 1997. This device, about the size of a personal digital assistant (PDA), enabled users to connect their *keitai* device with a cable and send and receive messages. Within two years of its release, over 700,000 Pocket Boards were sold. The user survey research for this product (September 1998) indicated that 80 percent of the users were women, of whom 70 percent were in their twenties. Sixty percent of the users reported that they "had no prior experience with data communication for personal or business use." The Pocket Board, as a media type, could be considered a link between pagers and short messages (NTT DoCoMo 1999).

3 | A Decade in the Development of Mobile Communications in Japan (1993–2002)

Kenji Kohiyama

In the past ten years the Japanese communications industry has seen enormous change. It has transformed itself from a monopoly into a competitive industry and has seen the establishment of a new wireless industry dominated by NTT DoCoMo. This chapter examines these developments, drawing from business literature (NTT DoCoMo 2002; Tachikawa, Kohiyama, and Tokunaga 1995; TCA) as well as my own experiences engineering the Personal Handyphone System (PHS) at NTT during this period. I focus in particular on key turning points in the development of contemporary Japanese mobile communications, historical junctures where social conditions and business decisions shaped the technology. These junctures include the privatization of the telecommunications industry in the mid-1980s; the competition between pager, Personal Handyphone, and cellular phone in the mid-1990s; the establishment of a successful NTT DoCoMo business model in the late 1990s; and current developments in location-based services, *keitai* cameras, iconic communications, and next-generation (3G) infrastructures. I conclude with a discussion of personalization in communication technology as an organizing trend that is likely to affect future developments.

Privatization and the Birth of the Mobile Communications Industry in Japan

Until the mid-1980s, Japanese telecommunications was a monopoly run by a public corporation, Nippon Telegraph and Telephone (NTT). In this section I describe the privatization of the industry and the origins of the wireless communications industry. Of particular interest are the differences between the development of mobile communications and fixed-line communications.

Wireless Communications in the NTT Era

1985 was a turning point in Japanese telecommunications with the privatization of NTT and the entrance of new common carriers into the communications industry. Until then, the Japanese communications market was under the complete monopoly control of NTT. It would not be an exaggeration to say the NTT family of companies virtually controlled the entire telecommunications industry in Japan.

White Paper on Communications in Japan (1986) shows that wireless was a minor player within NTT in this era. Before the introduction of pagers and mobile phones, wireless communications were one element within hybrid wired/wireless long-distance relay infrastructures. Using a microwave relay system, wireless communications constituted approximately 50 percent of the connections for long-distance calling. However, this ratio was not fixed, and the competition in both cost and transmission capacity between these two types of communication was the site of an industrywide struggle. At that time, the wireless spectrum was primarily used for broadcast media. Japanese defense forces, police, local governments, electric companies, and railways all had their own wireless communication facilities. In other words, from the point of view of radio systems, NTT's wireless system was by no means a monopoly. Furthermore, communication over radio wave systems required government authorization, and wireless operators needed access to the spectrum in order to conduct their business. The government's influence in regulating the spectrum was considerably stronger than its influence in regulating a monopolistic public corporation like NTT. In short, faced with various competitors as well as government regulation, NTT was not a major player in the wireless communication market even in its monopolistic days.

The Birth of NTT DoCoMo and New Common Carriers

Regarding the advent of new common carriers, mobile communications also has a different history than fixed-line communications. New companies entered the fixed-line communications market by setting up long-distance relays. Historically, charges were set high for long-distance calls, and charges for local calls were kept low by internal subsidization (see Nagai 1994). This suggests that customers were being charged much more than the actual cost for long-distance calls. This long-distance market thus held the promise of profit for new companies entering the market. New operators rapidly increased their market share, focusing on the dominant communications market connecting the urban hubs of Tokyo, Nagoya, and Osaka. The competition introduced by the new carriers drastically reduced charges for long-distance calls. However, fifteen years after new common carriers were approved, most subscription lines[1] are still owned by NTT (see Tsuyama 2000, 152). In other words, there were no new operators that put serious effort into the more challenging business of building subscription lines.

In 1992, when new common carriers were accepted into the mobile communications industry, mobile communications consisted primarily of car phones. Although there were large hopes for the future, at the time there were only about 1 million subscribers.[2] Nobody anticipated the adoption rates that we see today. NTT DoCoMo was established as a spinoff of monopolist NTT; DoCoMo standing for "Do Communications on the Mobile," the subsidiary was dedicated to wireless communications, including car phones and the nascent pager market. NTT DoCoMo did have an advantage over new operators because of assets brought over from its parent, NTT. This advantage was not as large, however, as those advantages associated with fixed-line communications. In fact, NTT DoCoMo saw its market share drop sharply at one point after new operators started entering the market.

The new common carriers in mobile communications had a different strategy for entering the wireless market than they had for the fixed-line market. Their services involved subscriptions to networks, with service areas encompassing the entire country. The main objective of a mobile communications business is to build a base of subscribers. Mobile subscribers tended to desire access to the whole country, in contrast to fixed-line use focused on urban centers. For cellular services, this meant providers needed to create an independent network, and furthermore, this network had to be nationwide. It is not possible to focus only on profitable urban regions because the size of the service area is an important draw for potential customers. Given these conditions, the mobile communications market presented a more level playing field than the one encountered by the new common carriers in the fixed-line market.

Keitai Handset and Component Manufacturers

How did privatization and competition effect handset manufacturing? Before NTT's privatization, the specifications of NTT handsets were extremely rigid. They even included NTT-exclusive specifications that were incompatible with other standards anywhere else in the world. While it is true that such rigid standards helped maintain reliable service, they also raised cost and functioned as trade barriers for foreign manufacturers. Competition in the *keitai* handset industry is unlike that in fixed-line communications. The lifespan of the *keitai* handset is short, and while it takes advanced technology to develop a product, price wars push prices down. Because of such conditions, handset manufacturers collect low-price, high-quality, and high-performance components from around the world, producing only key parts internally. In the overall landscape, major device manufacturers hold onto stable global market share, while assembly manufacturers remain locked in fierce competition. This situation is not specific to Japan; low-price, high-performance components can now be purchased from anywhere in the world. Costs are kept low through larger-scale production.

The Shift from Pagers to *Keitai* in the Mid-1990s

Competition between Pager, *Keitai*, and PHS

The mid-1990s was a pivotal period in Japanese mobile communications. When *keitai* was introduced in 1990, business use dominated. The mobile phone was a luxury item, an executive tool supplied by the corporation to a select few. Subscribers in 1992, the year NTT DoCoMo was established, numbered approximately one million (TCA). By this time, the pager was already a widely used communications device. At first, the pager's only function was to signal the receipt of a transmission with a ringing bell. Like the *keitai*, it was initially used almost exclusively for business purposes. The introduction of the display pager in 1990 changed this. This type of pager was designed to enable the caller's number to be displayed. These pagers were quickly adopted as an essential means of communication by high school students, who relayed messages in their own pager code (see chapter 2). This youth-driven creation of a new communications culture was a rare occurrence, even when viewed internationally, and has left a lasting legacy in the history of communications. The number of pager subscribers in 1992 was approximately seven million (TCA). Subsequently, the Personal Handyphone System (PHS),[3] extended the range of a cordless phone to be used outdoors.

Competition between these three product lines was intense from 1994 to 1996. At the time, pager developers were considering new functionality for pagers, including the ability to display Roman alphabet characters in addition to numbers, and to send as well as receive messages. The spread of pagers seemed unstoppable. They were becoming such a force among teens that some high schools prohibited students from bringing them on campus.

In the meantime, *keitai* providers were working to become more competitive by reducing costs. Initially, users leased *keitai* handsets from NTT, requiring a deposit as well as an activation fee. The deposit fee was gradually eliminated, as was the activation fee in 1996. Probably most significant was the shift from a lease to a purchase model, which began in April 1994 and became standard in 1996. Combined with the reduction in subscription costs, this last change was particularly effective in boosting adoption. Telecommunications Carriers Association statistics show that for the month of June 1994, two months after these changes, growth of new *keitai* subscribers surpassed the growth of new subscribers to fixed-line telephone services.

The path was clearing for *keitai*'s ascendancy. The PHS was its only obstacle. In fact, the *keitai* industry plotted its business strategy based on studies of the PHS market. The intense advance publicity for PHS pushed *keitai* operators to take price reduction measures. The number of PHS subscribers passed the 1 million mark within a year after the start of service in October 1994. During the intensifying competition between PHS and *keitai*, the pager steadily lost ground. Many of the youth who had driven pager use switched over to PHS. PHS was an instant hit among high school girls in particular and was given the nickname *picchi*. Despite its small service area, low cost and long battery life made PHS a huge success. Its adoption curve within the first year of its introduction was significantly steeper than the *keitai*'s. In 1995 PHS and *keitai* held practically identical market shares. However, after this point, PHS suddenly lost momentum. In 1996 *keitai* emerged as the mobile communications industry leader, and it has since maintained its lead.

The Context of PHS and the Reasons for Its Defeat

The forerunner to all these personal communications systems was London's Telepoint system, established in 1989. Telepoint is a system for setting up small antennae on the streets to connect to "cordless public telephones" for sending calls only. This project inspired Japan to first begin considering personal wireless communications systems. The cordless telephone for the home was popular, with annual sales of three million units. Since they were analog, however, they were easily tapped into by outsiders, and this was considered a social problem. This problem hastened development of the digital cordless phone.

PHS has several unique features, clarified by Kohiyama and Satoshi Kurihara (2000): (1) It is a digital cordless phone that makes it possible to make a call from the house, office, and outdoors from one phone; (2) connection costs are from one-half to one-third those of standard *keitai*; (3) the handset is small (100 cubic centimeters) and light (100 grams), with a long battery life (eight hours per charge); (4) audio quality is high enough to send music; (5) communication can be maintained up to walking speed; (6) it can receive incoming calls and make outgoing calls.

PHS was successful at the time of its launch but soon lost out to *keitai*. The limited service area was also a persistent problem. PHS had weak connections and could not be used inside most homes. Further, handsets had been virtually given away in order to gain new subscribers, and these price wars had reduced the cost of PHS handsets to such an extent that they were perceived as cheap goods. The effects of these price wars were more detrimental to PHS than to *keitai*. PHS relied on the NTT fixed-line network, and operators had to pay NTT for network charges. It is difficult to say whether these network fees were set too high, but the reality is that it was not possible to reduce these fees to offset handset price cutting. And *keitai* would soon approach the size, weight, and battery life of PHS. Ironically enough, PHS was a victim of its own success as the *keitai* came to emulate its appeal. Youth made a transition from pager to *keitai* via PHS (Okada and Matsuda 2000, 23–41).

The principle difference between *keitai* and PHS was the cost of communication. Since PHS is essentially a cordless phone, using the extension did not incur charges. The system was based on the three-million unit-per-year cordless phone market. However, the unexpected success of PHS immediately after its launch created an operator-driven environment. As a result, the focus turned to competing against *keitai* rather than capitalizing on the unique features of PHS. As mentioned earlier, terminals were given away at no charge, and consequently cordless phone manufacturers turned their back on the PHS business. Even now, the market share of cordless phones combined with a PHS handset is extremely low. This is in part because the small size of a PHS terminal is considered inconvenient for home use. PHS is able to support a wireless private branch exchange (PBX) telephone system,[4] where users can create local phone systems that do not incur telephone charges. The development of this PBX capability

Table 3.1
International PHS Adoption, 2000–2003 (thousands of units)

Country	2000	2001	2002	2003
China	1,200	5,010	11,820	30,660
Japan	5,880	5,700	5,570	5,200
Taiwan	0	160	500	570
Thailand	390	630	600	540
Other	0	0	10	90
Total	7470	11500	18500	37060

Source: PHS MoU group. Reproduced with permission.

into a business possibility is only recently being realized. There are various factors behind the defeat of PHS, but the major reason is the diversion from its fundamental concept and strengths.

A comparison with the situation overseas should provide a useful reference point. The Digitally Enhanced Cordless Telecommunications (DECT) system in Europe is similar to PHS. Although DECT and PHS rely on similar systems, DECT is applied to cordless phones and wireless PBX, and public use is restricted to networks that specifically tie together a home and a network in a Wireless Local Loop (WLL).[5] In the United States, Personal Communication Services (PCS) have a similar name to PHS but are actually based on a digital cellular system. In other words, while these systems may appear similar to PHS, there are actually no other systems like PHS that create a mobile phone network based on cordless phones integrated with a land line network. There are some PHS systems in use outside of Japan (table 3.1). The estimated adoption rates of PHS around the world by the end of 2003 stand at 37 million units. In contrast to China, where adoption rates have been high, Japan, Thailand, and Taiwan have passed the peak of adoption. Other countries planning to adopt PHS include India, Indonesia, Vietnam, the Philippines, the United Arab Emirates, and Ethiopia.

China, with its rapid adoption rates, provides an interesting case because of some key features: (1) In China, PHS could be considered a form of fixed-line communication because the PHS providers and backbone network providers are the same. By contrast, in Japan, PHS is considered part of the wireless telephony industry and is operated by a different business than the backbone network provider (NTT). (2) In China, tolls for mobile phone calls are split between the caller and the receiver. In Japan, for PHS, only the sender is billed. The cheaper handset cost (on average, over 1,500 RMB for a mobile phone for average monthly salaries of 800 RMB in contrast to on average 750 RMB for a PHS) combined with this billing mechanism has been a draw for Chinese subscribers. (3) Since China is so much larger than Japan, there are fewer consumer expectations for national coverage. (4) China's current economic development prioritizes short-term cash flow and favors the low capital outlays required for establishing PHS systems.

China's deployment of PHS favors public access and is not necessarily following the same developmental trajectory as in Japan. Among the differences just cited, the fact that PHS is provided by backbone network providers is probably the most fundamental difference in contrast to the situation in Japan.

NTT DoCoMo's Success

The Japanese *keitai* market is shared by three companies: NTT DoCoMo, KDDI's au, and Vodafone. Their shares are roughly 60 percent, 20 percent, and 20 percent, respectively. The difference between the financial standing of NTT DoCoMo and the other two companies is even greater than the difference in their market shares because of how the companies deal with sales and distribution of *keitai* terminals in Japan. J-Phone and au work through independent retailers and pay incentives on top of the retail price for the sales of the terminals. This system benefited the retailers selling PHS terminals, as they could make money off incentives even with free handsets. The business strategy that prioritized gaining market share over profit ultimately weakened the two companies' financial condition. In contrast, NTT DoCoMo established its own DoCoMo Shops, a retailing operation managed jointly with trading houses and other companies that exclusively sold DoCoMo terminals. Although I do not have access to the exact numbers, it appears that the incentives for these stores are significantly lower than those of their competitors. The DoCoMo brand image has also been enhanced because the handsets are not discounted. This situation is clearly reflected in the bottom line. In the midterm accounting of 2002, all three companies turned a profit, but the profits of au and J-Phone were less than one-tenth of DoCoMo's.[6]

Even DoCoMo, however, has faced business difficulties. In 1996 its share of the *keitai* market dropped below 50 percent. Although this is not widely known, streamlining operations helped salvage the situation. During this period, DoCoMo made a number of changes to its business operations. One was the elimination of the leasing system in favor of outright sale of its handsets. The other was the development of operations software that enabled the company to quickly process new subscribers, a major innovation at the time. As a result, DoCoMo could deliver better customer service, quick turn-around between application and start of service, efficient retail operations, better control of the complex distribution process, real-time management, and synchronization of the flow of goods and money.

New mobile technology networks have set the stage for a restructuring of the mobile industry in Japan. In introducing third-generation mobile networks, au utilized cdma2000, gradually building high-speed capability by relying on existing facilities. This strategy was a success, and the company has been rapidly increasing market share. NTT DoCoMo has also started third-generation mobile services, called FOMA, but these have experienced slower growth than au's. Although it is too early to judge, it appears that the successes and failures of third-generation services will have a large impact on the Japanese mobile communications landscape to come.

NTT DoCoMo's phenomenal success with the i-mode service is a notable example among recent Japanese business successes.[7] i-mode not only made Internet access an established element of *keitai* ownership but also established a system for paying for Internet contents. The *keitai* Internet network is built on a telecommunications industry model. The Internet is merely connected through gateways. In other words, from the Internet point of view, on the mobile communications network each mobile service company is just one narrow gateway to another network. For Internet people, the open architecture of the Internet is its central appeal. However, for at least some *keitai* users, the more limited and easier-to-use network has proved to be a selling point. This is an example of an Internet constructed by telecommunications people.

What is the secret to i-mode's success? One is that DoCoMo already had a national network for packet communications. Before i-mode, this national packet network was a corporate liability, and management was looking for ways to capitalize on it. A new division board was organized, which included people brought in from the outside to develop new competitive services such as e-mail, which fell outside the bounds of the telephony market. Their primary mission was to find ways to effectively utilize the packet network.

Another innovation of i-mode was its micropayment system. i-mode is probably one of the most successful examples of a business built on subscription-based content. This success is based on the micropayment scheme, which is derived from a telecommunications model. Telephone calls have always been charged in 10-yen units and can just as easily be charged in packet units. DoCoMo built on this existing structure and took the position of agent for the content providers. Users pay for content together with their monthly i-mode fees, the service provider passes these payments on to the content provider, and the content provider in turn pays a commission to the service provider. The ¥300 cap on monthly subscription fees was also a clever tactic. The amount is small enough (the equivalent of approximately US$3) that users show little hesitation in signing up for the content with a few taps on their *keitai* keypad. This micropayment system for content subscriptions provided a radical new model for Internet business that had previously tried using advertising as the primary source of revenue for content services.

The primary use of the *keitai* Internet is for transmitting e-mail; Web access is a growing but still relatively small share of *keitai* Internet traffic. Prior to the advent of i-mode, the transmission of messages between *keitai* was restricted to short text messages between handsets from the same service provider. The advent of the *keitai* Internet enabled users to send and receive e-mail with other *keitai* and PC users regardless of service provider and regardless of terminal device. The success of i-mode was thus also built on the existing practices of mobile messaging that were well established in the youth market at the time of its launch.

Current Developments

Location-Based Services

Unlike the fixed telephone, *keitai* is not limited to one specific location; it is a medium active in specific and wide-ranging locations. Because of this, location-based services hold much promise. These services have been designed to pinpoint the location of *keitai*; until recently, positioning was based on the location of cell stations. Currently, there are *keitai* terminals with Global Positioning Services (GPS) built in, offering greatly improved accuracy. There are a variety of positioning services that can be categorized by function: basic locational services, communication services, navigation services, and location-sensitive information services.

Some services notify the user of their own location; other services notify people of a *keitai* user's location. Some examples of the latter type of service are companies seeking to locate mobile workers such as sales reps or service technicians, or the elderly carrying a handset in the event that they get lost or disabled. Because of the problem of privacy, these services have been used only in limited instances.

There has been persistent concern over privacy with location-based services. Positioning services were first implemented with PHS. Since PHS is operated through many small cell stations, when someone knows the station being used for a given transmission, they can get relatively accurate positioning of the handset. The public reacted negatively to services that capitalized on this technical capability. The media wrote extensively about the privacy problems such services would involve. More recently, car navigation systems played a significant role in opening up a market for positioning services, and the privacy concerns have become somewhat muted while still remaining unresolved. The more surveillance-oriented systems are still unpopular. In general, the tendency has been for people to see privacy issues as an individual responsibility rather than something that requires public standards.

In reality, the communication type of service can entail a form of surveillance. Some dating sites locate nearby facilities and facilitate communication based on proximity of users. These services get around privacy concerns by informing users of their relative position vis-à-vis other users rather than disclosing absolute position.

Navigation systems build on basic positional services by providing transit information concerning roads and traffic lights, information about speed and direction, and activity indexes that predict time to destination. These kinds of high-performance car navigation systems are also used by corporate operations that deal with mobile workers.

The final type of service, location-sensitive information services, help users access local information based on positional data. The most popular services of this type display stores and restaurants in a particular area. Other services provide information on local history and culture and environmental information about weather, animals, plants, and landscape.

Third-Generation Mobile Communications and Wi-Fi

Third-generation (3G)[8] mobile communication devices were introduced with great fanfare in 2002 but are facing difficulties. au entered the market with a 2.5G approach that accelerated existing 2G networks, and it has stayed ahead of its competitors. In the personal computer market the expectation is that old technologies are upwardly compatible with new, more advanced technologies. By contrast, with 3G, new handsets fell behind the second generation in terms of certain key features such as service area, battery life, and cost. If there had been compelling applications and content, consumers might have upgraded to 3G. At the moment, however, interest is low, in large part because the video phone has not captured consumer interest. Au's success could be related in part to the fact that its new system maintained backward compatibility with its 2.5G system. When features and costs of 3G devices are up to par with those of 2G, this dynamic is likely to shift.

In a somewhat different domain, wireless local area networks based on IEEE 802.11 specifications (wi-fi) have recently been seeing dramatic growth in Japan. Many new wi-fi subscription services have been launched. One example is wi-fi hot-spot services in transit stations, hotels, fast food chains, and cafés. The success of wi-fi will have an important effect on the adoption of 3G and 4G mobile communications. While a number of companies are entering the wi-fi market, none have yet seen business success, and one, Mobile Internet Service in Sangenjaya, has already closed down. That company apparently lost its user base when it began to charge connection fees. It is unclear whether Japan will adopt the grassroots free wi-fi approach that has begun to take hold in some U.S. urban areas, such as the San Francisco Bay Area. Such a grassroots effort in Japan may also integrate 3G and 4G mobile communications.

Personalization in Communication

The central trends in mobile communication of the last decade can be summarized by the term *personalization* (Tachikawa, Kohiyama, and Tokunaga 1995; Kohiyama 1996). Okada, in chapter 2, has described some of the historical processes of personalization in youth communications media. In conclusion, I describe a conceptual model for understanding personalization in communications, relating these historical processes to future trends in mobile communications in terms of the personalization of devices, connectivity, and services.

Defining the Who, What, When, Where, and How of Personal Communication

A blank canvas for communication "whenever, wherever, with whomever, and with whatever" has been held as the ideal. At the same time, it is also critical to enable individuals to color this canvas to their own taste. A communications system must have a basic platform that is maximally open and unrestricted, combined with the ability of individuals to restrict and specify. Personalization involves enabling access to the broadest possible range of who, what, where, when, and how (4W1H) as well as enabling people to specify and restrict access based on their individual needs.

In terms of "who," the optimal system enables access to as diverse and far-flung a set of potential partners as possible while enabling users to target specific people within that network. In terms of "what," ideally any device should be able to access any type of information. At the same time, the system should allow users to categorize, edit, and filter information for individual needs. The "when" of communications means a network that is accessible at all times and can weather highs and lows in volume. In terms of personalization and restriction, we might consider services that deliver information at a particular time of day or that block certain types of calls at certain times. In terms of "where," the ideal basic platform is universal service in all locations, something that is achieving finer granularity with the advent of mobile phones. Personalization could mean the ability to access office communications infrastructures from home, or a service that restricts or enables particular communications at particular locations. "How" involves enabling a broader range of quality, volume, and type in communication through higher bandwidth connections and compression technologies. Restricting and specification here means enabling users to select quality of image and media type. Personalization also represents the ability of users to communicate across cultural, national, and language boundaries. Translation systems are one example of movement in this direction.

Personalization of Devices

The personalization of communication devices involves the development of personal consumer electronics such as the personal computer, or other personal electronics such as radio, TV, and cassette players. We can also see a trend in Japan towards *jo-ho kaden* (smart appliances) (see Yano Research Institute 2002), and we can expect that these informational devices will become increasingly common in the Japanese household. For communication devices, miniaturization has been central to this trend toward personalization, and we will likely see *keitai* functionality built into objects such as credit cards and wristwatches. Low pricing is also a key factor because the amount of money individuals are willing to spend on electrical appliances has been in steady decline.

The personalization of mobile communications in Japan began with the pager. The pager was small in size to begin with and had much potential for personalized communication. *Keitai* was originally developed as a car phone, and thus miniaturization of the device was not the most critical issue. However, since mobile phones became the prime market, the focus has been on making handsets smaller and lighter and extending battery life. The goal in size was 100 cubic centimeters, the size of a pack of cigarettes. Motorola's MicroTAC, released in 1989, was a breakthrough in achieving a 200 cc handset, a major contrast to the 600 cc handsets being used in Japan at the time. Two years later, after NTT lifted its control on handset specifications, Japan developed a handset lighter and smaller than the MicroTAC.

The first 100 cc mobile phone was a PHS, introduced in 1995. Immediately after the launch of the service, a 100 cc, 100 g PHS terminal was developed as one of the measures to differentiate itself from *keitai*. It received an unexpected review—people thought it was too light and cheap-looking. The current *keitai* and PHS are in fact bigger and heavier. It took some experimentation for manufacturers to arrive at the ideal size and weight, and this has remained relatively constant in recent years.

It was not only *keitai* that were being miniaturized. Computers, personal digital assistants (PDAs), and digital cameras were also being reduced in size and weight. Currently, PDAs and computers cater to different markets. Digital cameras went in two directions, either becoming extremely small and being integrated into other devices like *keitai*, or working toward achieving higher image quality at the expense of size. It is interesting that PDA and digital cameras that followed the path to miniaturization and simplification are now being integrated into *keitai*. In other words, *keitai* defined the future of both PDA and the digital camera by integrating their respective functionalities with communications functions.

Personalization of Connectivity

Telephones have not traditionally discerned the caller and the exact recipient. This is changing because of the personal nature of mobile phones and caller ID functions. One could also imagine other technologies that enable users to personalize the "who" of communication, such as password protection applied to mobile devices to protect against access by others, or services to screen certain calls or e-mails.

Keitai service providers aim to gain better access to the user's wallet, and systems for integrating *keitai* with credit cards and electronic currencies are well under way. Personal identification and other security-related functions become key elements in this effort. While there were efforts even with the launch of PHS to link devices to personal identifiers, it was unclear then what purpose this would serve. Now, with the connection of *keitai* to commerce, the motivation has become clear. With personal computing, personal identification is traditionally handled with passwords. With telephony, this was achieved through voice. Formal authentication through voice was considered an inhibitor to communication, and there are still no formal authenticating procedures in everyday voice telephony.

Concerns are arising as to whether a password is sufficient to authenticate *keitai* access once it becomes a device for financial transactions. In experiments currently under way, there is no need to input the number each time the *keitai* handset is used to make a purchase. It is treated like a credit card or a prepaid card in that there is no mechanism to identify the user of the device. If there were a method of simple personal identification that could be applied to even spontaneous calls, we could imagine a variety of applications. There is technology now to incorporate a fingerprint ID mechanism in *keitai* to identify the user. Implementation of these new technologies is contingent on a compelling application.

Personalization of Service

What does it mean to have your own communications service? Users would be able to select and arrange by themselves the various types of services and charges available on the network, as they would shop at a supermarket. It suggests the free manipulation of the personalization factors of 4W1H.

Personalization of service has been a major issue in the telephone era. Switchboards were focused on safe and stable delivery of information, and they could not offer a variety of services in a timely manner. The same can be said for telephone charges. It was only recently that the charges were diversified, brought about in part by changes in policies of the Ministry of Public Management, Home Affairs, Posts and Telecommunications (the former Ministry of Posts and Telecommunications). The personalization of services is at last becoming a reality in the diversification of service plans available in the *keitai* market. As *keitai* takes on expanded functionality, the ability to set difference service preferences will be absolutely critical.

Now there are many new communications services, and the user's freedom of choice is expanding. However, at this stage, selection can only be made from ready-made services provided by the communications operators. By contrast, in the field of personal computers, new applications are constantly being introduced, and users can combine and use them as they please. The range of services and applications currently available in mobile communications do not even come close to those available in personal computing. Unlike computers, high stability is a prerequisite for communication. As multimedia becomes standard in mobile communications, we may see the emergence of an alternative, more flexible mobile network that is separate from the basic communications lifeline. If this happens, we may acquire the ability to construct our own personalized operating environments with our mobile communications, just as we do with our personal computers.

Notes

1. Subscription lines are the lines that connect consumers to a telephone switching center, in contrast to the lines that connect the centers.

2. See NTT DoCoMo (2002) for details about the number of subscribers.

3. The PHS system relied on the existing long-distance and fixed-line infrastructure in connection with wireless access handsets. This was initially a cheaper way of creating a mobile phone network, in contrast to cellular networks, which had to be created from scratch.

4. Private branch exchange telephone systems are used to support organization-specific telephone systems such as those used internally by universities, where users can dial by internal extension number.

5. Refer to DECT Forum ⟨http://www.dect.ch/⟩.

6. These ratios are derived from corporate statistics published on the Web sites of the respective companies: ⟨http://www.nttDoCoMo.co.jp/corporate/ir/⟩, ⟨http:www.kddi.corporate/ir/index.html⟩, ⟨http://www.telecom-holdings.co.jp/ir/report/index.html⟩. Or see *Johou Tsushin Handbook 2004* (*Information and Telecommunications Handbook 2004*).

7. See Matsunaga (2000) for an account of the earliest stages of i-mode development.

8. Second-generation systems rely on digital GSM networks. 3G relies on IMT 2000 and delivers higher bandwidth digital connections capable of supporting video transmission. See Senju Kobayashi (2001).

II | Cultures and Imaginaries

4 | The Third-Stage Paradigm: Territory Machines from the Girls' Pager Revolution to Mobile Aesthetics

Kenichi Fujimoto

Imagine Hegel, Marx, and McLuhan encountering the *keitai* of the twenty-first century. Georg Hegel is astonished at seeing the spirit of the era dwelling persistently in our palms. Karl Marx complains that it is an alienating fetish object. Marshall McLuhan, his eyes sparkling, chimes in that it will turn the whole world into a village—no, a house. But in the next moment, he comes upon a realization that appalls him. "But wait!," he exclaims. "My wife and children will have the equivalent of a private room with a twenty-four-hour doorway to the outside world, fully equipped with a TV, a bed, and even a bathroom. Where would my place be in such a house?"

A Paradigm as an Aggregate of Dissimilar Cultures

When theorizing *keitai*, the sociology of knowledge represented by Karl Mannheim's (1929) theory of "existential connectedness" by class[1] is a key entry point. One might be skeptical of the utility of classical framings of ideology and class in addressing the high-tech realities of *keitai* knowledge. I would argue, however, that Mannheim is relevant to the topics at hand, particularly if we also bring in Thomas Kuhn's (1962) concept of paradigm in understanding knowledge societies.[2]

We might start from the idea that a given population (previously known as a class) holds "sharedness in common" (previously known as connectedness), a certain form of science and technology (previously known as knowledge, ideology, consciousness), in other words, a paradigm. Given this framing, what kinds of paradigms are brought by what kinds of social groups at moments of *keitai* use? Does this group's unique "perspective structure" (Mannheim 1929) deserve to be called mobile media literacy, or more materialistically, *keitai* ideology? Before rushing to such conclusions, I would first like to forge links with a concept of culture, reframing the idea of paradigm to align with approaches from sociology of knowledge and culture.

Keitai use in Japan holds much in common with its use in other countries, but there are several crucial differences. These could be glossed as differences in culture.[3]

By being limited to perspectives from the humanities and social sciences, however, the term *culture* has overemphasized historical invariance and ethnic idiosyncrasies, a bias embedded in terms such as "traditional culture" and "folk culture." Kuhn's (1962) celebrated "scientific revolution" theory proposed an idea that fundamentally corrected this bias. His paradigm theory introduced the following theoretical innovations:

• New, extraordinary ideas and discoveries made by pioneers are first taken up by a group of scientific specialists and then adopted by the general public. Only then are they accepted and diffused as new scientific theories.
• Therefore, the transition from old to new scientific theories actually occurs as a revolution driven by the power of the supporting population rather than by definitive proof or disproof.
• During events of this sort, the ways of thinking (behavior patterns) shared among a particular group is called a paradigm.
• A paradigm includes not only pure scientific theories but the values, ideologies, images, biases, and all dimensions of the ways of thinking that go along with these theories.
• In that sense, there is no intrinsic distinction between literary, political, and societal ideas, nor between the transition or diffusion of scientific theories and trends in literature, politics, and society, nor between scientific theories about nature and discourses of literature, politics, and society.

An addition can be made to these five points based on a philosophical foundation that is not explicit in Kuhn's writings: the idea of "asymmetry or asynchronicity in duties of the speaker and rights of the listener in language games," which comes from the later writings of Ludwig Wittgenstein (1953) and from Hans-Georg Gadamer (1960). In other words, the speaker becomes a speaker only by taking on the obligation to speak. By contrast, the listener can take on the existential privilege of listener (receiver, chooser, and reactor) while retaining the prior optional right to listen to or ignore the speech.[4]

What is being described as speech in this framework can be applied to all manner of forms of expression and objects, such as theory, art, media, commodities, and artificial environments. Kuhn's work was innovative in challenging the myth of discovery and invention by the lone genius speaker's subjectivity, the common belief of scientific groups as well as the general public at the time. Kuhn, by contrast revealed the importance of the anonymous supporter listener's subjectivity, in listening, receiving, choosing, and reacting.

I will not go into depth here on the subject-object dialectic reversal of presentation and the dominance of the parasite, about which I have written more extensively elsewhere.[5] Based on Hegel's (1808) "the master-slave dialectic reversal" and Michel

Serres's (1980) "the host-guest dialectic reversal,"[6] I would like to propose the following sixth point, "the consumer revolution in scientific theory."

- A scientific theory (or an utterance, speech, or discourse) first becomes a scientific theory when it is accepted, chosen, used, and disseminated by a social group (language community). Because of this, a scientist (speaker) discharges a unilateral obligation to the listener by presenting a scientific theory as a gift. And then, depending on whether it is chosen or ignored, he becomes parasitic on a supporting group, which is not only guest to his presentation but also master of the recursive language games again and again, through reception, acceptance, and choice.

Incidentally, this concept of paradigm (which includes the sixth point, "consumer revolution in scientific theory") resembles Kuhn's later theorizing, particularly in relation to his "disciplinary matrix" (1977), which drew from the holism of "Duhem's Thesis" (Feyerabend 1975) and Paul Feyerabend's "anything goes" philosophy. This approach does not, however, imply the intellectual defeatism of the "indeterminacy of translation thesis" put forth by another holist, Willard van Quine (1981) but is significant because of its orientation to active historical, ethnological, and contemporary fieldwork and description. On this philosophical foundation, following Kuhn's descriptive and concrete scientific and historical explanations, I would like to portray a contemporary ethnography of *keitai*.

By scaling down from a holistic notion of culture (which is both important and misleading), under a holistic theory of paradigm, we can salvage the analytic utility of a culture concept that often suffers from being framed too broadly. In other words, all the categories of cultures, for example, "alien culture," "counterculture," "hegemonic culture," "dominant culture," "subculture," and "personal culture" (including "up 'n' down," "groove 'n' edge," and various "modes"), I consider unreferrable—explicanda, in terms of logic—in cause-effect explanations.

Based on this framing, I describe certain paradigms associated with wireless communications, primarily *keitai*. These paradigms are tied to conflict and tension between the social groups of street-savvy teenage girls, or *kogyaru*, and older men/fathers, or *oyaji*. Through the case of what I call the girls' pager revolution,[7] I begin with a period of coexistence of the two groups, then focus on the critical decade of the 1990s, which was characterized by a fierce struggle for hegemony and consequently resulted in a scientific revolution for mobile media literacy. I then turn to the twenty-first century, where the girls' paradigm dominates, setting a unified de facto standard.

My secondary objective is to test out a model of how this paradigm shift in East Asia articulates on an international stage. I do not see this recent scientific revolution as exclusive to contemporary Japanese society but rather as embedded in a broader global and historical context, particularly paradigm shifts in the past century: the shift from military (soldier) to business (businessmen) to socializing (youth) and the merging and transformation of various "alien cultures" that accompanied these shifts. For example, kites start as a military reconnaissance weapon and then become a business communication tool, and finally, a children's toy. Similarly, cars, airplanes, and telecommunications each have three-stage paradigm histories.

This three-stage paradigm theory is a hypothetical model for mobilizing historical facts for theory building, and is one thread for understanding the backdrop to the girls' pager revolution (Fujimoto 2003a).

Nagara Mobilism and Japanese Tradition

Currently, a *keitai* paradigm is beginning to blanket the world. I have identified the girls' pager revolution as a key transformation in a pivotal critical period (Kuhn 1962) before the formation of a new paradigm. I define the most distinctive feature of the third-stage paradigm by my concept of *"nagara* mobilism," which differs in important ways from the established concept of "mobility." As I have discussed in more detail elsewhere (Fujimoto 1999a), mobility has tended to refer to functional dimensions of portability and freedom from social and geographic constraint. By contrast, mobilism emphasizes broader cultural and social dimensions such as malleability, fluctuation, and mobilization, particularly in the context of the Japanese aesthetics of everyday life. Japanese pager users, with their *nagara* mobilism, prefigured the current global *keitai* paradigm and its thought and action patterns. *Nagara* (while-doing-something-else) refers to the state of multitasking separately, in parallel, and asynchronously while walking, moving, or playing.

The features of the paradigm tying together the girls' pager revolution and the widespread *keitai* phenomenon can be summarized as follows: (1) ceaseless mobilism that takes pedestrianism as a base transitioning between walking, bicycling, taking trains, and riding in cars driven by others; (2) a *nagara* philosophy and practice involving dispersed parallel processing and multitasking of multimedia information (text, sound, and graphics) while walking; (3) economic mobilization, where scarce personal resources are invested almost exclusively towards the costs of owning and using personal media, in addition to mobilizing of surplus resources of society by, for example, becoming parasites to parents and boyfriends. In addition to these three key features, I add (4) this is a global and local paradigm with no Euro-American precedent.

In Japan this paradigm of *nagara* mobilism was incubated among a population of high school girls in their middle and late teens, called *kogyaru* (small gals), girls just younger than the college students and office ladies in their twenties called *gyaru* (gals). In the early 1990s this paradigm was viewed by adults as an alien culture (a strangely inferior culture, an oppositional culture, an anticulture, a nonculture) and was vigorously opposed by older men/fathers, the *oyaji*, who were socially hegemonic in Japan. I have described elsewhere (Tomita et al. 1997) the points of conflict between *kogyaru* and *oyaji*, and the specific processes of conflict and resistance that played out in homes, schools, workplaces, trains, and on the street. The basic comparisons are summarized in table 4.1.

Emerging from these conflicts, the current *keitai* paradigm is heir to *kogyaru* culture. In the wake of the triumph of the girls' pager revolution, *nagara* mobilism has attained hegemony on the streets, crossing gender lines and even extending to users in their thirties and forties. Although media use is no longer distinctively gendered, particular fashions, energies, and options attached to these practices still retain a distinctive cultural cast, attesting to the driving role that *kogyaru* have had in defining the *keitai* paradigm. Let me describe the streets of urban Japan at this moment in 2004.

What do foreign tourists see when they first visit Japan and leave the airport, approaching the heart of the city? Whether they are visiting a large city like Tokyo or Osaka, or a local city with a smaller population, if they come across high school students after school, between 3:00 p.m. and 4:30 p.m. before rush hour traffic, they would almost certainly notice their distinctive fashions and practices.

Considering the country's size, the climate by season and region in Japan varies greatly, but for some reason, in all seasons and regions, school uniforms for both male and female students are very similar throughout the country. Students often choose schools depending on their preference in uniforms. Since they wear uniforms by choice and are able to adapt them to their own tastes regardless of school dress code, their fashion is half-compulsory and half-voluntary (the origin of the ancient *kanji* for *clothes* also indicates this double meaning).

Around town, couples and mixed-sex groups are rare (perhaps reflecting the tradition of *wakashu yado* [young men's houses] and *musume yado* [young women's houses] in folk society). Japanese tend to hang out with people of the same sex (Fujimoto 1997 2000).

The male students, mostly wearing black, navy, or gray long pants, walk in groups along the streets, cutting a monochromatic silhouette against the late afternoon sky. Some dress austerely in traditional school uniforms buttoned up to the chin high-collared (*tsume-eri*), like the army garb of the writer Yukio Mishima, who committed suicide by hara-kiri.

In contrast, the basic three garments of female students are off-white long vests (reminiscent of the animated movie *Princess Mononoke*), supershort skirts (like the Russian pop duo t.A.T.u.), and navy or white knee-high loose socks (no stockings even in sub-zero weather).

The difference between the male students' austere fashion and the female students' skin-exposing, free-spirited fashion is a typical gender gap observed not only in Japan but in other countries as well. In contrast, their information behavior appears to be truly unisex. In fact, over the past ten years, the "Mishima" boys have been following

Table 4.1
From Second to Third Paradigm (Focused on Girls' Pager Revolution)

	~1980s	1990–2000	2000–
Stages of scientific revolution (Kuhn)	Normal (stable) period	Critical (paradigm-shifting) period	Normal (stable) period
Three-stage paradigm theory (Fujimoto)	Stage 2 (business) final period	Stage 3 (socializing) germination period	Stage 3 (socializing) equilibrium and diffusion period
Paradigm bearer	Businessmen (*oyaji*)	Young women (*kogyaru*)	General populace (excl. infants and elderly)
Patterns of thought and action	Diligent, punctual, top-down governance	Always loose; horizontal solidarity	Selective engagement/disengagement; ostensible compliance with orders
Symbolic media	Watch, fixed-line phone, desktop (reserve) PC	Pager, payphone	*Keitai* (incl. avatars, cameras, wallets, etc.)
Symbolic informational practices	Being beeped by the company	Friendships with anonymous *bell-tomo* during passing free time	Continual *nagara* e-mail
Symbolic icons	Kinjiro Ninomiya, the farmer-sage	*Kogyaru* sitting cross-legged on the street, like *jibetarian*	Bikers reading e-mail on device held in one hand
Foundational culture	*Nagara*	*Nagara* + mobilism	*Nagara* + mobilism

Figure 4.1
Young girl using *keitai* e-mail in front of convenience store. Photo by author.

the "t.A.T.u." girls' lead in mobile media literacy, both behaviors gradually merging to become unisex.

After their school clothes, what catches our attention about high school students is their *nagara* culture, created by the combination of incessant *keitai* e-mail transmissions and slow traveling speed (figures 4.1 and 4.2). Since even high school seniors are not allowed to commute to school by motorbike, moped, or car, they walk or ride bicycles. They laugh and joke around, strolling at a snail-like speed to and from school. We often see them pushing their bikes, walking side by side and blocking the way of other pedestrians. Or they might be riding their bikes on the sidewalk at a speed not much faster than walking.

Oddly, in Japan, even when there is a bicycle lane next to the pedestrian lane, people prefer to ride their bikes on the pedestrian lane. This blatant violation of the law is committed by men and women, young and old, and is grudgingly accepted by pedestrians and baby carriage users (young mothers with their children; the aged, who use them as walkers). Many point out the dangers and lack of consideration of sidewalk bicyclists, but there is no movement on a national level to enforce the rules for riding bikes. This leniency is similar to the "norm of noninterference" on trains, where people shut their eyes to the use of *keitai*, which is supposed to be turned off as a general rule, a *tatemae* (social facade) (figure 4.3).

While walking or riding their bicycles on the sidewalk, the students talk to each other and are engaged in a constant exchange of *keitai* text and photo e-mail with absent friends (*mail-tomo*). *Keitai* e-mails tend to be more prevalent than *keitai* voice calls.

Figure 4.2
Young girl using *keitai* e-mail on a bicycle. Photo by author.

Figure 4.3
Tatemae poster speaks while smiling: "Don't use *keitai* in trains." Photo by author.

Figure 4.4
A statue of Kinjiro Ninomiya reading while walking with a bundle of firewood on his back. Photo by author.

There are those who perform stuntlike feats: they ride their bicycles one-handed while reading and writing e-mail or talking on the *keitai*, pause at a vending machine to buy a canned drink or cigarette pack without dismounting, and then ride off.

This example of *nagara* culture—reading and writing e-mail while walking or bicycling—is an ironic twist on the legend of Sontoku (Kinjiro) Ninomiya (1787–1856), an inspirational figure and national role model whose statue can often be seen in front of bookstores (figure 4.4) or primary schools. Born into a poor farming family in the Edo period, he became a teacher, businessman, and government official through self-education and hard work. "He advocated diligence, cooperation, deference to authority, and thrift as ways of improving Japan's rural economy at the end of the feudal era … and established the *hotoku* movement to promote morality, industry, and economy."[8] As a young man, going about his daily work of gathering firewood, he would read as he walked. The image of high schoolers with their *nagara* e-mail is indeed the opposite side of the coin of Kinjiro Ninomiya's image.

In this sense, Kinjiro Ninomiya symbolizes the multilayered quality of contemporary Japan's *nagara* mobilism. Underlying the conflict in the 1990s between the *kogyaru* and the *oyaji* is a cultural current of *nagara* mobilism that predates the modern era. In the second (business) stage of the paradigm, carried by businessmen *oyaji* (see table 4.1), this cultural stream gives rise to the diligent Kinjiro Ninomiya image. This same *nagara* mobilism, in the third (socializing) stage, carried by *kogyaru*, leads to the image of *nagara* e-mail sent while loitering on street corners. In the Edo period, too, the haiku poet and secret ninja agent Basho Matsuo (1664–1694) traveled by foot around Japan, conducting *ginko* poetry sessions involving group improvisation of short poetry verses while walking with *yatate*, a portable ink-and-brush case. Even this refined work, in addition to being an art and an amusement, could be seen from the point of view of the first (military) stage of the paradigm as a camouflaged form of espionage. Basho's haiku poems and letters represented not only artistic exemplars and instructions to a disciplinary community across the country but also secret intelligence to distant espionage networkers about resisting feudal domination.

The young people on the streets with their *nagara* e-mails actually call to mind the image of Basho's *ginko* poetry. As the day turns into night, young people, walking in the dark, holding a *keitai* in one hand and peering at the LCD monitor that glows like a lantern, resemble the silhouette of Basho walking the streets in the Edo period with his portable ink-and-brush case. The riders of lightless bicycles hold the glowing light of an LCD in one hand. The bicycle light was the first hit product marketed by the founder of Panasonic, Konosuke Matsushita, who was once called the "father of electrical appliances," but most bike riders have all but forgotten this beautiful tradition and ride their bikes recklessly with no illumination.

To the foreign visitor in the big city, these ordinary scenes of Japanese daily life may seem more odd and more mysterious than *komuso* (zen priests wearing bamboo mesh hoods while playing bamboo clarinets on the street). Japanese adults of an older generation are also taken aback by this "alien culture" of high school students. Still, these youth cultures are rapidly spreading to people in their thirties and forties, creating conflict between the new and old cultures over hegemony of the streets. In the background of this conflict is a significant paradigm shift.

In this chapter I examine the state of this alien culture spreading through Japan, especially among young people, framing it tentatively in terms of mobile media literacy, a viewpoint on a culture of street walking, *nagara*, continual motion—asynchronous, constantly interactive, giving/taking tactical intelligence culture of quasi-military street mobilization. I also analyze the underlying paradigm shifts and discuss a range of related cultural issues, such as norms for rules and manners, public preferences, appreciation of beauty, and the philosophy of *nagara* mobilism.

I would like to return to high school uniforms as representing, like *keitai*, the alien culture that we see on the streets in Japan. Uniforms are generally designated by the school, and the students are required to wear them. However, the declining birthrate has created competition among schools, and in order to attract students, many schools have started introducing more fashionable designer uniforms.

The traditional uniforms, *tsume-eri* (high-collared) for boys and *sailaa fuku* (sailor outfits) for girls, were adapted from European military attire. Most people recognize that the newer blazer uniforms were derived from a preppy Ivy League look. The girls' skirts, adapted from the original referent styles, are excessively short and raise the eyebrows of the older generations, to whom they appear "shameless like prostitutes." This superminiskirt look, always complemented by socks on bare legs, remains the same throughout the seasons, from Hokkaido in the north to Okinawa in the south, and has been a popular style since the 1990s.

There is a clear difference between the miniskirt fashion that spread from England in the late 1960s and the early 1970s and the miniskirt fashion in contemporary Japan, which mirrors the difference between the Western-origin PC culture and the *keitai* culture that developed uniquely in Japan. The British miniskirt look conjures names like Twiggy, the Beatles, Courrèges, and Mary Quant. This Western fashion had a powerful influence on young people all over the world. It was, in a sense, the last Western fashion that the Japanese accepted completely and unconditionally. In contrast, the Japanese superminiskirt came to be the "uniform of Shibuya street" among high school girls in the early 1990s along with the pager. Eventually it became firmly established as a fashion throughout Japan, and it even turned streets around the country into mini-Shibuyas. The *kogyaru* miniskirt of the 1990s and the Twiggy miniskirt of the 1960s are two separate phenomena, and the *kogyaru* miniskirt is a native alien culture fashion unique to Japan. Like the loose socks fashion, it has no direct relation to fashions originating in Euro-American settings.

Many aspects of the pager and *keitai* as information and communication devices were developed by AT&T and Motorola (U.S.), BT (Britain), and Nokia and Eriksson (northern Europe), but their evolution in Japan has been unique. Japan is now a major exporter of *keitai* culture, which includes i-mode, icons, photo e-mail, movie e-mail, ring tones (including sampling voices), *keitai* straps (figure 4.5), wallpaper for the LCD monitor, and digital font styles that rely on unconventional combinations of existing characters and symbols, named *gyaru-moji* (gals' alphabet) or *heta-moji* (awkward alphabet).

The *keitai* is a locus that integrates Japanese subcultures. Young people use the latest J-Pop (Japanese pop music) for ring tones and Japanimation (Japanese anime and anime characters) like Pokémon, Doraemon, and Sen/Chihiro from the movie *Spirited Away* as motifs for wallpaper and *keitai* straps. A *keitai* is for them more than just a tool—it is something they are highly motivated to animate and to customize as a dreamcatcher, a good luck charm, an alter ego, or a pet. I might even call it an idol or a fetish and regard it as an animistic handy object (*tokko*) stretching a spiritual barricade (*kekkai*) around the body.

Keitai culture, together with high school girls' culture, anime culture, and character culture, has spread among the youth of Seoul, Taipei, Shanghai, and Bangkok. Western media are also predicting its penetration into the United States and Europe, invoking a version of "yellow peril" theory or Orientalist prejudice (Said 1978).

Japanese culture has been called a hybrid culture, of which the wireless communication technologies of the pager and *keitai* are representatives. At the same time, Japanese pictographic and iconic culture, manifest in *nagara* behavior, e-mail texting culture,

a

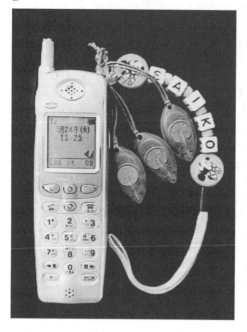

b

c

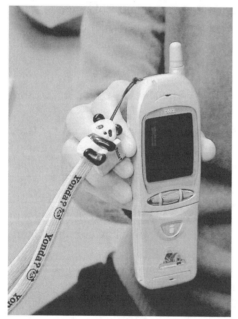

Figure 4.5
Keitai straps. Photos by author.

picture messaging, and recent multimedia messaging, has some historical roots in Japanese aesthetics and text presentation traditions.

While Japanese are often described as being poor at self-expression, the graceful and ethereal portrayal of inner landscapes with words and illustrations—as opposed to asserting the self—has been a traditional forte, manifest in the *Tale of Genji* and the *Pillow Book*, as well as wood block prints (*ukiyoe*) and tile engravings of the Edo period.

Ways of communicating with loquacious words and well-turned images evolved into new haiku arts and mass culture, such as the *etegami* (picture-letters) favored by the elderly or *keitai* photo e-mail among youth. The girls' pager revolution was an expressive literature movement—the *syasei* (sketching actually) movement, initiated by Shiki Masaoka, a nineteenth-century poet—that grafted haiku art onto the business tool of the pager. Japan's *keitai* culture has become a modern representative of a culture of haiku-like verse, infusing *syasei* e-mail with the energies of a popular artistic form and taking on a leadership role in global *keitai* culture.

Keitai's Tranformation to a Refreshing Favorite and Possibilities for *Keitai* Aesthetics

I am currently involved in research on new possibilities for *shikohin* (refreshing favorites) as part of the Shikohin Study Project.[9] In this project, we define *shikohin* as items that bring about sensory pleasures and refresh feelings. One member of this project, Masatoshi Takada, describes how human civilization has traversed three stages: agricultural revolution, industrial revolution, and information revolution. He describes the spread of *shikohin* as a social phenomenon as follows: (1) *shikohin* first appear in society as medicine then become accepted as commonplace items; (2) they are narcotics with psychoactive effects, broadly conceived; (3) their qualities are related to specific historical and social contexts; (4) they create new cultures and information industries; (5) *shikohin* industries become central to an information industry society (Takada 2003; Fujimoto 2002a).

Building on this framework, I would suggest that new amusements such as *nagara* e-mail are the same as *shikohin* and have a narcotic effect on the body. To what extent can we discipline these effects and elevate them to an aesthetic form? As it matures, *keitai* has the potential to become a *shikohin* to rival the big four of alcohol, cigarettes, coffee, and tea.

In the sixteenth century Rikyu Sen took a simple *shikohin* of tea and elevated it to the art of the way of tea (*chado*) through aesthetics of everyday life. Using the objects and media of *chado* as a starting point, Rikyu applied his exacting sensibilities not only to the actual substance of tea but to epoch-making culinary forms such as *kaiseki* (food eaten during tea ceremony) and *bento* (lunch boxes), interior and exterior architectural elements in the *chashitsu* (tearoom), and an expansive vision of a way of life for the *chajin* (person of tea) or *sukisha* (aesthete or connoisseur). *Chado* is an interesting case in that it came to have broad cultural influence in Japan in the aesthetics of everyday life, seasonal events, festivals, manners, and even fashions.

The historian Sakae Tsunoyama, who wrote *Cha no Sekai-shi* (*World History of Tea*) (1980) and *Tokei no Syakaishi* (*Social History of Watches*) (1984) claims that the West's tea culture was imported from Japan after 1610. Regardless, it appears that, concurrent with Japan's attention to tea, *shikohin* also had a large influence on the taste and aesthetics of the West, in the form of tobacco, herbs, and perfumes, alcohol such as wine and whiskey, beverages such as coffee and tea, and sweets such as chocolate.

In contrast to Japan, however, *shikohin* in the West constituted an inverse of traditional cultural structures associated with Greek philosophy and art, Christianity, and in the modern era, certain marginal cultures. In Japan tea went beyond its status as *shikohin* in everyday life, becoming a cultural reference point for worldviews, views on life and death, philosophy, and religion. The foundation of Japanese aesthetics has been rooted in encounters and conversations over tea. This deserves to be noted as a culturally unique development that did not occur elsewhere in the world.

Four centuries after Rikyu's death, contemporary Japanese aesthetics are being transformed in relation to developments in information and media. *Chaki* (tea implements) to this day are collectibles, and young people's practice of *chado* is still strong. Are we, however, seeing an emergent new gadget aesthetics and *keitai shikohin* culture emerging among young people?

For example, *keitai* straps have come to match the popularity of *netsuke*, which are luxury accessories and straps tying a cigarette case, a medicine case, or a wallet to a waist belt.

The expanded palette of *shikohin*—recreational consumer products—includes refreshments, cigarettes, chewing gum, comic books, and karaoke, but marketing analysts are blaming the slump in gum, magazine, and karaoke sales on *keitai*. In other words, they believe that *shikohin* and *keitai* compete in the same market arena. What they fear is that a variety of "refreshing experiences" or pleasures, will all become mediated and displaced by *keitai*. If this is the case, is there no place for *shikohin* (particularly refreshments, cigarettes, and chewing gum) in our future information society?

Historically, Europeans used modern *shikohin* such as coffee, tea, tobacco, sweets, and perfume as stand-ins for the exotic pleasures of the New World. The raw experiences of travel and adventure were reconstituted into *shikohin* through processes of mediation, mobilization, objectifying, packaging, and personalization. This was the first stage in the modern mobile-ization of *shikohin* delocalized and decontextualized from primitive pleasures and souvenirs.

Now we are confronted with the second stage of the modern mobile-ization of relocalized and recontextualized *shikohin*, that is, *keitai*'s replacing *shikohin* as media.

It is widely recognized that international products such as tobacco, coffee, and tea have been sophisticated informational and media commodities for the past four hundred years, since the emergence of commodity markets in modern Europe. While taking new forms, these basic products have remained resilient through global shifts toward urbanization and information flows.

I do not see contemporary mobile media technologies like *keitai* as vessels for insipid audiovisual information. Rather, I see *keitai* as providing ongoing access to real-time exchanges with others of photos, moving pictures, e-mail, live conversation, and vibrations. *Keitai* are less like books, which tend to be decontextualized, de-localized, and escapist media, and more like *shikohin*, as objects of recontextualization, relocalization, and actual media objects.

Of course there are negative dimensions to the trend toward *keitai*'s and mobile media technologies' taking on characteristics of *shikohin*. Not only does *keitai* erode the integrity of public spaces with the intrusion of private conversations but there are also reports that the electromagnetic waves can have negative health effects, and young people become ensnared by psychological dependence that is labeled *keitai* addiction.

The social philosopher Jean Baudrillard (1986) proposes using the term *gadget* to describe objects of any genre, whether *shikohin* or electronics, that are valued not for practical function but for communication, play, fun, and ideas. As world cultures increasingly turn toward play, art, and aesthetic sensibility, and as gadgetlike media with audiovisual dimensions become more prominent, one outcome is that our experiences with and relationships to objects become more intimate.

Japan has had a tendency to enjoy *shikohin* not in isolation but within the totality of associated objects, tools, and media. These objects, tools, and media often exist separate from the *shikohin* themselves and take on their own independent gadget identities in the form of *kawaii* (little, pretty, cute) stationary, fashion accessories, and character goods (Fujimoto 2002c; 2003b). These objects get exchanged in arts-and-crafts markets as they become popular and take on the status of collectors' items, antiques, and premium goods with artistic value. This flow from *shikohin* to accessory to art object happens spontaneously as part of fashion trends, but *chado* was the most successful Japanese example of this flow as artistic revolution, comparable to what happened with Art Nouveau or the arts-and-crafts movement.

Chado can also be viewed as having started with the taste and smell of tea, developing this into an artistic sensibility for taste and smell and extending it to the other senses, gradually expanding its artistic environments. *Keitai*, like tea, may have the potential to expand to include all five senses. Will we see a *keitai-do* (way of *keitai*) for the twenty-first century? Will there be a Basho Matsuo or Rikyu Sen of *keitai* (Fujimoto 2002a; 2002b)?

In this sense, today *keitai* competes with chewing gum and tobacco to be a successor medium to both tea and watches, as a distinctive *shikohin* that is not orally induced and that does not originate from plant material. The establishment of the four major *shikohin* of alcohol, tobacco, coffee, and tea was a global conversion of Central and

South American resources by the matured consumption revolution in Europe. I also see *keitai* as an example of a twenty-first-century consumption revolution.

Manufacturers and suppliers predetermine the design of all products, including *shikohin*, and they have a responsibility toward users of the product. This was a fundamental principle of consumer society. However, before *keitai* use, teenagers' pager use appropriated an existing business tool, resulting in an "invention" that disrupted the industry's marketing strategy and went on to become a huge business success through the industry's subsequent mimesis of the young consumers' invention. As a result, after the innovation, both manufacturers and carriers went on to ignore their traditional responsibilities and indiscriminately launched new products into the market without a clear product plan. This led to a trend toward using the market as a site for natural selection for survival, depending on the creativity of the users to determine how a product would be used. The manufacturers and carriers that did not follow this new trend and tried to impose their own ideas about design and use on young people were eliminated from the market. In the end it is up to consumers to determine these corporate fates.

This recognition of user innovation was an epoch-making phenomenon that reversed the roles of manufacturers and consumers and the roles of planned design and incidental selection. The years from 1990 to the present were marked by events where consumers subverted the intent of producers and challenged fundamental principles of manufacturing and design. I have called this immense pager and *keitai* craze the girls' pager revolution.

There are too many technologies to mention here that were enthusiastically accepted into postwar Japan and that dramatically changed people's lives, like the television, refrigerator, and car. However, there had never been a *shikohin* like the pager, which subverted the intent of the manufacturer to become a hit consumer product.

There are, of course, cases like the *bosozoku* (motorcycle/car gangs) racing cars at high speeds, ignoring traffic rules, or illegally remodeling their vehicles. But even though speeding is illegal, cars are built at the outset to run at speeds beyond the legal limit, so uses of the car by the *bosozoku* should not come as a total surprise to manufacturers.

By contrast, the pager and *keitai* were subject to legal but new uses invented by consumers and not anticipated by manufacturers. These devices and services, designed as emergency communication tools for adults, were reinvented by youth to serve as tools for *nagara* playing and socializing. This reinvention is a truly rare case of *shikohin* that came to be used by all generations in Japan as adults were won over in later years.

The Anti-Ubiquitous Territory Machine

I have described *keitai* from a variety of perspectives, in relation to *nagara* mobilism, popular trends, and *shikohin* cultures. This view of *keitai* in contemporary Japan suggests Japan's resistance to cultural globalization as well as Japan's leadership role in certain aspects of globalization. At an even more general level, these countervailing tendencies rest on the shifting plate tectonics of the third-stage paradigm shift (from *oyaji* to *kogyaru*), although the future is uncertain.

I would like to stress that paradigms are not Hegelian *Zeitgeists*, invisible spirits of the times, or abstract folk spirits passed along through intangible transmission. Rather, I see paradigm norms and structures as including both the mental and the material. A paradigm is an encompassing conceptual structure that integrates a wide range of factors like literature; social customs; laws; implicit standards; political, economic, social systems; and dynamics of science, technology, and industry. Although I cannot cover all these factors here, I discuss paradigm shifts through the concepts of time and space, which have been regarded as important schemes of epistemology since Kant (1781). I focused on the transformation of media related to self and to territorial awareness. I have developed a 3 (military, business, socializing) × 2 (on/off) × 8 (media, space, norm, time, self, aesthetics, literacy, physical transformation) matrix (table 4.2).

What I present is merely a hypothesis; providing empirical evidence is my future challenge. What I aim to demonstrate here is that even one paradigm shift related to media involves an intertwined set of factors (different cultures) that cannot be explained one-dimensionally by theories of technological progress, economic development, cyclic trends, or cultural genetics. For instance, there is a constant stream of abstract theories, such as technical determinism vs. social constructionism, which are misleading unless accompanied by detailed descriptions. Multifarious manifestations of culture and civilization related to *keitai* are spread across the globe, changing over time; these may one day appear in retrospect as a generalized paradigm shift. Similarly, to theorize that the three-stage paradigm transitions of wireless mobile communication are necessarily leading to "global economic development," "standardization of technology," and "a ubiquitous society" would be to perpetuate a short-sighted myth, an ideology. While I respect the efforts by companies and engineers that celebrate trends read off certain surface indicators (sales, improved performance) of the *keitai* phenomenon, human sciences must first grapple with the structure of basic theoretical concepts.

Throughout the 1990s I studied the wristwatch (especially Swatch and G-Shock), the pager, and *keitai* as media, and I have discussed their position in society as embodiments of alien cultures. Since then, I have drawn from media theory to write about issues such as *naka-shoku* (mobile eating and drinking), *jibetarian* (young people sitting on the street), café spaces, sleep environments and the information society, and *shiko-hin* culture and globalization. I have focused on Japan, East Asia, and Eastern Europe and have made it my personal quest to salvage *keitai*, which appears to be a fetish and a gadget corrupted by information capitalism, by viewing it from the more general viewpoint of paradigm theory.

Table 4.2
Paradigms (Explicantia) and Cultural Forms (Explicanda)

	The succession of different cultural forms		Patterns of paradigm shift
Basic order of succession			
Basic paradigm	I. Military (including disaster prevention)	II. Business (office work + domestic work)	III. Socializing (including romance, social life, and play)
Population	Soldiers (supplemented by civilians at home)	Businessmen (supplemented by wife and child)	Youth (mature children led by girls)
World epoch	1905 (sea battle of *Tsushima*)	1958 (first pager service, 1968 in Japan)	Early 1990s (girls' pager revolution)
▲ ON type work	International wars	Corporate business competition	Interpersonal dating wars
▼ off type work	Communal living	Household reproduction	Solidarity among friends (including *mail-tomo*)
▲ ON type territory	Nation-state affiliation	A person's business circle and post (self-employed or employed)	Friendship circle with telecom network
▼ off type territory	A person's village	A person's family	A room of one's own or a favorite place to be
Examples of media types			
△ ON media types	Telegraph (for commanders and commissioned officers) Bugle (for privates)	Sound-only pagers (for managers and field personnel) Wrist watches/sirens/chimes (for everyone)	Text pagers/*keitai* (for everybody)
▽ off media types	Village bells (churches and temples)	Sirens/chimes (for everyone)	Text pagers/*keitai* (for everybody)
△ ON media functionality	Boosting morale and relaying secret commands	Standardization of Greenwich Mean Time	Elevating personal flair and sharing secret intelligence among friends (including *mail-tomo*)
▽ off media functionality	Sharing life rhythms	Family sharing of leisure time/having personal time	Having personal time/seizing spiritual territory (*kekkai*) on the fly

Table 4.2
(continued)

Examples of technologies of place			
△ ON type technologies of place	Battleground (encampments)	Company/factory/school	Street
▽ off type technologies of place	Native land/homeland	Home/coffee shop	Bedroom transformed into a "room paradise" or favorite café space
Examples of social norms			
△ ON type social norms	Mobilization/commanding and regrouping/attacking	Fulfilling contracts (customer comes first)	*Deai* (encounter) and dating (women are in control of romance)
▽ off type social norms	Idyllic village	The happy family	Loose-relaxed, reciprocal relationships (friendships)
Examples of perceptions of time			
△ ON type perceptions of time	Action regulated by time in battle	World Business Time (WBT)	Spontaneous time (groovy, riding, "on" time, syntony)
▽ off type perceptions of time	Village seasonal events scheduled by folkcustom	Leisure time with family	Spontaneous time (edged, cutting "off" time, distony)
Examples of identities			
△ ON type identities	Citizens	Gainfully employed members of society	Sociable person blessed with friends and lovers
▽ off type identities	Villagers	Household members who love their family	Connoisseur surrounded by refreshing favorite objects
Examples of aesthetics			
△ ON type aesthetics	Courage!	Gumption!	Cute (*Kawaii*)! Cool! Popular!
▽ off type aesthetics	Simple & pure!	Sweet to family and subordinates!	Natural! (Self-absorbed but lovable)

Table 4.2
(continued)

The succession of different cultural forms	Patterns of paradigm shift		
Basic order of succession			
Basic paradigm	I. Military (including disaster prevention)	II. Business (office work + domestic work)	III. Socializing (including romance, social life, and play)
Examples of literacy			
△ ON type literacy	Receptive listening and weapon operations	Reading and writing/calculation/conversation/driving	Reading and writing/*keitai* operations/conversation
▽ off type literacy	Bracketing thought, emotion (shutting off fear)	Bracketing literary skills (shutting off fear of others)	Bracketing attachment to older favorite technologies (Shutting off fear of both techno-loving *otaku* and techno-hating Luddite)
Types of bodily transformation (automated mechanization of sensation)			
△ ON type bodily transformation	War machines	Business machines	Machines for socialization and dating (*kogyaru*)
△ off type bodily transformation	Machines for manual labor/machines for cooperative systems	Machines for family bonding	Machines for accumulation (*otaku*)/literature (haiku)

As summarized in table 4.2, *keitai* is a late modern *shikohin* in that it summons up animated pulsatory rhythms like *nori* (groove or syntony) and *kire* (edge or distony) from a flat Kant-like dimension, and can be regarded as an anti-ubiquitous media armament that instantly creates a personal and spiritual territory (*kekkai*) in opposition to trends toward ubiquity.

Keitai are still evolving as direct descendants of certain intellectual and fashionable information-gathering tools like the newspaper and coffee of Western Europe, which subtly maintain distance between strangers. Values inclusive of time-space awareness, skin sensation, and aesthetics are changing, not because of the singular effect caused by *keitai* but in an interactive and co-constitutive constellation with *keitai*.

The masters of the old paradigms scorn change. The *oyaji*, "raspy and thick-voiced," want to turn their gaze away from the paradigm shift as they heap blame for social ills on *keitai* and feel self-righteous in the comfort of their own territories. But in reality the high-pitched voices (or "yellow voices") of women and children are conquering the streets, trains, offices, and homes. Following is an excerpt from one of my earlier papers (Fujimoto 2002d), an ethnographic sketch of the fight for the hegemony of the streets with the advent of the third-stage paradigm shift.

So as not to be mistaken for a "raspy and thick-voiced" *oyaji*, young men (including wanna-bes and self-made men) are already trying to differentiate themselves by practicing and imitating the behaviors of women and children. Naturally, among women and children, the true privilege of emitting the carefree, high-pitched voices belong to the figures clothed in those symbols of the high school girl, the superminis and loose socks. Their high-pitched voices travel back and forth through space as e-mail on the prized territory-generating, vocalizing apparatus known as *keitai*.

With a *keitai*, a girl can turn any space into her own room and personal paradise (*kekkai*), whether that be her favorite café or her own stall in a flea market. The *keitai* is a jamming machine that instantly creates a territory—a personal *keitai* space—around oneself with an invisible, minimal barricade.

Even when signals aren't sent out as voice or text, carrying a "cute" *keitai* is itself an effective visual anti-*oyaji* signal. But sadly, when an *oyaji* plays with his *keitai* on the train instead of reading a newspaper, the *keitai* turns into an *oyaji keitai*. The same cute *keitai* in the hands of a young attractive man becomes a greasy, phallic object when held by a dirty *oyaji* even though its shape and size remain the same. When an *oyaji* has command of the *keitai*'s vibrating signal, the tremors evoke perverted images. When his *keitai* is a camera phone or a third-generation multimedia phone, he could be taken for a Peeping Tom, an up-the-skirt photographer, or a stalker. It is the dominant public (women, children) who decides whether that anonymous body on the train is an *oyaji*.

Some theories call *keitai* a ubiquitous medium, but this represents a shallow viewpoint that addresses only the functional level. As Shigeru Kashima (2000) suggests, the communal horse-and-buggy and the train stole from people their raw voice as primitive territory machines, or the wild and natural community embodied in village life under the pretext of civilization. Displacing voices of conversation and recitation, silent reading of books and newspapers became a new territory-generating apparatus for maintaining a comfortable personal distance between oneself and other passengers. This was when the raspy and thick voice was modernized and turned into the newspaper. Its dry rustling sound creates a partition, a personal living room, a shield, providing men with a safe haven from which to peep at women's bodies. It also became a sexually molesting paper tentacle with which they could even cop a feel, if lucky.

This modernization brings about the materialized alienation process. Like a woman's scream formerly, the male cough—ahem—a degrading utterance with no verbal meaning, pathetically asserts the authority of the raspy and thick voice. The active vocalization of the raspy and thick voice, which used to be so uninhibited and oppressively present on the streets, has been pitifully reduced to the passive reception of information in the enclosed space of the train, the silent reading of a newspaper, coupled with the weak broadcast of information with the empty cough in the implosion of media exchange.

At the beginning of the 1990s, as the girls' pager revolution proceeded, the streets and trains changed from paternalistic communities to private territories of women and children. By switching media from the raspy and thick voice to the newspaper and empty cough, aging men tried their best to create a home, to find peace and quiet inside the trains. But the *kogyaru*, clutching their new territory-generating apparatuses, the pager and *keitai*, freely broadcasting their high-pitched voices, wearing loose socks and munching on snacks, started spilling over from the streets onto the trains. The space of pleasure that had allowed men to hide behind newspapers and peek at the bodies of submissive female passengers was transformed into a humiliating slaughterhouse where *oyaji*s are mercilessly butchered and demeaned by *kogyaru*'s violent demonstrations, rendering the former utterly invisible and insignificant. *Kogyaru* couldn't care less if a subhuman *oyaji* peeked at their underwear or eavesdropped on their conversations. To the *oyaji*s' ears, all ring tones sound like *Dona Dona*.

On the streets and in the trains, there is a quiet but decisive change in whose voices are hegemonic.

The Segway, an electronic pedestrian mover that glides on two wheels balanced on gyroscopes, initially appeared on the streets as an alien object in an alien landscape, but is gradually becoming a part of everyday life in some U.S. cities. It appears to have qualities similar to the illegal, lightless bicycles in Japan in that it runs on the sidewalks and roads at a slow speed, upsetting traffic systems and rules. Even Americans, who have relied heavily on car phones tethered to four-wheel vehicles, might start getting into the habit of writing and reading *keitai* e-mail if they are riding a Segway.

On the streets in modern Japan, where the third-stage paradigm has reached its prime, *keitai* and humans have merged to create a powerful base station for a territory machine. When the bicycle is added to this combination, the unit becomes even more powerful. But the bicycle, stripped of all nonessential attachments such bell, luggage carrier, light, and gear lever—except for the front basket, which has managed to stay—is turning into a purely secondary organ of the *keitai*/human unit. Bags, too, can change in shape and size; the wristwatch, the Walkman, and other personal belongings can appear and disappear; attitudes and behaviors are also malleable and optional elements of a paradigm that takes the *keitai*/human unit as the stable core.

Transportation systems and wireless communication devices have previously been too big and heavy to deserve the label of tools integrated with the human body. In an earlier era, people used books, newspapers, umbrellas, and canes as territory machines on the train, cigarettes while walking on the street or in the office, coffee in coffee shops, and the car on the road. In both the West and the East, in the era of feudalism, people saw an identical phenomenon where the ruling class, riding horses and carrying swords, achieved hegemony of the streets by forming a supreme platform unit of human-horse-sword, which they used as their territory machine.

Inside the house, the rice paddle and the kitchen furnace, and later electrical home appliances, formed the base unit for the "women's place" (aka stockade, ghetto) of the kitchen or the living room. By contrast, merged with the car phone and radio, cars used to constitute a base unit for the "men's place" (aka stockade, ghetto) of work space between the office and the client.

As long as they conform to the demands of a larger paradigm, the bicycle, *keitai*, the Segway, and the linear motorcar are able to form a wide range of combinations (although I don't expect to see Segways in Japan because the Road Traffic Law prohibits them from being driven on public roads; one of these days, though, they may claim the sidewalks and bicycle lanes).

In a sense, the process of merging self and *keitai* into a territory machine and the process of globalizing ubiquity and *keitai* both represent alien cultures. Despite their conflicts, they represent two cultural extremes of the same third-stage paradigm. They have the same potential to prognosticate the coming, soon-to-be-birthed fourth-stage paradigm, which still lacks definitive shape.

Before casually preaching the myth of harmony to come that we call multiculturalism, it is crucial to examine first the specific realities of cross-cultural conflicts and probe the paradigms that lie in the background of these conflicts. High school girls' superminis and *keitai* are just the starting point. Unconscious behaviors and minutiae of everyday life, from waking to sleep, the surrounding trinkets and habits, are all intimately tied to these cross-cultural conflicts and emergent paradigms.

We must listen more carefully to our inner voices as part of an effort to deepen our understanding of strange cultures around our familiar world. The Japanese idiom *"i no naka no kawazu"* (frog in a well) is a warning against the arrogance of the ignorant. But in order to recognize those alien cultures closest to us, we must dig deeper into the well of our own internal aesthetic and cultural sensibilities.

In *The Dialectic of Enlightenment*, Max Horkheimer and Theodor Adorno (1947) discussed the dialectic reversal of civilization (logos) and savagery (mythos). They concluded that the more civilization advanced, the more savagery exercised hegemony. For example, they describe the fascist movement in Germany and the uniformity of culture industries in United States as inevitable results of the enlightenment movement.

Lastly, adopting their dialectic philosophy,[10] I would like to suggest the dialectic reversal of territorial and ubiquitous principles in advanced information technology society. Territory anticipates ubiquity from the outset in terms of tending to spread to the surroundings the comfortable time and space that satisfies only one's own desire for refreshing favorites. As the ubiquity of media environments in public spheres advances, the inclination of individuals to establish their own spiritual territory "right in the middle" increases. Ubiquity and territorialization are two sides of mutually connected dialectic processes. It follows that the more features of a ubiquitous machine *keitai*—the universal platform of the times—acquires, the more omnipotent power it exhibits as a territory machine.

In the second-stage paradigm, the powerful civilization of business overwhelmed weaker groups with PCs and pagers as ubiquitous machines. But the more information technology spread out agelessly and sexlessly all over the country, the more the weaker groups, including children and aged people, were driven to create minimal, spiritual territories (*kekkai*), driven by necessity.

In the third-stage paradigm, unconsciously, against the enclosures of business, the powerless, the helpless, and innocent school girls came out from the closet, acquired pagers as violent weapons with adult powers, and screamed out like modern barbarians on the street.

Perhaps in the future, both principles (civilization-savagery, subjectivity-objectivity, mastery-slavery, majority-minority, and territory-ubiquity) will be vibrating with pitch and roll in the dialectic processes.

Notes

1. Methodologies for the sociology of knowledge begin with Mannheim (1929). Successors are described in, for example, Berger and Luckmann (1966). For a discussion of contemporary culture in terms of fields and objects of study (rather than issues and methodologies), see Fujimoto (1999b), and Katz and Aakhus (2002). My use of terms such as *knowledge* and *culture*, and more specifically, *mobile media literacy* should not be confused with a narrow version of information literacy. Rather, my work spans a broad range of interdisciplinary studies from linguistics, anthropology, and management studies.

First, I draw from Chomsky's concept of "linguistic knowledge," particularly the debate over "local and universal grammar" and "linguistic competence and performance" (Chomsky 1975; 1986; van Quine 1981; Fujimoto 1985).

For my discussion of spontaneous and indigenous generation of "dialect," "groupware," "local knowledge" of mobile media literacy (emergent skill from media-user evolution, at subverting the meanings of the pager and *keitai*) of the marginal social group of young women (*kogyaru*), which challenged and overthrew the restrictive and antagonistic hegemonic social group of aging men (*oyaji*), now beginning to connect and merge with a globally shared corporate culture and global knowledge, I use as my basic concept Clifford Geertz's (1983) concept of local knowledge. Further,

I follow studies in organizational culture that critically adapted this theory of local knowledge to shared organizational cultures, knowledge resources, the construction of global standards, and corporate cultures (Deal and Kennedy 1982) as well as the more recent extensions of this work in knowledge management (Dixon 2000; Krogh, Ichijo, and Nonaka 2000).

2. For paradigm theory, see Kuhn (1962). See Popper (1994) for the criticism of adaptation of paradigm theory to sociological methods.

3. For international comparisons of *keitai* cultures, see Katz and Aakhus (2002). For more on contemporary Japanese mass culture, see Ueda (1994), Ukai, Nagai, and Fujimoto (2000) and Fujimoto (2002a).

4. For a discussion of "presentation" as logically and pragmatically superior to "communication," see Kadono et al. (1994).

5. For Serres's (1980) concept of "parasite" adapted to social theory of *keitai*, see Fujimoto (1999b; 2002b; 2004).

6. For Hegel's (1808) philosophical concept of *Geist* adapted to sociological study of *keitai*, see Katz and Aakhus (2002). But this renewal *Apparatgeist* is not explicit yet.

7. For more discussion of the *kogyaru* girls' pager revolution and the 1990s paradigm shift, see Fujimoto (1997; 1999a; 2002c; 2003a) and Tomita et al. (1997).

8. Michael Several, Los Angeles, November 1997. ⟨http://www.usc.edu/isd/archives/la/pubart/Downtown/Little_Tokyo/kinjiro.html⟩.

9. For more on our *shikohin* study project, see ⟨http://www.cdij.org/shikohin/⟩.

10. Myerson (2001) attempted to adapt German critical theory for *keitai* study, but this attempt only explained two types of "communication" concepts. I expect him to mount a broader challenge (Fujimoto 2004).

5 | Japanese Youth and the Imagining of *Keitai*

Haruhiro Kato

Keitai have thoroughly penetrated daily life in Japan, making it difficult to think of it as a special medium. This chapter analyzes the ways in which *keitai* have become utterly unremarkable and ubiquitous presences by examining the patterns and rhetorics of "*keitai* communication" themed video productions made by Japanese college students in a course I taught.

These college students are heavy *keitai* users. Their stories reveal their sense of *keitai*—the reality and perception of their communication and social relationships. There are several distinct patterns to the stories that they produced through which I identify the way they imagine *keitai*, their subconscious associations with *keitai*, and the limits to their imaginings of media. In addition to informing our understanding of college students' relationships to *keitai*, these stories provide insights into the spread of *keitai* communication in modern society more generally.

Directly reading students' sense of *keitai* from class-assigned videos is a risky enterprise. Certainly video productions are not a direct reflection of their shared impressions or culture. However, when I created this assignment, I did not specify any content guidelines. The students creating these productions were majoring in sociology or media studies and had a more sophisticated vocabulary for describing media than the average college student. They were not studying *keitai* per se bur rather participating in the class because of an interest in video production. In this sense, these works can be considered a reflection of a typical level of consciousness regarding *keitai* held by media-literate students in a private college.

Project Background

As part of a broad media literacy curriculum and a studio-based education effort, I asked students from various universities to produce dramatic visual or radio stories on the theme of communication, in particular, *keitai* communication. They decided on the content of the stories themselves and were encouraged to develop their works through critique by their peers. The productions had to be completed in about six weeks of weekly ninety-minute lessons, although some had to be completed in just two or three weeks. The videos are ten to twenty minutes long. The radio dramas, lasting between four and ten minutes, were performed live. The class produced eight videos and thirty radio shows.

The stories demonstrate that the *keitai* is an indispensable tool for Japanese students, who use it enthusiastically and on a regular basis. There are certain patterns to the *keitai* stories they produce, illustrating the imagination and perceptions of *keitai* held by Japanese students as well as illustrating their views on the properties and limits of communication. In other words, this program enables us to study how today's pager and *keitai* generation portrays scenes and dramas of *keitai* use, and to observe the images they have of *keitai*.

Keitai are no longer just youth media. My expectation has been that students' stories would reflect a particular view of *keitai* based on their lack of social experience and immaturity of social imagination. Furthermore, students generally accept the negative image of encounter sites presented by the mass media. Online dating and anonymous encounter sites have generally been depicted as dangerous and unwholesome. Because most Japanese students see online interaction as distinct from ordinary relationships, they imagine these dating and encounter sites as full of falsity and deceit. They can't comprehend why people would use these sites and can't accept that a marriage made through online dating could be truly happy and loving. I have not expected much critical insight regarding this issue to appear in their narratives, and my goal has not been to solicit such insights. Rather, the hope has been to encourage a unique narrative framing that reflects the students' awareness, which in turn describes some of the characteristics of contemporary *keitai* communication.

Keitai and Everyday Cyberspaces

One thing that became clear through this program is that it is becoming increasingly difficult to produce stories about *keitai* and discuss *keitai* itself as a narrative object. The students comment on this difficulty themselves. I would even say that it is becoming unreasonable to use *keitai* as a theme because it has become such a transparent and self-explanatory medium that productions on this theme lack clarity. *Keitai* are embedded in so many contexts and functions that they cannot be defined or described in a single way. For Japanese youth, *keitai* are just that—*keitai*. During critiques of the productions students often described how difficult it was to create stories about *keitai* as special media.

The *keitai* is really convenient. It's a voice medium for people who want to talk, a text medium for people who want to write, and a graphic medium for photographs. It accommodates users' tastes. [girl, junior]

[After analyzing a popular TV drama series, *Kitano Kuni Kara 2002, Yuigon* (*From the North Country-side 2002*)]: A lifestyle with *keitai* is so natural that one without *keitai*, or one from which *keitai* is taken away, sounds unreal. [boy, junior]

In other words, a lifestyle incorporating *keitai* communication is firmly established in these students' lives, and they continuously and simultaneously inhabit both the "real world" and this alternative space of connections. "In college, the structures of groups and relations that lead to bullying no longer exist, but in exchange there is no space where different social groups coexist. Consequently, *keitai* and other media that foster co-presence become increasingly vital in college" (Kashimura 2002, 211–249). As discussion about *keitai* is continued, students describe the connections enabled by *keitai* as simply a matter of personal convenience.

One male senior explains, "It's not that I want to be connected, it's that I want to be able to make a connection." A female senior describes a similar sentiment: "I want to be able to keep in touch, but I turn it off when I am on a date with another guy because I don't want to be tied down." After an experiment when his *keitai* was taken away for a week, a male junior said, "It didn't bother me too much, but my friends had a hard time." These statements reflect a medium that is most convenient for the person who is doing the contacting rather than the person being contacted.

The students who participated in these productions felt that "maybe this is the last year" when they could make stories that focused specifically on *keitai* communication. In other words, it is becoming impossible for students to question *keitai*'s existence and regard it as a new medium. Their interests lie in communication and social relationships rather than in *keitai* as a particular tool or piece of technology.

These students' comments point to an underlying set of conceptual issues in their understanding of *keitai*. Cyberspace studies, dominated by Internet research, have often considered the ways in which online interaction provides liberation from everyday relationships and institutional relationships through anonymous encounter sites and other forms of online participation chosen by individuals. These social forms can be contrasted with "systematic" relationships based on status-role relationships of everyday institutions, which are often not an outcome of individual choice. This dichotomy between everyday identities and institutions and cyberspace identities and communities has been subject to critique based on studies of how "real-life" identities and institutional power relations operate in cyberspace (see Ito's introduction and chapter 9).

Keitai represent another set of dynamics in this debate about the relation between online and offline practices and identities. Unlike the Internet accessed through the PC, *keitai* communications are not tied to a particular location and can be accessed in a wider range of times and places. Recent *keitai* also include communication types, such as *keitai* e-mail, that can be used in unobtrusive ways and that do not interfere with the ongoing stream of "real-life" activity, whether in a classroom setting or a public space such as public transportation or a performance. In other words, activity can be layered, such as when a *keitai* user listens to a lecture while sending text messages to a friend. Finally, with the advent of additional *keitai* communication functions such as picture messaging, online and offline contexts become complementary. Camera phones enable even more sharing of the "real-life" setting as part of the content of online communication.

In short, *keitai* communications do not occur in a cyberspace set off from everyday institutional identities and offline settings and practices; rather, they integrate what have often been considered two separate and antagonistic worlds of real and virtual social contact. The unremarkable and taken-for-granted status of *keitai* and the primacy of the "real world" are themes that surface in student discussion as well as in the video narratives themselves. The student productions consistently reject a radical separation between nonsystematic relationships and systematic relationships and between online life, mediated by *keitai*, and offline social life. Students see even contact with anonymous others in *keitai* encounter sites as eventually returning to relationships fostered in traditional institutions such as the local peer group and school. The struggle to conceptualize what is special about *keitai* communication is another indicator of the thorough integration of *keitai* with the everyday institutions, relationships, and identities of Japanese youth.

Production Case Studies

Despite these struggles in creating *keitai*-themed video works, many productions were made under this rubric, and certain unmistakable patterns or descriptive styles emerged.

- Pattern 1 Tales of *keitai* removal, creating a world where there are no *keitai*, no *keitai* reception, or where *keitai* signals cannot reach.
- Pattern 2 Tales of encounters, where "new" relationships are created but in a surprise ending the other person was actually somebody they had already known.
- Pattern 3 Tales of communication from another land, for instance, a call from the world of the dead.
- Pattern 4 Tales of everyday breakdown, where characters get into a fight because one forgot a *keitai*, but eventually reconcile, through *keitai*.

Patterns 3 and 4 will not be analyzed here; these works are focused on rhetorical ingenuity and don't display much significance at the level of student impressions and the commonly held meaning of *keitai*. By contrast, patterns 1 and 2 are clearly significant for the topic of this chapter; in these works there is a narrative of reversal against the social order of inescapable *keitai* connection.

The first pattern is based on a fantasy premise of constructing a world without *keitai*. The second pattern sets up a contrast between *keitai* and face-to-face communication, concluding that face-to-face encounters are the contexts where "true" communication occurs. In other words, this pattern creates dramatic tension from the fact that face-to-face communication is the prior and more essential form.

Whether about PC e-mail or *keitai* e-mail, movies and television films construct dramatic narrative by suggesting that e-mail initially creates a sense of freedom from the constraints of the household or workplace. This kind of "grammar of drama" and "grammar of fantasy" could be considered representative of the first phase of deployment of e-mail communication. A society where *keitai* is well established demands a different source of dramatic tension or "grammar of drama." Students who see *keitai* as a well-established cornerstone of their social lives create the fiction of a world without *keitai* or a world of "pure" face-to-face communication as a source of dramatic rhetorics and grammars. This is a reversal of the prior rhetorical strategy. In the first and second patterns, we can see a symbolic representation of the *keitai* imagination of Japanese students at this moment in the early twenty-first century.

Tales of *Keitai* Removal

The most typical plot of *keitai* stories dismisses the two worlds created by *keitai* and reconstructs a world in which *keitai* does not exist. In other words, the tales portray a society before the existence of the two worlds. Rejecting the media communications space that *keitai* brings about, they stress the importance of face-to-face relations, sounding a cautionary note about the convenience and excessive dependence on *keitai*. *Keitai* society's characteristic dependence on constancy and pervasiveness of the two worlds is eliminated in these narratives. Emphasizing the value of place, space, and social contexts that cannot be disrupted by calls and e-mail from remote worlds, they defend the intrinsic context of the here and now in which we are situated. Face-to-face encounters are emphasized.

Constructing a world without *keitai* by taking it away or disconnecting it from the network is the most straightforward plot strategy. At the same time, this type of plot is difficult to adequately frame and explain. Students struggle to depict persuasive reasons for *keitai* being absent or being rejected. One commonality across the productions is the difficulty of conveying a clear message about *keitai* communication. Without this clarity, the intent of the production is difficult to grasp, and audiences have a difficult time relating. Common feedback from the student viewers was that they didn't understand the message of the stories.

A Man without a *Keitai* This 2001 film is thirty-nine minutes long. The main character is a man who does not have a *keitai*. He is late to the party that he has organized and is berated by his friends because it is difficult to get in touch with him. While feeling pressured to get a *keitai*, he reflects on what it means to have one. In the end, his reason for not getting a *keitai* is simply, "It is a pain. I don't need one really." In other words, he is simply reacting against friends who have criticized and nagged him about not carrying a *keitai*. The students who created this drama may have only been able come up with this reason for not carrying a *keitai*. Today it is difficult to conceive of a positive reason or social significance for not carrying a *keitai*. Not owning one has been reduced to a personal choice or mark of individuality.

The story is conveyed by the following scenes:

Five male students are chatting on a university campus. The protagonist is among them and is suggesting organizing a party with a group of girls. All the boys say they will participate.

On the day of the party, four boys are waiting at the train station for the protagonist who organized the party.

Meanwhile, the protagonist is feeling ill and repeatedly visits the toilet. After emerging from the toilet, he calls a friend's *keitai*, only to hear the message, "The mobile phone subscriber you are trying to reach is currently outside of the coverage area or has turned off the phone."

The following week, the five students are gathered again on campus, talking. Giving the excuse that he tried to call to tell them he was late, the protagonist apologizes repeatedly.

Protagonist: I had to run an errand on campus and was running late, so I tried to call. I couldn't get through.
Friend 1:   Were you faking being sick?
Friend 2:   Were you faking being sick?
Friend 3:   Were you faking being sick?
Friend 4:   Were you faking being sick?
Friend 1:   We were waiting at a place with reception. We tried calling you, too, but could only call your home number. Why don't you get a *keitai*?
Friend 2:   You have a lot of nerve not carrying a *keitai* in this day and age.
Friend 3:   Yeah, it's weird.
Protagonist: I have a phone at home, and I am usually at home, so I think the home phone should be fine.
Friend 3:   But you weren't home.
Friend 1:   Not being able to reach you when we are trying to reach you ...
Protagonist: That happened to me, too!
Friend 1:   Yeah, and we don't even really know if you tried to call us.
Protagonist: I'm sorry. But I don't have any desire to carry a *keitai*.
[The four friends leave.]
Protagonist (*to himself*):   What should I do? Do I really need a *keitai*? But they all sound like idiots, *keitai, keitai, keitai.*
[Two girls are talking at a corner of the campus.]
Girl 1:   I got on the bus while I was talking on my *keitai* and an old guy yelled at me.

Figure 5.1
Rejected at a pay phone. From *A Man without a Keitai*. Reproduced with permission.

Girl 2:   But you didn't hang up because you were talking to your boyfriend.
Male Voice: Nobody goes around without a *keitai* nowadays. You are stuck in the Stone Age. I don't even want to talk to you.
[The protagonist goes to a pay phone, tries to make a call, and gets a parody message: "This telephone card is no longer in use. Please try again from a *keitai*."]
Voice 1:   He's so uncool.
Voice 2:   He smells.
Voice 3:   He's an idiot.
Protagonist (*to himself*):   *Keitai, keitai.* Everyone is so annoying. Ugh. But I don't really need a *keitai*. I don't need it!
[As the protagonist watches TV in his room, he gets a number of strange phone calls. He screams.]

I have had numerous opportunities to show this work in class and to have students write about their impressions of it. The viewers, who are also students like the creators of the work, seem to feel that the title *A Man without a Keitai* has an appealing fantasy-like ring that draws their attention. For the viewers, there is a heightened expectation before viewing the work, and they keep watching for a special revelation of why the protagonist doesn't have a *keitai*. In other words, the most crucial scene is where the protagonist mobilizes a vocabulary for not wanting a *keitai*. But the creators were not able to clearly and persuasively describe his reasons for rejecting a *keitai*. In this sense, the production could be considered a failure. On the other hand, constructing a fantasy out of some discourse and the simple fact of not having a *keitai*, this work could be considered a success. Of interest for us here is the difficulty of articulating a reason for having or not having a *keitai*. In other words, the difficulty of describing a special reason for not having a *keitai* is tied to the difficulty of describing a special reason for having one.

About Being Here   This 2001 film is twenty-nine minutes long. A student sends *keitai* e-mail to a broken-hearted female friend to say that he wants to go on a trip with her. She shares her plans to go somewhere that *keitai* signals cannot reach. The man says, "That is even better," and joins her. During their trip together, he confesses his love for her, and she accepts. The central point of the story is that the two characters go to a place that *keitai* cannot reach so they can be alone and away from others. The story places significance on the couple's going where they will not be disturbed by *keitai* so that they have the freedom to talk about their affections.

The couple's going out of their way to seek a place with no *keitai* reception signifies a break from their social relations symbolized by the constant connection and tethering by *keitai*. The characters in this story mobilize a rhetoric that reverses the characteristics of distance-spanning telecommunications by (1) seeking a place that will not be interrupted by *keitai* contact from others, (2) going out of their way to travel across distance, and (3) engaging face-to-face. By taking these actions, they construct a stage for a dramatic romance that is in contrast to everyday reality. This is a stage that is appropriate for "communicating something important," and "communicating true feelings." By setting up a special situation, the message of "communicating love" takes on drama and reality. No *keitai* is needed here. That is clearly the creator's intention. The student who directed this production heads and performs in a theater group. This experience is brought to bear on this rhetoric.

This production portrays the importance of face-to-face interaction. The success and failure of the production hinges on whether the work is able to get viewers to identify with the reasons for going out of the way to go somewhere where there is no *keitai* reception.

The story unfolds in the following scenes:

Four friends, including Tomo, are gathered talking on a college campus. Another student, Saki, approaches. She says she has just been dumped and is going on a trip to nurse her wounds.

Tomo leaves the group and goes to his room. He sends an e-mail to Saki, saying that he wants to accompany her.

In her room, Saki wavers about what to say, then starts typing. The scene proceeds with e-mail messages between Tomo and Saki.

Figure 5.2
Sending a text message. From *About Being Here*. Reproduced with permission.

Tomo: It's about the trip you were talking about yesterday. I really want to go with you. Is that okay?
Saki: Why? What's up?
Tomo: There is something I want to tell you in private.
Saki: I'm going really far away. We might not even have *keitai* reception. Are you sure you want to go?
Tomo: It is better that way. What I have to tell you is really important.
Saki: I don't know what you are talking about, but sure.
Tomo: I think I am being a pest, sorry. But thanks.

This scene of the two exchanging messages is one that the student viewers can generally relate to. In their written responses to the work, students wrote that they could identify with Tomo's worry and his anxious expression as he waits for the e-mail from Saki. In the case of short text messages of this sort, the repeated exchange of e-mail is what builds and reaffirms the social context. Most students have a shared experience of worry, uncertainty, and impatience as they wonder if they are communicating effectively, whether their feelings are being understood, and whether they will get a reply. This shared experience is what builds identification with this scene.

Tomo's message, "It is better if there is no *keitai* reception. I have something important to tell you," is the most important theme of this work. Why is it not enough to simply turn off the *keitai*? All *keitai* users have heard the message, "The mobile phone

Figure 5.3
The moment of truth. From *About Being Here*. Reproduced with permission.

you are trying to reach is turned off or out of the coverage area." It is an announcement signaling that the subscriber is in a no-service area or the phone is off. Following are scenes from the story where the two discuss why Tomo wanted to go to a no-signal area, where this message would be in effect.

Tomo and Saki are using the Internet to research their destination, the last station on a private rail line slated to be shut down soon. Tomo looks like he wants to say something to Saki, but can't. She asks, "What's wrong?" "Oh, nothing," he replies.

Tomo and Saki are in a mountain village without even a convenience store. They go to see a famous old temple and then a historic thousand-year-old cherry tree. At the park with the cherry tree, the two have the following conversation.

Saki: So why did it have to be here?
Tomo: Because I didn't want anybody to interrupt us.
Saki: Then it could have been in Toyota [city].
Tomo: But there is the *keitai*.
Saki: Oh, right.
Tomo: People give you a hard time when it is turned off. It is better to go somewhere there isn't any reception.
Saki: Okay, I get it. But what did you want to say?

Tomo is too embarrassed to convey his feelings for her. But Saki tells him she has guessed his feelings. Saki agrees to start seeing him, and the story has a happy ending.

Because what he had to say was important, Tomo tells Saki in person at a place where *keitai* cannot disturb them. The setting of the story clearly places significance on the act of removing *keitai* from the social scene.

The characters say, "It is better to go somewhere there isn't any reception" and "People give you a hard time when it is turned off." Places out of reach of *keitai* signals can be divided into places outside the service area and areas inside the service area where there are weak signals, such as in the subway. In this story, there is symbolic significance attached to being outside the service area as well as to the seriousness and sincerity of Tomo's love for Saki. The viewers of the production were able to relate to Tomo's sincerity, but they could not see the necessity of conveying his feelings in a place outside the *keitai* service area.

At the same time, there are two underlying points of identification. The first is that the *keitai* is a media form that makes communications ubiquitous. In this sense, *keitai* connection means being confined by social relations. "I didn't want anybody to interrupt us." The second is the importance of speaking directly. Students viewing this production understood that being outside the service area creates a scene that epitomizes the importance of direct communication. Being outside the service area creates a special situation. At the same time, one of the student viewers commented: "If my boyfriend turns his *keitai* off when we're on a date, I get suspicious. I might guess, from casual clues like this, that he is cheating on me." The special significance given to someone deliberately turning off the *keitai* or going to a no-signal zone illustrates how established the constancy and pervasiveness of the two worlds has become.

A Big Dependence on Something Small This 2001 film is eighteen minutes long. It is a documentary-type production where fictional events are combined with an experiment in which a student who loses in "paper, scissors, rock" has to live without her *keitai* for a week. The student encounters many problems because of her lack of a *keitai*. For example,

- She cannot reach her boyfriend when she wanted to cancel their date because she had a fever.
- She could not get credit for a class because she didn't know the deadline for a report.
- She could not contact her employer of her part-time job and was fired after she had been given a final warning about being late.
- She had an appointment to celebrate a friend's birthday but had not planned anything in detail. Not being able to call her, she tried going to her friend's house, but she was not home.
- The log in her journal on the last day of the week without her *keitai* sounds alarmingly desperate.

This drama is seemingly something that could happen to anybody. However, in other stories where a character is placed in a similar situation, he or she is depicted as finding ways of overcoming the difficulty. Consequently, the viewers of this production felt that lack of a *keitai* shouldn't cause so much trouble. Their feedback included comments that people would get used to living without a *keitai* and that they would find ways of working around the problem. Many students believed that people get dependent on *keitai* because they have it, but if they did not, they would get used to the inconvenience.

When asked, "What is *keitai* for you?" many Japanese youth reply, "I could never live without my *keitai*," or "It's part of my body." However, a documentary like this one tells us that the necessity of *keitai* is a myth. Most of the students who created and viewed this production had been carrying a *keitai* for only a few years, since high school. These same students bought into the myth that *keitai* is absolutely essential to their lives, that "*keitai* is a necessity," and "It is part of my body." This attitude toward *keitai* is apparent in the ways that a life without *keitai* can only be imagined as exceptional and fantastic.

Swimming This 2002 film is eleven minutes long. Today it is taken for granted that everyone carries a *keitai*. This story rejects this social standard. When the protagonist wakes up one morning, he is the only one in the world who has a *keitai*, and everyone else is communicating by letters and land-line calls. Finding himself in this environment, he experiences nostalgia for this world without *keitai*. Then he wakes up again to find he is back in a world where everyone has *keitai*. In the last scene of the story, the protagonist throws away his *keitai*. Through the protagonist's confusion about using *keitai*, the story tries to remind us of communication in the pre-*keitai* era.

The story portrays things we have lost because of the adoption of *keitai* through familiar everyday scenes of student communication before the *keitai*:

- A scene of memos written on the blackboard: "I'll see you in the library."
- A scene of a mutual friend waiting with a message: "He's waiting for you in the cafeteria."
- A scene of writing a letter to parents: "You haven't called lately, but it's about the money."
- A scene of frequent use of a pay phone.
- A scene of asking for a private phone (for conversations parents cannot overhear?).
- A scene of going home "to wait for a phone call."
- A scene of making a date to make a phone call: "I want to talk to you about something. Can I call you at 9?"

These are all scenes that could be addressed with *keitai*. In other words, the story reverses settings where *keitai* functions as a useful tool that liberates people from the constraints of time and place.

In one of the stories, one day all *keitai* in Japan stop functioning for two weeks. We see people getting accustomed to not having *keitai*. After two weeks there are two types of users. One is the user who takes the opportunity to stop being dependent on *keitai*. The other is the user who starts using it again with fervor. In another story, *keitai* stop functioning for all students except one. At first, the student with the functioning *keitai* is happy, but in the end he stops carrying it around because there is no one to reach. He gradually loses the desire to use it.

These stories all have *keitai* removal themes. Rather than depicting problems with *keitai* per se, the stories portray alternatives to *keitai* communication by depicting face-to-face communication that is regulated by specific times and places. The stories do not, however, condemn *keitai*. They do illustrate nostalgia for certain inconveniences and remind us that we should not forget the importance of face-to-face communication and that there are things that cannot be relayed by *keitai*. If we were to look for a deeper message, it would be an awareness of the real, the unexpected, and the lost through the exploration of social life where *keitai* is not a standard item. Japanese students inhabit a country that has embraced *keitai*, but they remain faithful to the mythology of face-to-face communication. They believe they should talk about important things in person because the message may not be conveyed otherwise. Although not as strongly as with bulletin board and PC e-mail communication, students feel that *keitai* e-mail communication is "honest." At the same time, they feel a deeply rooted mistrust, a sense that feelings are not conveyed and are not real unless "you meet in person." Because of these perceptions, stories praising face-to-face communication and warning against *keitai* dependence are likely to be reproduced at any Japanese college campus.

Tales of Encounters

One-to-One   This is a story of people who meet on a *deai-kei* (anonymous encounter) site but find out when they meet in person that they are already friends. Miyuki, a college student, is the shy protagonist. She meets an e-mail friend on an encounter site. Her handle in this space is "Happy," and the other girl's handle is "Barbie."

Miyuki's inner voice speaks: "It might be because it is e-mail, but I can talk about anything really honestly to her. This might sound flakey, but I feel like she is a true friend." Miyuki confides in "Barbie" about her social awkwardness in communicating, and "Barbie" cheers her up. One day Miyuki is berated by her friend Sachi for showing up late for their shopping date. She is not good at communicating with the more outgoing Sachi. Miyuki calls Sachi to apologize but is yelled at again. Depressed, Miyuki sends e-mail to "Barbie" saying she wants to meet her. The two decide to get together. When Miyuki shows up at the meeting place, she finds that "Barbie" is Sachi. The two have a meal together and revive their friendship. After that, the previously withdrawn Miyuki is able to enjoy communicating with her friend in a more outgoing way. Miyuki's inner voice says, "I am glad I was Sachi's e-mail friend. Meeting Sachi in this way was both accidental and necessary. I gained some confidence in myself somehow."

These words are similar to the lyrics of the closing song of the 1998 Hollywood movie starting Tom Hanks and Meg Ryan, *You've Got Mail*. "I was so glad it was you."

In the last scene of the student production, the following message appears:

Recently, perverse crimes have been proliferating on *deai-kei* sites.... These two were able to see the true nature of a friend who was close to them in their real lives. Maybe you, too, can gain a precious friend or a precious something. In the world of e-mail you can become your ideal self. But don't shut yourselves into that world and forget your true self.

The most common stories are boy-meets-girl scenarios mediated by *keitai*. These are tales of men and women who meet in the message space of *keitai* communication and become *meru-tomo* (e-mail friends). They most often meet on a *deai-kei* site, the Internet Web sites accessible with *keitai*. Other stories are about men and women who

Figure 5.4
Two friends meet. From *One-to-One*. Reproduced with permission.

become close after one of them loses his or her *keitai* or wallet, creating an opportunity for them to start exchanging phone calls and e-mail. These are stories of *keitai* being used as a device to connect people and create new relationships. These kinds of stories typically make up about one third of both the visual and radio productions. The plots start with the characters exchanging e-mail and always end with them meeting in person.

One-to-One also suggests that being limited to the e-mail world means losing one's true self. Behind the stories that make use of *deai-kei* sites is a conviction that one's true self is realized in the world of face-to-face interaction by escaping the *keitai* e-mail world where you perform your ideal self.

The two worlds of face-to-face and cyberspace are often thought to oppose one another, but often in students' everyday media communication they complement one another. In these stories told by the students, however, the two are not seen as mutually reinforcing. Cyberspace is a merely a vehicle to return to the face-to-face. In other words, cyberspace is a prop to support face-to-face space, and therefore all these tales of encounter end with the characters meeting in person.

Many of the *meru-tomo* stories that the students produce have a surprise ending in the final meeting. Most of them find out that the other person was already an acquaintance or family member. In the early stages of the plot the emotional bond that creates intimacy is acknowledged in the e-mail interaction. Stories that describe encounters with strangers, however, are pervaded with a negative aura and sense of disgust. These stories end as tales of betrayal. In all cases, there are two patterns to the conclusions of stories. One preaches a lesson to beware of *deai-kei* sites, and the other is about not being discouraged by a mistake and using the sites to search for a new partner.

There are a number of ways to interpret the endings of these stories of encounter. If students were simply to celebrate *deai-kei* sites, they could create a happy story where people use the sites as devices to find a partner. In the case of happy endings, students always put in a surprise twist where the partner is actually an acquaintance or family member. Does this propensity to make the endings of these encounter stories into a negative reversal drama point to an impoverished imagination among Japanese students? Or is it simply a rhetorical limit? It is impossible to determine here. What is more important is that, as I have stated repeatedly, students hold tightly to the mythology of face-to-face encounters. Every production taking up *deai-kei* sites ends with a face-to-face encounter. None of the works tell a story where the characters don't meet and only exchange e-mail.

Conclusion: The Demise of *Keitai* Stories?

This chapter has examined typical patterns in student-produced narratives. It has also discussed how *keitai* has become such a mundane medium that it has become difficult to use as a noteworthy object that can drive a narrative forward. *Keitai* have become an indispensable, or rather a seemingly indispensable, medium for Japanese youth.

Each student in the group that wrote *A Man without a Keitai* gave his or her own view of *keitai*. Their remarks describe a ready-to-hand convenience deeply rooted in their everyday lives:

"Love and friendship"
"Available any place and any time"
"Wakes me up whenever I fall asleep"
"Everybody has one"
"Easy to get in touch with everybody"
"Convenient when meeting up with people"
"Lets me talk to people whenever I want"

In order to portray the *keitai*, an object as personal and mundane as the wristwatch or the wallet, students had little choice but to adopt an extreme scenario such as removing it.

My educational effort in media production will continue. What will the students produce if I set up another assignment a few years down the road on "stories about *keitai*"? In 2002 students found it difficult to create a story about *keitai* communication, and I concluded that "this year might be the last." For this reason, my theme for 2002 was "communication stories," and *Swimming* was the only work to feature *keitai*. My theme for 2003 was "social bonds." In order to encourage an imagination with a rich rhetoric, I found it necessary to reframe the project to allow students to explore totally fictional dramas or everyday communication dramas within which *keitai* might play one role.

I don't believe that Japanese students can easily change their assumption that face-to-face communication is primary. On the contrary, I expect that as *keitai* becomes further embedded in everyday life and become even more unremarkable, things that are not like *keitai*—live performance, face-to-face, embodied settings—will become even more the sites for seeking and imagining authenticity. If "stories about *keitai*" are possible even in the face of this trend, I would hope for two directions in developing and discovering new rhetorics. One is a rhetoric that concludes within the space of *keitai* communication. The other is a rhetoric that doesn't place face-to-face communication outside of *keitai* communication but sees both as primary and on an equal plane.

The daily communication space of Japanese students is extremely narrow. Their interpersonal relationships are developed in a small, intimate, and systematic framework established upon their entrance to college. These relationships are framed by their identities as students, organized extracurricular activities, and seminar groups. It is understandably difficult for them to imagine a space of social relationships that involves liberation from this core set of identities. Worlds such as *deai-kei* sites and a broader palette of identities outside of their known set of relationships are peripheral to their self-understanding and their understanding of others.

At the same time, current college students were adolescents when *keitai* became a fact of life. Even their early years were spent in households where *keitai* were being used. Media literacy is an essential part of them. This is not mere functional media literacy but a media literacy that involves being able to fluently differentiate and mobilize the unique characteristics of different media forms—"transit literacy." How will new generations with this transit literacy discover and develop the two new rhetorics that I have proposed, and what will the next generation of *keitai* narratives look like? I have great hope for these new imaginings.

III | **Social Networks and Relationships**

Misa Matsuda

The mid-1990s saw the rapid adoption of *keitai* in Japan. As with other countries and regions, observers initially focused on the disruptive effects of *keitai* in public spaces. Subsequently, attention turned to its effects on interpersonal relationships and society. The influences of *keitai* on the individual and society have been characterized in a variety of ways in public discourse: for example, users pay more attention to *keitai* interlocutors than to people they are with, whether friends, family, or somebody they happen to sit next to on the train; now one is expected to be available for contact anytime, anywhere, and waiting is not tolerated; parents are no longer aware of who their teenagers' friends are.

I have conducted research on a wide range of *keitai* uses since 1995.[1] One trend has emerged from this research as being central to *keitai*'s effect on interpersonal relations: the increase in individual selectivity. As Leopolinda Fortunati (2002, 51) points out, using *keitai* "is to be reachable not by everyone, but only by those with whom we want to communicate—intimate friends or selected others whom we want contact." In theory, *keitai* can be used anytime, anywhere, but in reality people foster relationships with those whom they choose to contact.

This chapter discusses certain trends in *keitai* use in Japan, focusing on the increase in individual selectivity in choosing relationships. I do not argue, however, that *keitai* alone determines the structure of relationships and social life. Rather, I locate *keitai* use within existing structures and trends in Japanese society. *Keitai*'s rapid adoption in the 1990s grew out of, and was driven forward by, these existing contexts, and in turn the device accentuates and accelerates the social trends within which it was born.

My discussion is based primarily on research conducted by the Mobile Communication Research Group in November and December 2001, which sampled people between the ages of 12 and 69 in Japan (1,878 valid responses, collection rate of 62.6 percent). The next section focuses on *keitai* Internet access, with an emphasis on youth. Following that, I examine the original function of *keitai*: voice telephony and the effect on overall relationships with family and friends. Next, I highlight the increasingly common practice of *ban-tsuu-sentaku* (call screening)—using caller ID, to select which calls to answer—and discuss marital status as a factor determining this practice. Finally, I consider the facilitation of choice in interpersonal relationships as a central effect of *keitai* use.

E-mail-Centered *Keitai* Internet

Based on the discussion in the introduction and in chapters 1, 2, and 3 we can see that Japan has led the world in pioneering the *keitai* Internet. When viewed from the position of the user, however, a somewhat different picture emerges. Only a small percentage of users use the *keitai* Internet for Web access. According to a study by the Mobile Communication Research Group (2002), out of all *keitai* owners (64.6 percent of the population), 36.9 percent access the Web, 36.1 percent do not access even though the function is available, and 27.0 percent do not have access. Moreover, the ratio of teenagers (including children aged 12) who own a *keitai* Internet terminal is 84.5 percent, of which just under 70 percent actually use the connection. In contrast, the ratio of people in their sixties who own an Internet-enabled *keitai* is slightly over 50 percent, of which only 6.0 percent use the connection.

In the same study, we surveyed rates of information site access by type (table 6.1). The most frequently accessed sites were ring tone download sites (67.7 percent) and wallpaper download sites (35.0 percent). These are both services for customizing *keitai* terminals. Other popular sites included weather forecasts (23.1 percent), search engines (20.6 percent), and news (19.1 percent). Access rates to transactional sites such as ticket reservation sites, stock exchange, and banking services were low, only 1–3 percent. The most popular Web sites provide decorative and customizing services similar to those described in chapter 4: stickers, print club pictures, and decorative handstraps for pagers and *keitai*.

We can make an additional connection with the fact that only 0.2 percent of respondents reported building a home page with *keitai*. (In the same survey, the ratio of people doing this with a computer was 10.8 percent.) The Internet has often been touted as a medium that puts publishing in the hands of the masses and makes one-way mass media obsolete. Based on the current use of the *keitai* Internet, it would be difficult to make such a claim.

The *keitai* Internet's primary use is for e-mail. According to the same survey, among *keitai* users, 57.7 percent use the e-mail function. By age, the percentages are 89.2 for teens (ages 12–19), 82.7 for those in their twenties, 66.8 for those in their thirties, and up to 19.5 for those in their sixties.[2] This means that even though the Internet can be accessed with *keitai*, most users merely exchange messages, as they did with text pagers. Hence we can say that the *keitai* Internet is substantially different from that accessed by personal computers; it is an extension of individual ownership and

Table 6.1
Rates of Use for *Keitai* Internet Sites

Information Site Type	User Ratio (%)
Ring tone download	67.7
Wallpaper download	35.0
Weather forecast	23.1
Search engines	20.6
News	19.1
Games/fortune telling	18.9
Transportation info	16.1
Music	13.9
Sports	13.2
Leisure/travel	7.7
Movie (theaters)	6.2
Maps	5.7
TV programs	5.0
Ticket reservation	3.2
Deai/friends	2.7
Stock exchange	2.5
Cooking/recipes	2.2
Online shopping	2.0
Restaurant reservation	1.5
Banking services	1.2
Job information	0.2
Other	3.2

Source: Mobile Communication Research Group (2002).

personal uses of a youth mobile communications medium that has transitioned from pagers to *keitai*.

Just as in Norway (Ling and Yttri 2002) and Finland (Kasesniemi and Rautiainen 2002), the user demographic that most commonly uses mobile e-mail is young and female. According to a time use survey conducted in October 2001 by the NHK Broadcasting Culture Research Institute, female teenagers (including 10–12-year-olds) use mobile e-mail on average 1 hour 58 minutes each day. This is an average across all e-mail users; 31 percent of teenagers used e-mail on the day this survey was conducted (Mitsuya, Aramaki, and Nakano 2002). So, with whom do these users exchange e-mails during these two hours every day?

The most common partner for mobile e-mail is a "spouse/lover" (figure 6.1). And compared to *keitai* telephony (figure 6.2), e-mail is used more often in contacting friends not seen on a regular basis.

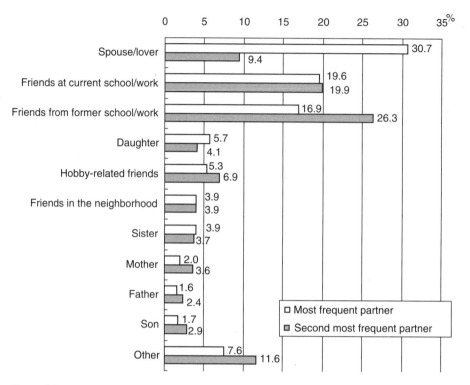

Figure 6.1
Keitai e-mail partners. From Mobile Communication Research Group (2002).

In the interview survey of youths conducted by the Mobile Media Research Group in the summer of 1996 in the amusement and shopping districts of Tokyo and Osaka (Matsuda et al. 1998), there was widespread use of several slang terms; *jimo-tomo*, *chu-tomo* and *ona-chu* were three of them. The first is a contraction of *jimoto no tomodachi* (friends from the neighborhood who one is not in school with), and the latter two derive from *onaji chugaku ni kayotta tomodachi* (friends from junior high school who one is not currently in school with). Behind the newly routine use of such terms is the presence of the pager, a popular device among youth at that time.[3] Terms such as *chu-tomo*, *ona-chu*, and *jimo-tomo* signal a growing consciousness of the category "prior friends," which grew hand in hand with the adoption of the pager. Although the category of "friend from middle school" existed before the pager, the buzzword did not. Now this category of friend is both more common and more notable as a category of relationship.

Before the widespread adoption of pagers, even close friends found it difficult to stay in touch when they moved on to different schools. Their lifestyles changed through

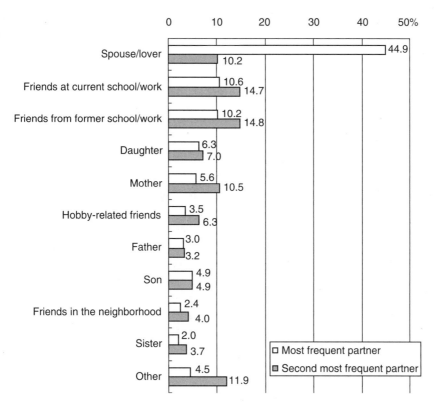

Figure 6.2
Keitai phone call partners. From Mobile Communication Research Group (2002).

their attendance at the new schools, and it became a challenge to maintain a relationship.[4] However, as an asynchronous form of communication, pagers allowed people to send messages at convenient and flexible times. Keeping in regular contact by pager messages also facilitated face-to-face meetings. In the years that they were popular among youth, pagers enabled the easy maintenance of friendships with people in distant environments. Now *keitai* e-mail has taken on this role.

By supporting ongoing relationships with former friends, *keitai* e-mail makes it possible to be friends with people in different places and under different circumstances. In this sense, types of friendships are on the increase, but this does not necessarily translate to the diversification of friendships. Pagers and *keitai* do not increase the number of friends that young people meet with and associate with outside of school.[5] They primarily provide opportunities to *maintain* relationships with friends who used to go to the same school. Furthermore, a category such as *chu-tomo* does not include every friend from junior high. It refers more specifically to "friends who were especially close during junior high" with whom one maintains a friendship after graduating. Through the pager before, and now through *keitai*, young people are maintaining associations with close friends who have mutually selected each other. These relationships are actually extremely homogeneous.

Kaeru Calls and Appointments

Telephone Calls

I turn now to telephone calls made with *keitai*. As mentioned earlier, 64.6 percent of the sample of the Mobile Communication Research Group (2002) survey owned a *keitai*. The rate of use is the highest among people in their twenties, at 86.8 percent. Although the rate drops as age increases, now approximately 30 percent of people in their sixties own one. More men than women use *keitai* (70.1 percent and 59.3 percent, respectively), but this is an effect of a large disparity between men and women age 40 and above, largely an outcome of occupation: 76.3 percent of full-time employed persons use *keitai* compared with 42.7 percent of full-time homemakers.[6] Frequency of use and the ratio of personal versus professional use differ between men and women; men are heavier users than women, and more women use *keitai* for personal purposes. Only 1.8 percent of respondents said they did not use the *keitai* for telephony.

Who are they calling, and for what purposes? As shown in figure 6.2, the most common partner is "spouse/lover," and the next most common is "friends at current school/work." Figure 6.3 shows the purposes of *keitai* phone calls. The most common purpose for calling friends is "rendezvous/appointments." For calls to family, the most common purpose is to confirm "time of arrival at home."

In summary, the most common forms of *keitai* telephone calls are *kaeru* calls (going-home calls) and appointments. In other words, *keitai* telephony is most often used to make a *kaeru* call to the family, to make a rendezvous with friends seen regularly, and to make appointments.

In an interview survey conducted in 1995, when business *keitai* users still dominated, respondents would note that "my wife started to call often to ask me to buy toilet paper on the way home." Wives had previously never called the office even when their husbands had a dedicated line. However, it was reported that wives found it easier to call with *keitai*, which the husbands "always carry on them." These calls are not communicating personal and professional emergencies, nor are they lightweight chatty calls; there is a specific bit of information that needs to be communicated. Communication of this kind is indicative of a new form of contact enabled by mobile communications, which I call casual business. Elsewhere I have made the argument that *keitai* has become a tool for conducting this kind of casual business (Matsuda 1996a).

The paradox of the Internet is that it increases opportunities to create and maintain social ties but tends to reduce in-person social contact. Some are concerned that con-

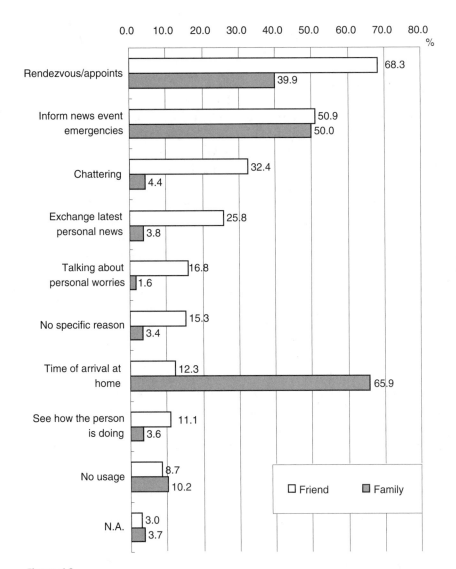

Figure 6.3
Keitai phone call purposes. From Mobile Communication Research Group (2002).

tact made available by *keitai* would decrease face-to-face communication. In reality, most feel that "contact with others increases through owning *keitai*."

Changes in the Family and Increases in *Kaeru* Calls

In public discourse, we have seen certain fears associated with youth *keitai* adoption. Parents of junior high and high school students have reported that they are no longer aware who their children's friends are and that they are worried about their children's having started going out more often after owning *keitai*. A common conclusion is that "*keitai* is disintegrating family bonds." When taking a broader historical perspective, however, it is uncertain whether it is *keitai* that is causing youth to spend more time out of the home. Based on time use surveys of the Japanese people, Okada (2002) notes that the amount of time people in their late teens to twenties spend at home has been declining steadily for the past twenty years, well before *keitai* became widespread. In fact, there are indicators that *keitai* has increased contact with family members who were frequently out of the home and had few occasions to call. We have documented many cases of family members' making new forms of contact through mobile phones, such as a mother and daughter making frequent *keitai* contact or a grandfather and grandson who exchange *keitai* e-mail. The question of whether *keitai* severs or enhances family bonds must be contextualized by broader changes in the contemporary Japanese family.

Yoriko Meguro (1987, iv) characterizes changes in the Japanese family over the last forty years as a shift toward individualization: "A shared family life is no longer universal or taken for granted. Rather, it is increasingly seen as something achieved at particular times by individuals creating unique personal bonds." Prior assumptions that family is a natural outcome of being together or sharing blood ties have been displaced by this notion of family as something to be created and maintained through the ongoing efforts of individuals. Following on Meguro's work, others have argued that certain changes in the Japanese family can be understood by terms such as *individuation, privatization*, or *individualization*. As Shinji Shimizu (2001, 98) has argued, "The phenomenon indicated by these terms is casting a shadow both directly and indirectly on trends in the family."

Kaeru call became a buzzword a few years before Meguro proposed these shifts in the experience of family. The *kaeru* call, shorthand for "a call home to inform the family what time you will be home," was a part of a 1985 campaign by the newly privatized NTT national telephone carrier to promote telephone calls. The term was popularized in part through humorous TV commercials that punned on the other meaning of *kaeru* (frog).[7] The spots featured close-ups of thoughtful husbands calling loving wives. As the camera zoomed out, they were revealed to be sitting on the back of a huge frog. Since then, calling home before one's return from work has become an icon of thoughtfulness and attention to a spouse. This symbolic status of phone contact with the family is an important backdrop to the current practice of making frequent *keitai* contact between family members.

Behind the image of loving phone contact is an emergent social equation: if family bonds are embodied in these routine communications, without these forms of contact the family could disintegrate. This pervasive conception of the family, popular now, was unthinkable a mere fifty years ago.

Drawing from ethnographic research, Hidetoshi Kato (2002) describes an earlier, more taciturn Japanese society and criticizes the more communicative contemporary norm. In a 1955 study that he was involved in, conducted at Nikaido Village of Nara Prefecture, the members of a three-generation household made 195 utterances during the three hours between dinner and bedtime. For this family of eight adults, this meant only eight utterances per hour per member. Furthermore, most of the utterances were "yeah" or "uh-huh" and other such responses or acknowledgments (Kato 1958).[8] Kato concludes that "small groups of people bound together in the intimacy of family or village didn't require those interactions that we now glorify as 'conversations'" (2002, 176).

So how much conversation is there in today's families? Since there has been no similar ethnographic research with contemporary families, I draw from the results of a survey on the length of the time spent in conversation among family members.

According to a survey conducted by Nomura Securities Company Ltd. (2001) with 700 homemakers with children in high school or younger living in the Tokyo (350) and the Kyoto, Osaka, and Kobe areas (350), the average amount of time a mother and child engaged in conversations each day was 2.3 hours, and a father and child 0.5 hours. In an earlier study of Tokyo homemakers conducted by an insurance company and reported in the *Asahi* newspaper on September 6, 1988, the average conversation time of spouses on a weekday was 73.3 minutes. There was wide variation, with 10.3 percent reporting over three hours and 38.2 percent reporting less than thirty minutes. This article also referred to other research conducted during the same period by a consumer appliance manufacturer that concluded that wives wanted more conversations with their husbands.

Two points are worthy of note. The first is that there is a growing discourse and consciousness about talk between family members; it has become possible to ask questions not only about length of conversation time but also about the amount of time spent in conversation with spouse or family. The time use survey by the NHK Broadcasting Culture Research Institute (2002) illustrates an average day of a family consisting of father and mother in their forties and a high school daughter. The study depicts all family members present in the house for approximately one hour from 7:30 p.m. on. This is the time for the so-called *kazoku danran* (family togetherness), a time when family members relax together watching TV or eating a meal. This time represents conversational engagement between family members, but there are other activities going on simultaneously. It is difficult to determine how much actual talk is involved; perhaps only the same amount as there was in 1955. In any event, there appears to be at least research and broader popular support for the belief that the volume of conversation within a family is indicative of certain social trends. Many other studies cover similar ground.

This research, as well as many other attitude surveys, also shows that many people see value in extending the length of conversations. Conversation between parents and children as well as between spouses has become desirable. According to Mami Iwagami (2003), the lack of communication between Japanese parents and children came to be viewed as a social problem beginning in the 1980s.[9] "The lack of family communication" was "discovered," and increased conversation became a new social necessity.

This is an extension of societal changes between the end of the Meiji period and into the Taisho period, characterized by the increase in family conversation. Hidetoshi Kato (2002) quotes folklore researcher Kunio Yanagita ([1931] 1990, 513): "Western influence has made men have to talk to their wives and sisters. [Previously] if conversation was thought of as a refined art, then it was in fact a courtesy to not make conversation with those who are intimate."

In short, the image of "the family that frequently contacts each other through *keitai*" emerged during a long process of Japanese society's becoming more communicative, in particular, the growth of family conversation. Of course, we cannot read changes in the Japanese family from simple tallies of changes in the volume of family communication. For example, in the Nikaido family study, family members might have felt that they had a shared understanding despite the fact that they weren't exchanging a lot of words. Quality rather than quantity could certainly be a more important factor. What I want to stress here is the trend toward a consciousness that it is important for families to frequently exchange words. This consciousness is one foundation upon which current *keitai* communication has flourished. In fact, I might argue that *keitai* has increased quantity through communication of everyday business rather than quality through facilitating understanding among family members. Why is this increase in the quantity of conversation so welcomed? If we take into account the broader historical context of *keitai*, we can avoid the technological determinist conclusion that *keitai* increased or decreased family communication.

Making contact with family has become an individual's responsibility and choice. In order to maintain family in a larger sense, beyond the husband-wife or parent-child relationship, people have to stay in close touch with one another on a regular basis through *keitai*. If they didn't feel this necessity, they would not communicate with family but would only contact close friends outside the family with *keitai*. In reality, mother-child communication has increased through *keitai*, but not father-child communication. For relationships like the latter, where there was little conversation to begin with, there is also little reason to go out of the way to make *keitai* contact. That is why *keitai* is seen both to sunder and to bind the family.

Appointments and Choice of People to Meet

For many users, *keitai* is a tool to arrange regular gatherings with friends and to make appointments. If we look to youth in particular (most notably, college students), these practices can be characterized by the term "full-time intimate community" (Nakajima, Himeno, and Yoshii 1999) or "telecocoon" (see chapter 8).

In the course of our research, we often meet youths whose *keitai* address books list several hundred people. When they meet new people, they exchange *keitai* numbers as though they were exchanging business cards (Kawaura 2002). In a 1999 study of college students, when asked to whom they didn't want to give their *keitai* number, 40.2 percent said "a person of the opposite sex whom they meet for the first time," and 38.1 percent said "a person whose age is very different from theirs whom they meet for the first time"; 27.2 percent said there was "no kind of person to whom they wouldn't give their number" (Okada, Matsuda, and Habuchi 2000). In other words, one out of four college students would give their *keitai* number to just about anybody.

On their own, these data would give the impression that youths contact extremely large numbers of people through *keitai*, but many say they do so regularly with only about ten people. The number of registered *keitai* numbers decreases the older a user is, but there is little variation in the number of people contacted on a regular basis (Matsuda 2001a). This suggests that youths have many "superfluous" numbers in their *keitai* address books. Though there is the appearance that youths have expanded their circle of acquaintances with *keitai*, judging by their communication patterns they are mostly making frequent contact with a select few. Upon returning home, they make calls and exchange e-mails with the same friends whom they just saw at school. This round-the-clock set of relationships with an exclusive group of friends is what Nakajima, Himeno, and Yoshii (1999) characterize as a "full-time intimate community."

These trends are also tied to an expanded sense of selectivity and choice. *Keitai* are used most often when choosing a gathering point or making an appointment. Shin Dong Kim (2002, 70) describes the situation prior to *keitai*: "If a worker left the office without making any appointments, it was almost impossible to track the person down and bring him or her to a gathering place." Now, with *keitai*, he or she can be tracked down any time, from any place.

The social patterns of Japanese university students, who have a lot of free time, have been undergoing phenomenal change. Before the penetration of *keitai* during the early 1990s, students would gather at a hangout such as the cafeteria or a coffee shop barring any other prearranged appointments. They would spend time there or decide to go elsewhere. College students still do this, but it is becoming less and less common.

Now they make direct calls on *keitai* to people in their *keitai* address books and make appointments in advance. As soon as class ends, students pull out their *keitai* and begin to make calls, contacting those friends who are part of their everyday friendship networks.

Keitai connect users with close friends through phone calls and e-mail. This frequent communication also provides opportunities for them to meet in person. Of course, the same could be said for fixed-line telephones. Before the advent of the telephone, visitors used to show up at homes unannounced, but after the wide adoption of the telephone, it became impolite to make a visit without advance notice (Wakabayashi 1992). Through *keitai*, in a similar fashion, one fixes appointments and reduces chance encounters and spontaneous gatherings. Now people must always make conscious choices of whom to call and meet. *Keitai*, or more accurately, the *keitai* use of Japanese youth, increases opportunities for individuals to choose their relationships.

Ban-tsuu Sentaku

Restoring the Rights of the Call Recipient, and the Screening of Personal Relations

Telephones are intrusive because callers can initiate communication whenever it suits them, but recipients do not know who the caller is until they pick up the phone. Telephones gave the upper hand to the caller. Now caller ID has changed this situation: recipients know who the caller is before they answer the phone and they can ignore the call. Caller ID has restored the rights of the call recipient.

In Japan, household telephones started carrying caller ID in February 1998, but it was earlier for *keitai*; the first terminal that displayed the number of the caller was introduced in 1994. Caller ID became a standard service because there were no additional charges and because of users' tendency to frequently upgrade their *keitai*.[10] Matsuda et al. (1998), Matsuda (2000a), and Okada, Matsuda, and Habuchi (2000) examined call screening with caller ID, or *ban-tsuu sentaku*, as a common practice among youth. Compared with married users in the same age group, single users tend to do *ban-tsuu sentaku* more often, demonstrating how the individual's personal network regulates use (Matsuda 2001a). This is discussed in more detail later, but first I describe the practice of *ban-tsuu sentaku* based on a survey by the Mobile Communication Research Group (2002).

One question in our survey asked, "Have you ever not answered your *keitai* after seeing the caller's number or name?" The answers were "very often," 6.4 percent, "sometimes," 30.5 percent, "hardly ever," 37.7 percent, and "never," 23.1 percent.

There are several reasons for doing *ban-tsuu sentaku*. For example, if we consider age as a regulating factor, *ban-tsuu sentaku* strengthens the connection between youth and new media. It is easy for young people to get acclimated with new media, but as they get older, people find it more and more difficult to familiarize themselves with new

Table 6.2

Rates of Call Screening

	Married (%)	Single (%)
Very often	6.6	6.6
Sometimes	25.7	40.1
Hardly ever	39.0	37.9
Never	28.6	15.4

Source: Mobile Communication Research Group (2002).

$p \leq .000$

practices of this sort. Another regulating factor is the range of people to whom individuals give out their *keitai* numbers. As indicated earlier, youths' address books generally contain a large number of entries, and their *keitai* numbers have also been given out to many people. It would be reasonable to assume that they would need to do *ban-tsuu sentaku* to screen callers and find out who, among these many potential communicators, is calling.

The Mobile Communication Research Group's data show that younger people tend to give out their *keitai* numbers more often than older people do and that this tendency is stronger in men than in women. However, the rate of execution of *ban-tsuu sentaku* has no correlation to age or sex. The only correlation is marital status: single people are more inclined to do *ban-tsuu sentaku* than married people (table 6.2). The variation was especially evident among people in their thirties.

Personal Phone/Mobile Phone/Private Phone

Why do single people do *ban-tsuu sentaku* more than married people? This can be explained by the difference in the style of *keitai* use between single and married people described in one of my earlier papers (Matsuda 2001a).

For singles, *keitai* is a "personal phone." They decided to have their own *keitai*. They select many people with whom to make contact via *keitai*, but they also use *ban-tsuu sentaku* to screen incoming calls. They are also used to having long conversations on *keitai*. The difference between the sexes is that single men tend to adopt *keitai* as their sole means of making contact, whereas single women are more likely to use the household telephone. But the difference is small. These styles of use are similar to those of high school and college students (Matsuda et al. 1998; Okada, Matsuda, and Habuchi 2000). Singles proactively adopt *keitai* as their primary means of contact.

By contrast, married people orient to *keitai* as a "mobile phone." Compared to singles, they contact relatively few people via *keitai*, and they rarely do *ban-tsuu sentaku* with incoming calls. They generally do not have long conversations on *keitai*. Based on the fact that the length of calls from their household phones is constant, they appear to be selective about whom they give their *keitai* numbers, making *keitai* contact with those persons only when a fixed-line telephone is unavailable.

The people selected for inclusion in married people's *keitai* network include men and women. When we asked employed people about their private uses of *keitai*, men referred to their business associates more often than women did. Questioned as to why they started carrying a *keitai*, many men said, "It was required by the company." (Women also said, "My family urged me to have a *keitai*.") Thus business use regulates much of *keitai* use for married men outside of family. In contrast, women, even if employed, have more contact with acquaintances outside of workplace associates and family, and are more inclined to use *keitai* for personal purposes. While married men tend to see *keitai* as a "mobile phone," for married women, often using *keitai* for social uses, *keitai* is regarded as a "private phone."

Jukka-Pekka Puro (2002, 23) writes that in Finland "women use their mobiles for all types of social purposes, such as keeping in contact with children, relatives or friends, in addition to non-household work-related uses." Therefore, "70% of women keep their mobiles on all the time, whereas half the men turn their phones off at night." Japanese women use *keitai* in a similar fashion. The percentages of users who "don't turn *keitai* off at home" are 81.0 for married men and 87.7 for married women (based on analysis of the data from Mobile Communication Research Group 2002).

Changes in the Personal Network from Marriage

These differences in styles of *keitai* use between married and single people are based in different individual networks. The difference between singles' "personal phone" and the marrieds' "mobile phone/private phone" is tied to the general tendency of the singles' personal networks to concentrate more on "just friends" rather than on family and neighbors (Otani 1995). In contrast, married people more often socialize with relatives and neighbors. This same "marital effect" was evident in my own research (Matsuda 2001a) when people were asked specifically whom they met and contacted frequently.

Relationships with relatives and neighbors rest on externally given conditions; relationships with friends are based on a higher degree of individual choice. Taking this into account, it is understandable that married people, who have comparatively more ties with relatives and neighbors, do not often take advantage of *ban-tsuu sentaku* for screening callers, because there is generally little choice but to maintain these relationships, and it is difficult to pick and choose when to engage. We could also consider a different interpretation. One reason why married people tend to use *keitai* and household phones for different purposes could be because they differentiate media use depending on relationship. The household phone is used for long conversations with relatives and family or to make an appointment with neighbors, which is conveniently done from the telephone at home. Contact with "just friends" at varying locations is done more often by *keitai*.

I turn now to the difference between married men's "mobile phones" and married women's "private phones" from a personal network point of view. According to Shinsuke Otani (1995), when respondents were questioned about what person is closest to them, the most significant difference between the sexes was that men more often replied "co-worker" and women, "neighbor." Similarly, in an analysis of workplace relations of self-employed and employed workers, men reported more relationships than women with co-workers and friends at the office (Tanaka 2000). In other words, the fact that men use *keitai* to contact co-workers or business associates more often than women is merely a reflection of their personal networks rather than a reflection of more business-oriented use.

Japanese *keitai* adoption in the late 1990s was driven by a generation of people called *dankai junior* (Generation Y, or second-wave baby boomers, born between 1970 and 1974). Considering the changes that tend to happen to personal networks after marriage, now that this generation has reached marrying age,[11] *keitai* use should be on a decrease. Respondents report that while they do not get rid of their *keitai*, they no longer make long phone calls or use it as often as they did before marriage.[12] *Keitai* does not by itself impact communication or interpersonal relationships; its uses are integrated into each individual's lifestyle.

Selective Interpersonal Relationships

Urban Interpersonal Relationships

This chapter has examined *keitai* use in contemporary Japan with a focus on the expansion of individual choice. In particular, *keitai* has made it more convenient for youth and singles to choose their interpersonal relationships. It is important, however, to consider influences other than *keitai* that deliver similar choices in interpersonal relationships.

The surveys on Japanese value orientation conducted by NHK Broadcasting Culture Research Institute every five years since 1973 provide interesting data regarding interpersonal relationships. Over the twenty years since 1973, there has been a decrease in people of all generations who want "comprehensive" relationships with relatives, neighbors, and co-workers, and an increase in people who want "partial" or "formal" relationships (Akiyama 1998). According to a study by Shinsuke Otani (1995) in 1989, given the choice between "being a good friend but not too intimate" and "being completely open and intimate with friends," the larger the population of the area, the more people tended to answer that they preferred the former. This question focused on desired rather than actual relationships with relatives and friends, and the responses suggested that the preference for "selective interpersonal relationships" over "completely intimate interpersonal relationships" was related not only to *keitai* adoption but also to a broader context of concurrent generational change and urbanization trends.

Beginning with work on *Gemeinschaft/Gesellschaft* (association/community), the contrast between flexible interpersonal relationships, which people can enter and leave as they please, and binding relationships, which are often systemized and organized, has been studied by many researchers. For example, Chizuko Ueno (1994) characterizes *ketsu-en* (kinship), *chi-en* (community), and *sha-en* (sodality)[13] as "obligatory relationships" and *Sentaku-en*, the flexible and pluralistic interpersonal relationships in which people choose their friends, as "selective relationships." What is important here is her characterization of *sentaku-en* as "relationships born from urbanization."

Claude Fischer (1976; 1982), in his research in northern California, proposes several hypotheses about the influence of urbanism on interpersonal relationships. He theorizes that urbanism does not destroy friendships but that, on the contrary, city dwellers have a slightly greater number of friendships. The frequency of their social contact is not substantially different from that of non-urban residents. The difference is that urban respondents "are involved with more 'just friends'—the large category of associates called 'friends' who shared no other social context with the respondents—than were small-town respondents" (1976, 258).[14]

Fischer defines an urban area based on the "number of people living in and near a community." Yasushi Matsumoto (1992), examining this issue from a personal network point of view, says that a definition should be based not solely on population but also on the "number of people with whom contact can be made on a day-to-day basis from a particular location." The youth population is characterized by growing frequency in contact with others as well as by a substantial minority (one quarter) who freely give out their *keitai* numbers.[15] If we adopt Matsumoto's definition, youth could be described as increasing the "number of people with whom contact can be made on a day-to-day basis" and thus engaging in "urban interpersonal relationships." In this vein, youths' *keitai*-supported selective interpersonal relationships illustrate an intersection of the theories of Fischer and Matsumoto. It is not so much that the growth of youths' selective interpersonal relationships are a result of *keitai* ownership or their identity as youth, but rather that the growth is the result more generally of an expanded social network or increase in the number of people with whom contact can be made.[16]

Sidney Aronson (1971, 162) uses the term "psychological neighborhood" to describe changes in social networks of city dwellers due to the adoption of the telephone:

With the spread of the telephone a person's network of social relationships was no longer confined to his physical area of residence (his neighborhood, in its original meaning); one could develop intimate social networks based on personal attraction and shared interests that transcended the boundaries of residence areas. It is customary to speak of "dispersed" social networks to denote that many urban dwellers form primary groups with others who live physically scattered throughout a metropolitan area, groups which interact as much via the telephone as in face-to-face meetings.

Much research has criticized this sense of relationships transcending the boundaries of residence areas. Arguments include, "Those with whom people have close relationships actually do not live that far away" and "Telephone calls are most often made within the same region." However, the relationships of *chu-tomo* and *jimo-tomo*, discussed earlier, are important cases. Even if people live in the same neighborhood, if they cannot see each other on a regular basis, they are in a sense far apart. With *keitai*, it becomes possible to maintain connection with people who have become socially distant because of changes in lifestyles. With the telephone and *keitai*, people can develop an intimate social network based on personal attraction and mutual interests. We might describe this not in terms of "transcending distance" but as a small excursion from the patterns of everyday life.

Homophily in the Future

Whether or not *keitai* is the sole facilitator of greater choice of communication partners, what consequences can we expect from the shift toward selective rather than given or serendipitous relationality?

Fischer (1976; 1982) discusses homophily as an increase in the homogeneity of networks due to the expansion of individual choice in relationships. As indicated earlier, *chu-tomo* and *jimo-tomo* are highly homogeneous relationships. They have other characteristics as well. As pointed out by Chantal de Gournay (2002), *keitai* encourage the scaling down of relationships from diffuse social network to a group of close friends. People use *keitai* to maintain only comfortable relationships and associations with those with common interests. *Keitai* facilitate an insular life with little attention to the public and "the other."

From a somewhat different perspective, Hidenori Fujita (1991) has coined the term *shumi-en* (special-interest relationships) to describe selective interpersonal relationships. He suggests that while necessarily grounded in a foundation built through obligatory and binding *sha-en* (sodality)-type relationships organized by schools and companies, "The ideal society of the future is perhaps a '*shumi-en*-type crossover society' in which permissive, open *shumi-en* relationships increase and intersect with one another" (30). His belief is that selective interpersonal relationships "could possibly provide a basis for people to build individual identities, a plane that is different in kind from the definition and order of the school and workplace, the typical places of activity in modern industrialized society" (31). I also share this hope for an ideal future.

This relates back to the concepts of *chu-tomo* and *jimo-tomo*, which could be considered associations that inhibit efforts to proactively make new friends in the new environment of high school. To the school, they can be problematic relationships that resist institutional surveillance. We could, however, also consider these ties as means for making friendships outside of school (see note 4). There have been cases where *chu-tomo* came to the rescue for students who could not make friends or were being bullied at a new school. This does not constitute an example of Fujita's increasing and intersecting *shumi-en*-type associations and could be perceived as a limiting of relationships to solely *chu-tomo*. However, it is not the teachers' or parents' obligation to prohibit or control friendships with *chu-tomo* or *jimo-tomo*, nor would they be able to. We need to consider the broader social shift toward privatization and our perceptions of this shift without singling out the youth population as a site of concern. If this broader shift is overwhelmingly negative, then we need to address avenues for intervention.

Interpersonal relationships mediated by *keitai* are not solely the result of *keitai*; the same can be said of societal changes related to the adoption of *keitai*. If we truly want to understand the influence of *keitai* in the context of Japanese social life, we must examine conditions in place well beyond the ten years since *keitai* has made its appearance. This chapter has worked to provide this broader social and historical backdrop, contextualizing *keitai* use and adoption within long-standing trends and changes in Japanese society. Only after considering this historical flow of events can we begin to understand the changes brought about by *keitai* and the possibilities for the future.

Notes

1. My primary *keitai* research is as follows:

• Telephone survey of twenty-one *keitai* users conducted between November 15 and December 11, 1995 (Matsuda 1996a).

• Interview survey of youths in entertainment and shopping districts in Tokyo and Osaka by the Mobile Media Research Group. These were conducted in August and September 1996 in Tokyo's Shibuya and Harajuku districts; in January 1997 in Minami and Osaka; and in August 1998 in the same areas of Tokyo and Osaka (Matsuda et al. 1998; Okada and Tomita 1999).

• Questionnaire survey of college students in the Tokyo and Kansai areas, conducted by the Mobile Media Research Group. Surveys resulted in 590 valid responses and were conducted from May to June 1999 (Okada, Matsuda, and Habuchi 2000).

• Interview survey of seniors conducted by the Mobile Media Research Group in 2000.

• Questionnaire survey of employed persons aged 20 to 59 living within 30 kilometers of Tokyo, conducted by Misa Matsuda and NTT Ado at the end of January 2001. Surveys resulted in 400 valid responses (Matsuda 2001a).

• Questionnaire survey of college students in the Tokyo and Kansai areas conducted by the Mobile Media Research Group. Surveys resulted in 586 valid responses and were conducted from May to June 2001 (Matsuda 2001b).

2. Other data also suggest that the *keitai* Internet is mainly used for e-mail transmission (Cabinet Office 2002a).

3. In Jiyukokuminsha's yearly dictionary of new key words, *Gendaiyougo no Kisochishiki* (*Fundamental Knowledge of Contemporary Terms*), published since 1948, there is a section devoted to "Interpretation of Young People's Terms." *Jimo-tomo* and *ona-chu* appear for the first time in 1999, and *chu-tomo* in 2002.

4. Schools provide most of the opportunities for Japanese youth to make new friends. According to a study on youth (aged 18 to 24) of eleven countries including Japan and the United States (Ministry of Public Management 1999), the percentages of youth responses to "How did you get to know your close friends?" were, in Japan and in the United States, respectively, "at school/ university" 92.2 and 82.4; "at work" 39.6 and 47.9; "at a club or group other than at school/ university" 15.0 and 27.4; "in the neighborhood where you live" 12.6 and 44.7; "grew up in same area" 5.4 and 27.1; "in a public place such as bus, subway, restaurant, café, park" 4.8 and 26.2.

The lifestyle of Japanese students is also worth noting. According to the time use survey conducted in 2000 by the NHK Broadcasting Culture Research Institute (2002), a large portion of high school students' time is spent on academics (particularly school-related) activities. On weekdays, an average of 8 hours 18 minutes is spent on academics (out of which classes and other school activities take up 6 hours 24 minutes) and 1 hour 17 minutes on average for commuting. The amount of time on weekdays for "social obligations," which includes volunteer activities and local events, is zero. Students have 3 hours 55 minutes for "free-time activities," as distinguished from "obligatory activities" (work, school work, etc.) and "necessary activities" (sleeping, eating, etc.), which is less time than the 4 hours 44 minutes available for "free-time activities" to "all adults" and not much different from such time available to "employed people" (3 hours 33 minutes). In other words, the life of high school students revolves around school. (This tendency is strongest for junior high school students, who are still in compulsory education. By contrast, college students are extremely free of scholastic commitments). See Mitsuya and Nakano (2001) for an overview of the survey in English.

5. The study by the Youth Affairs Administration (Ministry of Public Management 1999) has been conducted every five years since 1972. Since the second study, there has been no change in the trend for the response of "at school/university" to the question of how youths met their close friends.

6. In light of the fact that a larger percentage of people in their thirties are "full-time homemakers" than in other age groups, some sort of age-related gender norm could be involved.

7. *Ie ni kaeru* ("I'm going home") and *kaeru* ("frog") are pronounced the same in Japanese.

8. This Nikaido Village case is not exceptional. Kato (2002) indicated similar research and results conducted by the National Institute for Japanese Language in the Tohoku district in 1949–50.

9. In research conducted by the NHK Broadcasting Culture Research Institute in June 1998 (Matsumiya 1999), among fifth through ninth graders the most common answer to the question "What should be done to prevent problematic behaviors at school, such as bullying and violence?" was "Parents should make frequent communication with their children" (43.3%). A prevalent lack of conversation between the parents and children is thought to be the cause of problematic behaviors.

10. According to data from Japan's Cabinet Office, the average length of time a *keitai* phone is owned before being replaced is 2.3 years. 〈http://www.esri.cao.go.jp/jp/stat/shouhi/0309shouhi.html〉.

11. The average age upon first marriage for men in 2001 was 29.0, and for women, 27.2. The average age is higher in urban areas (*Vital Statistics of the Ministry of Health, Labor, and Welfare*).

12. A similar "marital effect" was evident in the use of the household telephone before *keitai* adoption (Matsuda 2000b).

13. *Sha-en*, as proposed by Toshinao Yoneyama (1981), is largely an association that corresponds to *Gesellschaft*, but Chizuko Ueno (1994) equates it with obligatory relations such as *chi-en* and *ketsu-en* because of the lifelong employment and corporate housing arrangements that were in force up to the late twentieth century in Japan.

14. Much follow-up research has been conducted in Japan to verify this hypothesis (for example, Otani 1995; Matsumoto 1995).

15. Even in the Mobile Communication Research Group's (2002) data, 14.6 percent of students (junior high to college) said they would "give out their *keitai* number to anybody."

16. There is clearly still room for further debate regarding the appropriateness of equating people with whom contact can be made on a day-to-day basis as mediated by *keitai* and people with whom contact can be made on a day-to-day basis as brought about by urbanization.

7 | The Mobile-izing Japanese: Connecting to the Internet by PC and Webphone in Yamanashi

Kakuko Miyata, Jeffrey Boase, Barry Wellman, and Ken'ichi Ikeda

Internet Use in Japan

Is Internet communication in Japan different than in the United States? Does the proliferation in Japan of mobile phones that can connect to the Internet (we call them "webphones") affect who talks with whom online? In this chapter we compare the social relations in everyday life of Japanese users of mobile phones and personal computers (PCs).

Our investigation into the unique character of Japanese Internet relationships departs from the scholarly norm. Most accounts of the Internet have been universal ones. They have suggested that the global evolution of the Internet's population and use is following the course of American forebearers, with early non-American users being predominantly young, well-educated men exchanging e-mail over a wide range of people with whom they have strong and weak social ties (Chen, Boase, and Wellman 2002). Such accounts have assumed that the users and uses of the Internet will eventually be the same around the world and that current variations in Internet use exist only because other countries have not caught up to the United States or are too impoverished to do so.

Yet, societies often differ in their interpersonal relationships, socioeconomic systems, norms and values, and climate and geography. Not surprisingly, the users and the uses of the Internet often vary among societies (Miller and Slater 2000; Chen and Wellman 2004). The Internet is not a system floating ethereally above societies. It is embedded in the concrete realities of people, practices, and power (Wellman and Haythornthwaite 2002). For example, although a recent study of Catalonia described much less interpersonal e-mail use than in the United States, it also showed substantial use of the Web for dealing with institutions: obtaining information, booking theater tickets, finding plane schedules, and the like (Castells et al. 2003). East Asian use of the Internet is also different. Heavily broadbanded Koreans are immersed in multiplayer online games (Tkach-Kawaski 2003), and mobile-ized Japanese (and Scandinavians) often use both webphones and PCs (Akiyoshi 2004).

This chapter is the first published report of our study of the users and uses of the Internet in Japan. Survey data from 1,320 adult respondents were collected in late 2002 in the somewhat rural prefecture of Yamanashi, more than 100 kilometers west of Tokyo. The respondents were asked about their use of both webphones and PCs: how often they use them, what they use them for, and what sorts of relationships they have with the people they contact. The chapter focuses on the social characteristics and the social relationships of the users of Internet-connected webphones and PCs. We examine the Japanese context, different from the canonical U.S. context; compare the users of webphones and PCs; compare communication via mobile phones and PCs; analyze how webphones and PCs sustain strong and weak ties in social networks; and analyze how webphones and PCs are related to the provision of social support.

Our work addresses the ongoing debate about the effects of Internet use on community and social support. We conclude by discussing the implications of webphone use for the "mobile-ized" nature of Japanese society. We believe that our results are informative about the nature of Japanese communication and social networks, Japan-U.S. differences in Internet use, and the turn toward "networked individualism" that is happening in Japan as well as in North America (Meguro 1992; Nozawa 1996; Otani 1999; Wellman 2001; 2002).

The Shift from Solidary to Networked Communities

For over a century, the developed world has been experiencing a shift away from communities based in villages and neighborhoods and toward flexible partial communities based in networked households and individuals. A good deal of research is now showing that rather than destroying community, the Internet adds to existing relationships with community members: friends, acquaintances, relatives, and even neighbors (Wellman and Haythornthwaite 2002). This is not a static phenomenon. Communication through the Internet appears to be facilitating the turn away from bounded, holistic communities of kin and neighbors and toward far-flung, multiple, and partial communities.

Until now, thinking about the Internet has focused on its ability to enable communication across continents at nearly the speed of light. At the same time, Internet users have been "glocalized" (Wellman 2003), bound to their desktops by the wires connecting to the Internet even as their interactions range widely in space. It is time to consider a new era: how the peripatetic mobile users of the Internet communicate with the members of their social networks and communities (also see Rheingold 2002).

The Japanese have been at the forefront of this new era. Already experienced and active users of *keitai*—lightweight, feature-laden mobile phones—are using new models of mobile phones to communicate with social network members and to surf the Internet for information. Our study examines who is doing this and with what consequences for their social networks.

Mobile Culture in Japan and Around the World—An Emphasis on Youth

Accessing the Internet through mobile phones has already become integrated into daily life for a significant proportion of the Japanese population (Barnes and Huff 2003). By the end of May 2001, more than 40 million Japanese were able to access the Internet through their mobile phones, with the number rising 55 percent to more than 62 million by the end of March 2003 (Ministry of Public Management 2003). By contrast, only 4 million North Americans could use mobile phones to access the Internet in August 2000 (Funk 2001).[1] In fact, mobile phones that can access the Internet have been so rare in the English-speaking world that we had to coin a new word, webphone, to refer to them.

The four major Japanese webphone access providers are NTT DoCoMo, KDDI, Vodafone, and Tsu-ka (in order of number of subscribers, January 2004). Each uses a variety of Internet protocols: DoCoMo's i-mode is the most popular, followed by WAP (Wireless Application Protocol) and WAP2. Although WAP2 is the least popular at present, it is rapidly gaining a foothold in the market because it permits advanced 3G (third-generation) services that provide GPS (global positioning system), video clips, higher speed, and other advanced features (Kageyama 2003).

Japanese webphones have relatively large screens compared to all but the most recent American mobile phones. Sending e-mail through Japanese webphones is similar to sending e-mail through PCs, although users have to cope with less user-friendly telephone keypads. Users enter the e-mail address of the recipient, a subject line, and then the contents of their message. Moreover, webphones can send and receive e-mail to and from PCs as well as other webphones.

The percentage of young adults in Japan who use webphones to e-mail is much higher than in the United States and many parts of Europe, where webphone e-mail has failed to attract a majority of people from any age group. This difference is partly due to marketing strategies of Japanese mobile phone providers that catered to the desires of youth and young adults. As Okada describes in chapter 2, young people were probably predisposed to send e-mail via webphone because of their extensive use of pagers in the 1990s to contact friends and organize social activities. (Parents who wanted to keep tabs on their children's activities also spurred the use of pagers.) This incorporation of pagers into everyday routines set the stage for the adoption of webphones, with their advantage of smoothly integrating voice and message contact. After gaining a foothold in the youth market, webphone providers increased bandwidth and Web interfaces, making their services more attractive to a wider audience.

Ethnographic studies of webphone use indicate a concentrated, active use of mobile phones to expand and enhance contact with close friends and immediate family (see the introduction; chapter 2). The advantages that webphones offer Japanese youth are probably similar to the advantages that ordinary mobile phones (those that cannot be used to access the Internet) offer youth in other countries. Youth adopt mobile phones worldwide to increase their autonomy and the quality of their ties with friends. For example, European youth are more likely than their parents to use mobile phones to build their social networks and to tell parents their whereabouts (Ling 2001; 2004). Furthermore, mobile phones have become incorporated into youth culture to such an extent that text messages, air time, and even mobile phones themselves are heavily shared, binding youth closely together. Text message exchanges are often incorporated into face-to-face contact with peers during "hang-out time." When messages and phones are shared among the group, they add to the interaction of the entire group rather than only of their owners (Weilenmann and Larsson 2001; Taylor and Harper 2003).

Will young users continue to rely on webphones as they grow older? On the one hand, the desire to be in constant contact with friends may dissipate as young adults enter more instrumental relationships at the workplace and save their recreational time for contact with spouses and family at home. On the other hand, heavy habitual use of this technology among friends and family may continue as people age and continue to integrate webphones into their work and domestic relationships.

In contrast to young adults and youth, older adults first encountered the Internet by using PCs to e-mail and surf the Web (Miyata 2001). The mobile phones they first used were not able to access the Internet. Hence, it is possible that older adults have not developed the habit of using mobile phones to access the Internet even when their new webphones have this capability.

Internet Users in Yamanashi

Our study of Internet users is based on a random sample survey of 1,320 adults in Yamanashi prefecture, Japan. Yamanashi is a mixed rural and urban area located in the center of Japan. It is typical of Japan (outside of the Tokyo and Osaka urban agglomerations) in the characteristics of its population and its Internet users (table 7.1).

Within the Yamanashi prefecture, forty neighborhoods were randomly selected by postal code, with a random selection of thirty-three individuals within each of those neighborhoods. Potential respondents were chosen from a voters' list of people 20–65 years old. Surveys were in paper form, delivered in person and collected in person three weeks later. Three quarters (76 percent) of the selected individuals completed the survey, providing a total sample size of 1,002 respondents. Surveying took place between November 15 and December 5, 2002.

Table 7.1
Internet Activities in Yamanashi Prefecture

	Population (Feb. 2003)	Internet Users[a] (% of Population)		
Japan	127,450,000	46.4		
Yamanashi	888,210	44.5		

	Exchange Information (% of Users)	Send Information (% of Users)	Gather Information (% of Users)	Other (% of Users)
Japan	39.5	5.6	32.4	14.0
Yamanashi	37.8	5.7	30.4	13.4

a. Based on survey of time use and leisure activities, NHK Broadcasting Culture Research Institute (2001).

For our analysis, we divided respondents into three types: those who use both webphones and PCs, those who use only webphones, and those who use only PCs. Our rationale was that those using only webphones or only PCs might have different characteristics and patterns of use than those who use both media.

Age, Gender and Mobile/PC Use

Age Age marks the largest difference in use of e-mail by mobile phone and PC. The percentage of people using both webphones and PCs declines dramatically with age. A large majority of young adults in their twenties access e-mail through webphones (figure 7.1). About half (46 percent) of the respondents in their twenties use both the webphone and PC, while 46 percent only use webphones. In sum, 92 percent of all respondents in their twenties use webphones for e-mail.[2] The use of webphones by those in their twenties is so marked that these people make up 39 percent of all webphone users even though they constitute only 21 percent of the sample. They are 1.9 times as likely to be webphone users than the average respondents. Older adults in their thirties also are disproportionately high webphone users, constituting 29 percent of all webphone users though only 20 percent of the sample. They are 1.4 times as likely to be webphone users than the average respondents. The disproportionate use of webphones by young adults is so great that adults aged 20–39 own two thirds (68 percent) of all webphones though composing only two fifths (41 percent) of the sample.

Percentage of e-mail users

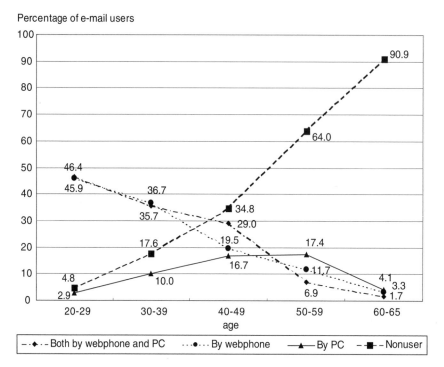

Figure 7.1
webphone and PC e-mail users, by age.

By contrast to low webphone e-mail use by adults over age 40, the percentage of respondents using only PCs to exchange e-mails increases until the age of 60. Moreover, older adults are more apt to use only PCs for e-mail, whereas younger adults are more apt to use both webphones and PCs. Not only do the ways of accessing the Internet vary by age, but so does the frequency of sending e-mail. Japanese over age 50 are much less likely to exchange e-mail (or use the Internet) at all, by contrast to North America, where those aged 50+ frequently exchange e-mails and use the Internet (UCLA Center for Communication Policy 2003).

Gender Gender is also related to the mode of accessing the Internet. For each age group, different percentages of women and men use webphones, PCs, or both (table 7.2). For example, more women (47 percent) than men (45 percent) in their twenties use only webphones. Gender differences change for older age groups. In general, men in the older age groups are somewhat more likely to use a combination of webphones and PCs: 38 percent of men in their thirties use both to e-mail, compared to 33 percent of women in their thirties. While 35 percent of men in their forties use

Table 7.2
Webphone and PC Users by Gender and Age

Age	Men (%)				Women (%)			
	Use web-phone and PC	Use web-phone	Use PC	Nonuser	Use web-phone and PC	Use web-phone	Use PC	Nonuser
20–29	44.7	44.7	1.9	8.7	48.1	47.2	3.8	0.9
30–39	38.1	32.0	11.3	18.6	33.3	41.2	8.8	16.7
40–49	35.2	17.2	20.5	27.0	21.2	22.2	12.1	44.4
50–59	6.5	11.3	23.4	58.9	7.3	12.2	11.4	69.1
60–65	3.6	5.4	8.9	82.1	0.0	1.5	0.0	98.5

both media, only 21 percent of women in their forties use both. Many middle-aged men make heavy use of their PCs at work and have less need for webphones. Moreover, Japanese women have historically used PCs less than men (Ono and Zavodny 2004).

Webphone Expertise

As we have seen, many young adults use both webphones and PCs to access the Internet. In addition, they are more likely to use only webphones to access the Internet. However, age is not the sole predictor. Yamanashi respondents with lower levels of technological skill are more likely to use only webphones to access the Internet (table 7.3). The lower the respondents rate their ability to use various kinds of technology, the more they tend to use only webphones.[3] By contrast, those using PCs (either alone or in combination with webphones) report themselves to be more skilled in using technology. This accords with a nationwide Japanese survey that found that people using both webphones and PCs had relatively higher levels of perceived skill using keyboards (Ikeda 2002).

Webphone and PC Contact with Social Networks

Do people maintain different kinds of networks when they use webphones, PCs, or both to e-mail? The differences could be because of the kinds of people using these different media or because of the nature of the media used. For example, Japanese businessmen are reluctant to use the more intimate, less formal webphones to send business-related e-mail.[4] To address this question, we examine how much e-mail is exchanged, and the extent to which e-mail is exchanged with local or distant ties, and with strong and weak ties.

Table 7.3
Multiple Regression Analysis: Demographic Variables and Perceived Ability to Use Technology

	Use Webphone and PC			Use Webphone Only		
	B	S.E.	Exp(B)	B	S.E.	Exp(B)
Gender (0 = female, 1 = male)	−0.495*	0.228	0.609	−0.120	0.201	0.887
Age (ref. 20–29)						
30–39	−0.714**	0.272	0.490	−0.296	0.245	0.744
40–49	−0.870**	0.299	0.419	−1.289**	0.277	0.275
50–59	−2.098**	0.374	0.123	−2.042**	0.324	0.130
Over 60	−2.259**	0.798	0.104	−3.340**	0.586	0.035
Education (reference = middle school)						
High school	0.344	0.608	1.411	0.694	0.362	2.001
Some college	0.157	0.620	1.170	0.266	0.392	1.305
Undergraduate degree or more	0.468	0.625	1.597	−0.623	0.431	0.536
Employment Status (ref. full-time)						
Working Part-time	−0.231	0.327	0.793	0.046	0.270	1.047
Self-employment	0.273	0.411	1.313	−1.075*	0.503	0.341
Student	1.180*	0.598	3.254	−0.615	0.569	0.541
Home Maker	−0.537	0.377	0.585	−0.380	0.315	0.684
Other type	−0.857*	0.424	0.424	−0.009	0.303	0.991
Unemployed	0.118	0.548	1.126	−0.630	0.480	0.532
Partner (0 = no, 1 = yes)	0.432	0.345	1.541	−0.564	0.305	0.569
Kids living together (0 = no, 1 = yes)	0.414	0.294	1.513	0.081	0.269	1.085
Perceived Ability to use technology	0.323**	0.032	1.381	−0.051*	0.023	0.950
Constant	−6.081	0.793	0.002	0.824	0.519	2.279
Cox & Snell R^2	0.311			0.174		

Notes: N = 969
* $p = .05$.
** $p = .01$.

Frequency of E-mail Contacts

Slightly more e-mails are sent by webphones than by PCs. Those who only use webphones send an average of about six e-mails per day, whereas those who only use PCs send two thirds fewer, an average of about two e-mails per day. The greatest number of e-mails are sent by those using both webphones and PCs, about six e-mails per day via webphone and about two e-mails per day via PC, for a total of eight e-mails per day.

Those who only use webphones and those who use both webphones and PCs send, on average, the same number of webphone e-mails per day: about six. Those who only use PCs and those who use both PCs and webphones also send, on average,

Table 7.3

(continued)

Use PC Only			Nonusers		
B	S.E.	Exp(B)	B	S.E.	Exp(B)
−0.053	0.315	0.949	0.692*	0.270	1.997
1.267*	0.514	3.550	1.637**	0.479	5.141
2.087**	0.517	8.062	2.390**	0.471	10.916
3.252**	0.541	25.843	2.882**	0.479	17.858
2.399**	0.780	11.014	3.738**	0.592	41.996
−0.081	0.676	0.922	−0.325	0.397	0.723
0.762	0.707	2.142	−0.242	0.448	0.785
1.363*	0.692	3.908	−0.713	0.480	0.490
−0.097	0.522	0.908	0.270	0.347	1.311
0.443	0.438	1.558	0.260	0.417	1.297
−18.122	8611.048	0.000	0.554	1.279	1.741
0.615	0.471	1.850	0.703	0.369	2.019
0.952*	0.423	2.592	−0.230	0.370	0.794
0.303	0.841	1.353	0.648	0.635	1.912
0.082	0.429	1.086	0.425	0.387	1.529
0.095	0.315	1.100	−0.494	0.277	0.610
0.297**	0.040	1.346	−0.386**	0.030	0.680
−9.808	1.039	0.000	2.474	0.646	11.865
0.178			0.504		

the same number of PC e-mails per day: about two.[5] In short, using both media adds to the number of e-mails sent: one medium does not replace the other. This finding—that using both a PC and a webphone is associated with more frequent e-mail use—is congruent with other studies' findings that the more media used, the greater the overall amount of communication (Haythornthwaite and Wellman 1998; Quan-Haase et al. 2002; Hogan 2003). It further suggests that different sorts of e-mail are being sent by webphones and by PCs.

Local and Long-Distance Contact

E-mails sent by webphones are more likely to be more local than those sent by PCs (table 7.4). People who use only webphones are especially likely to send e-mail to people who live nearby. People who use both media are more likely to use webphones to

Table 7.4

Webphone and PC E-mail Users by Distance between Communicators

Distance between Communicators	Use Webphone and PC		Use Webphone Only	Use PC Only
	E-mails by webphone (%)	E-mails by PC (%)	E-mails by webphone (%)	E-mails by PC (%)
Living together	19.1	11.5	18.1	0.0
less than 10 minutes away by car	12.7	4.9	13.5	3.8
less than 1 hour away by car	42.7	39.3	46.8	34.6
less than 5 hours away by car	19.7	34.4	18.1	38.5
more than 5 hours away by car	5.1	4.9	2.9	15.4
Living abroad	0.6	4.9	0.6	7.7

send messages to others living nearby (within one hour's travel by car) and PCs to send messages to others living further away. People who only use PCs are the most likely to communicate with others living further away.

These findings are consistent with ethnographic research discussed above showing that young Japanese use webphones to send quick e-mails to nearby friends. Their exchanges may simply be about trivial matters, used to maintain a feeling of connectedness or to arrange things, such as asking a spouse to pick up food on the way home from work. The mobile nature of webphones makes them perfect for arranging meetings or changing plans at the last minute (Ling and Yttri 2002; Smith 2000). We believe that users reserve PC e-mail for richer, in-depth contact because typing messages on a webphone is more difficult than typing them on a PC keyboard.

Contact with Weak and Strong Ties

Does the Internet help maintain social networks? To address this question, we examine three properties of social networks: the number of supportive ties, network diversity, and network size.

Number of Supportive Ties   Does the immediacy of portable webphones or the range of PCs facilitate the availability of social support? To measure the number of supportive ties, respondents were asked to report the number of network members who would give them words of encouragement (emotional support), provide them with a small amount of money (financial support), or aid them in tasks such as moving house or providing other goods and services (instrumental support). The number of network members in all three categories was summed and divided by 3 (table 7.5).

Table 7.5
Number and Diversity of Network

	Use Web-phone and PC	Use Web-phone Only	Use PC Only	Nonusers	F
N	251	244	111	378	
Number of supportive ties***	2.53	2.49	2.49	2.38	2.67*
Diversity of contancts	4.09[ab]	3.60[b]	4.48[a]	3.63[b]	3.48*
Number of new year's cards sent	60.70[b]	39.55[c]	90.77[a]	47.21[bc]	19.87**

Notes: * $p = 0.05$. ** $p = 0.01$.
Means that share the same letter do not indicate a significant difference by Scheffe test.
Only those means with different letter pairs are significant (i.e., a and b or b and c).
***Comprises informational, emotional, and financial support.

The data show that users of both modes of Internet access—webphones and PCs—have the highest number of supportive ties (2.53), followed closely by users of PCs only (2.49), users of webphones only (2.49), and nonusers (2.38). Are these small differences in the supportiveness of networks due to the mode of communication used or to some other factors? We used regression analysis to take into account intercorrelations among key variables. For example, respondents who use only PCs tend to be older than those who use webphones (alone, or with PCs). Social networks grow in size with age, greater time in the workforce, and experience.

Having more supportive ties is associated with older age, being a member of an organization, and sending greater amounts of e-mail by webphones (table 7.6).[6] Hence, the differing number of supportive ties for users of webphones and PCs reflects disparities in the age of users of each mode rather than the mode itself. People tend to add to their supportive ties as they grow older. People who are members of organizations tend to gather more supportive ties, especially when they are active members.[7] Frequent communication also matters. Those who send more than six e-mails by webphone have significantly larger supportive networks than those who do not send any e-mails by webphone. It may be that communication leads to more support; and it may be that those who need more support are in more communication (for similar pre-Internet Canadian data, see Wellman 1979; Wellman and Wortley 1990).

Network Diversity Diverse networks provide access to new sources of information and resources. The more different types of people one knows, the more social milieus one is likely to be connected with (Feld 1982). Network diversity has many different facets, such as occupational diversity, gender diversity, and ethnic diversity. We focus on occupational diversity because people who work in different occupations often

Table 7.6
Regression Analysis: All Respondents

Predictors	Supportive Networks	Diversity of Contacts	No. of New Year's Cards Sent
Gender (0 female, 1 male)	0.046	0.320	11.512*
Age (ref. 20–29)			
30–39	−0.232**	0.438	18.942**
40–49	−0.279**	0.307	14.804*
50–59	−0.182	1.045*	21.349**
60–65	−0.104	0.424	25.338**
Education (ref. middle school)			
High school	0.113	−0.377	9.116
Some college	0.061	0.008	13.511
Undergraduate degree or more	0.021	0.894	27.848**
Employment status (ref. full-time)			
Working part-time	0.042	0.049	−6.639
Self-employed	−0.071	1.055*	5.538
Student	0.115	−2.593**	−7.932
Homemaker	0.062	0.255	−15.172*
Other	0.180*	−0.446	−6.001
Unemployed	−0.054	−0.838	−0.090
Partner (0 no, 1 yes)	−0.070	−0.032	27.944**
Kids living together (0 no, 1 yes)	0.044	−0.369	−0.814
Associations	0.473**	3.618**	25.189**
Number of e-mails sent by webphone yesterday (ref. nonusers)			
1–5	0.065	0.374	3.694
6–10	0.255*	1.473**	−1.530
More than 11	0.293*	1.153	−1.332
Number of e-mails sent by PC yesterday (ref. nonusers)			
1–5	−0.010	−0.383	14.954*
6–10	−0.210	0.799	−12.636
Supportive networks		0.843**	8.613**
Diversity of contacts	0.036**		3.205**
Number of greeting cards	0.001**	0.011**	
Constant	1.628	−3.542	−68.098
Adjusted R^2	0.126	0.237	0.233
N	817	817	817

* $p = .05.$
** $p = .01.$

come from different social backgrounds (Lin 2001). Respondents were asked to indicate if they had any relatives, friends, or acquaintances in any of fifteen categories of diverse occupations. A count of the number of different occupation categories was made for each respondent, yielding a score from 0 to 15, with a higher score indicating greater diversity of contact.

Those who use only PCs for e-mail have more diverse networks (mean score = 4.5) than the users of both webphones and PCs (4.1), nonusers (3.6), and webphone-only users (3.6) (see table 7.5). As with our analysis of supportive ties, we used regression analysis to see if this difference is actually linked to the mode of communication used. The data show that, as with supportive ties, organizational involvement and frequent communication are positively associated with network diversity. In fact, involvement with formal and informal associations are the two strongest predictors of having occupationally diverse networks (see table 7.6). Not surprisingly, the more groups one joins and is active in, the more varied are the occupations encountered. The data show that PC-based e-mail facilitates political participation, whereas webphone–based e-mail does not (Ikeda, Kobayashi, and Miyata 2003).

Network Size We hypothesized that webphones (like other mobile phones) are especially important for maintaining strong ties (close friends and family) and that PCs help to maintain weak as well as strong social ties. Thus, we tried to discern the size of weak-tie networks. Yet, this is hard to do in a short, closed-ended survey. To estimate the size of weak-tie networks, we asked respondents how many New Year's greeting cards they had sent in the past year.[8] As the great majority of ties in a network are relatively weak (Bernard et al. 1990; Watts 2002), this is a crude estimate of the number of weak social ties that respondents have.

Those who send e-mails only by PC sent a significantly larger number of greeting cards, a mean of 91 (see table 7.5). Users of both webphones and PCs trail substantially, sending a mean of 61 cards. However, they send somewhat more than nonusers (47) and webphone–only users (40).

Regression analysis shows that older age is positively associated with sending New Year's cards, as is organizational involvement, being married, and having at least a university education (see table 7.6). All these variables provide contexts for meeting other people. Sending PC-based e-mail is also positively associated, supporting our hypothesis that sending more PC-based e-mail is associated with having larger networks and a larger number of weak ties.

As these associations between variables are similar to those found for the number of supportive ties and for the diversity of networks, it is not surprising that all three network variables are related. Larger numbers of weak ties are associated with more network support and more diverse networks. In general, those who use only PCs and those who use both PCs and webphones have more diverse networks and larger numbers of weak and strong ties than those who use only webphones and those who do not use either media. Their social networks lack richness in comparison to those who use PCs, whether in combination with webphones or not.

The Networks of Webphone and PC E-mail Users

The preceding results suggest that the natures of webphone and PC e-mail are different. The media are not fully substitutable, with the handiest one being used. For example, because PC e-mail is entered by keyboard and is read on bigger screens, it should contain longer messages than webphone e-mail, which must be typed on cumbersome thumbpads and read on small screens. Because weak social ties are usually contacted less frequently, they may likely be contacted with longer, PC-based e-mails that explain their purpose in more detail than the common webphone messages: "I love you," "Meet me in 10 minutes," and "Please bring home a bottle of beer." In this way, the medium affects not only the message but also the kind of person to whom the message is sent.

Network Characteristics and Modes of Communication

To investigate the relationship between network characteristics and modes of communication, we focused on those respondents who use both webphones and PCs to send e-mails (table 7.7). We have already seen that such dual users have more strong and more weak social ties. Regression analysis also shows that for dual users the frequency of sending PC e-mail is more strongly associated with being in diverse social networks than is the frequency of sending webphone e-mail. The more e-mails sent, the greater the statistical association, with the strongest association being among people who sent PC e-mail at least six times on a particular day. Moreover, moderate PC e-mail users (those who sent one to five messages yesterday) sent more New Year's cards, indicating a larger number of weak ties in their networks. There is no association between sending webphone e-mail and sending New Year's cards. Indeed, although they are not statistically significant, the data show that the more webphone messages sent, the smaller the greeting card network.

As another indicator of the size of networks, respondents were asked to report the number of e-mail addresses kept in their webphones or PCs. Those using both webphones and PCs keep an average of thirty-six addresses in their webphones and twenty-three in their PCs. Even if the addresses they keep in both devices overlapped completely, we would know they have exchanged e-mail with at least an average of thirty-six people. This number is much higher than the twenty-six addresses of those who only use a webphone and the seventeen addresses of those who use only a PC.

Most of the social ties in PC address books are weaker ties of acquaintanceship rather than stronger ties with close friends and family. This is because respondents report that

Table 7.7

Regression Analysis: Users of Both PCs and Webphones

Predictors	Supportive Networks	Diversity of Contacts	Number of New Year's Cards Sent
Gender (0 female, 1 male)	0.058	0.379	11.734
Age (ref. 20–29)			
30–39	−0.260*	−0.237	19.720*
40–49	−0.479**	−0.043	21.863*
50–59	−0.133	−0.302	21.192
60–65	−1.908**	−0.255	113.682*
Education (ref. middle school)			
High School	−0.193	−3.507*	12.985
Some college	−0.180	−3.619*	19.293
Undergraduate degree or more	−0.157	−2.877	23.376
Employment Status (ref. Full time)			
Working part-time	0.153	−0.447	−3.649
Self-employed	0.214	−0.253	−12.802
Student	−0.123	−1.843*	−13.582
Homemaker	0.263	−0.240	−10.120
Other	0.540*	−0.310	−16.096
Unemployed	0.083	−0.321	1.447
Partner (0 no, 1 yes)	−0.037	0.706	24.654*
Kids living together (0 no, 1 yes)	−0.049	−0.608	6.399
Associations	0.305	2.710**	21.726
Number of e-mails sent by webphone yesterday (ref. nonusers)			
1–5	0.020	0.780	−4.141
6–10	0.174	0.900	3.319
More than 11	0.292	0.497	−7.952
Number of e-mails sent by PC yesterday (ref. nonusers)			
1–5	−0.120	−0.367	21.631**
6–10	−0.420	3.727**	−13.229
Supportive networks		0.836**	17.040**
Diversity of contacts	0.045**		1.915
Number of greeting cards	0.003**	0.006**	
Constant	2.061	0.738	−77.142
Adjusted R^2	0.177	0.228	0.303
N	219	219	219

Notes: * $p = .05.$

** $p = .01.$

These entries are unstandardized multiple regression coefficients.

they only exchange supportive e-mail with between one and three network members, presumably those with whom they have strong ties. This may be a low estimate, as both earlier Japanese (Otani 1999) and Canadian data (Wellman 1979; Wellman and Wortley 1990) show a mean of five strong ties.[9] The inference is clear: only a few of the names in the PC and webphone address books represent strong supportive ties. Most represent weaker ties, but still significant enough to be entered into address books. Moreover, many weak ties are not in these address books. These correspondents are answered when they contact the respondents, or their infrequently used addresses are found in other ways.

The number of strong ties maintained through the use of e-mail is greatest for those using both a webphone and a PC. These users reported an average of 2.8 supportive people in their social networks; those using only a webphone reported 2.6, and those using only a PC reported 1.1 supportive people ($F = 11.33$, $p < .01$). Thus, not only do people who use both webphones and PCs have larger networks and more contact with network members, they also have a larger number of supportive ties in their networks. Here, too, using multiple media to communicate with network members is associated with being in larger, more communicative, and more supportive social networks.

Conclusion

Summary

The use of webphones and PCs varies by age and gender. Those who only use webphones tend to be in their twenties and thirties, and rate themselves low in their ability to use technology. In addition to their heavy use of webphones, people in their twenties and thirties are also heavy users of PCs, with many being dual users of both webphones and PCs. Between the ages of 30 and 59 there is an increase in the proportion of respondents who exchange e-mail through PC only.

Although gender differences in exchanging e-mail are not great among those in their twenties, they become more pronounced among older respondents in their thirties and forties. In these latter age groups, a greater percentage of men combine media, possibly because they are more likely to use PCs at work.

There is significant variation by medium in the amount and kinds of contact with social networks. People use webphones more than PCs to send e-mails, even when they have both webphones and PCs available. Yet webphone e-mail tends to be with people who are nearby, whereas PC e-mail tends to be with people who are further away as well as nearby. Webphones are most often used to exchange short, quick messages with close friends and family, keeping emotionally connected, and to organize meetings or facilitate arrangements of everyday activities. Those who send many webphone e-mails have more supportive ties. By contrast, PC-based e-mail messages are apt to be exchanged with weak as well as strong social ties, and are not as likely to be linked to imminent physical get-togethers. Those who send many e-mails have larger and more diverse networks.

Using two modes of communication supports larger social networks. To a certain extent, the two modes are complementary, with webphone e-mail used differently than PC e-mail for maintaining ties. Webphone e-mail is more closely associated with strong supportive ties. PC e-mail is used to maintain weak as well as strong ties. Webphones support intensive relationships with loved ones and other strong ties. They are interfaces for intimate contact and enable intimates to be accessible anywhere, anytime. Their personal nature and small screens afford more private e-mail conversations than the often-shared PCs at home and office. By contrast, PCs support more extensive networks. They are interfaces with the rest of the world as well as with intimates and allow users to be selective in whom they contact and when contact takes place.

It is probable that within a short time the great majority of Yamanashi residents will use both webphones and PCs to exchange messages. Hence, those respondents who use both webphones and PCs are a harbinger of this future. Not only do such dual-mode users have more strong and weak ties with whom they can exchange e-mail but they are in more frequent contact with them. Moreover, those who exchange move e-mails tend to have a greater number of strong ties.

Japan: A Mobile-ized Society

The proliferation of the Internet is facilitating social changes that have been developing for decades in the ways that people contact each other, interact, and obtain resources. Now the emphasis on connectivity in the developed world appears to be moving from transportation to communication: from airport terminals and road networks to computer terminals and networks. In this technologically changing world, the Yamanashi study indicates the interplay between communication modes and social networks in a society that is increasingly emphasizing computer-mediated communication. The study also shows that the North American Internet experience is not necessarily the norm for other developed societies.

There are important similarities, too. The Yamanashi evidence, taken together with research done in North America and Europe, shows that the Internet is not a self-contained transcendental world but is immanent in everyday life (Wellman and Hogan 2004). Thus, these Japanese data are consistent with North American data that show that people who use multiple media tend to have stronger and more frequently contacted ties (Wellman 1988; 1992; Haythornthwaite and Wellman 1988; Wellman, Boase, and Chen 2002; Quan-Haase et al. 2002; Castells et al. 2003). Rather than operating at the expense of the face-to-face world, the Internet is an extension of it, with people using all means of communication to connect with friends and relatives. The

Internet is another means of communication that is being integrated into the regular patterns of social life. Other research by our NetLab suggests that this integration of online and offline life is also true for communities of practice at work (Haythornthwaite and Wellman 1998; Koku, Nazer, and Wellman 2001; Koku and Wellman 2004).

The Yamanashi study highlights how different forms of computer-mediated communication are used for different purposes. Webphones are most often used to exchange short, quick messages with those who are physically nearby. They are less often used to access the Internet, and they are rarely used to gather information about social issues or to participate in online communities. It appears that webphones are useful to maintain strong ties with people who are socially or physically close (see also Rivière and Licoppe's (2003) French data). However, webphones are not used much to contact weaker ties or to develop more diverse networks. This may be because webphones are not well suited to accessing sites where weak-tie relationships may be formed, such as chat rooms and issue-oriented sites. Then, again, there may be a population cohort effect because webphone–only users are younger adults who may not be as interested in discussing issues as middle-aged and older Japanese adults are.

Our results call into question the traditional stereotype of Japan as a closed, bounded society (see Meguro 1992; Nozawa 1996; Otani 1999; White 2002). People are on the move, getting information, making arrangements, and contacting friends and relatives through wireless webphones and wired PCs. Young adults in their twenties are especially heavy users of webphones. Yet, the fact that those in their thirties also use webphones heavily suggests that this is not a twenty-something phase. Rather, it is a trend toward mobile connectivity, which we call mobile-ization, that is becoming increasingly prevalent throughout Japanese society. As those in their twenties and thirties grow into middle age, we expect their mobile communication to continue, although tempered by a heavier reliance on faster and more informative big-screen PCs at work and at home.

Toward Networked Individualism

The Yamanshi results support our NetLab's working assumption that developed societies worldwide are becoming less bound up in neighborhood and kinship solidarities and more constituted as networked individualism. The traditional criterion of "community" is of a neighborhood or village in which many of the residents know each other (Wellman and Leighton 1979). Yet, communities have become far-flung and privatized, shifting from groups interacting face-to-face in visible public spaces to individuals communicating privately over much larger distances. Communities and societies have been shifting toward networked societies where boundaries are more permeable, interactions are with diverse others, and linkages switch between multiple networks

(Wellman 1997; 1999; 2001; Castells 2000). Hence, many people communicate with others in ways that spread out across group boundaries. Rather than relating to one group, they cycle through interactions with a variety of others, at work or in the community. Their work and community networks are diffuse, sparsely knit, with vague, overlapping social and spatial boundaries.

Changes in the nature of computer-mediated communication both reflect and foster the development of networked individualism in networked societies. Internet and mobile phone connectivity is to persons and not, by contrast, to wired telephones that ring in a fixed place for anyone in the room or house to answer. The developing personalization, wireless portability, and ubiquitous connectivity of the Internet all facilitate networked individualism as the basis of community. The person has become the primary unit of connectivity, rather than the household, work unit, voluntary organization, or social group. Because connections are to people and not to places, the technology affords shifting of work and community ties from linking people-in-places to linking people anywhere. I alone am reachable wherever I am: at home, office, highway, shopping center, hotel, or airport. Computer-supported communication is everywhere but is situated nowhere. The person has become the portal.

This shift facilitates personal communities that supply the essentials of community separately to each individual: support, sociability, information, social identities, and a sense of belonging. The person rather than the household or group is the primary unit of connectivity. Just as continuous Internet availability makes people sitting in place at their desktop PCs easily reachable, the proliferation of webphones and wireless computing is making people more easily reachable without regard to place. Supportive convoys travel ethereally with each person (Castells 2000; Ling and Yttri 2002; Katz 2003; Katz and Aakhus 2002). Webphones allow instant access to intimates and imminent gatherings. PCs allow more leisurely messages to intimates as well as more formed messages to a much larger of set of correspondents. By contrast to the one-to-one nature of webphones, PCs are more easily portals to chats and message exchanges with large sets of people.

The technological development of computer networks and the societal flourishing of social networks are affording the rise of networked individualism in a positive feedback loop. Just as the flexibility of less bounded, spatially dispersed social networks creates demand for collaborative communication and information sharing, the rapid development of computer communications networks nourishes societal transitions from group-based societies to network-based societies (Castells 2000; Wellman 2002). Moreover, networked societies are themselves changing in character. Until quite recently, transportation and communication have fostered place-to-place community, with expressways and airplanes speeding people from one location to another (without much regard to what is in between). Telephone and postal communication have been delivered to specific, fixed locations. At present, communication is taking over many of the functions of transportation for the exchange of messages. Communication itself is becoming more mobile, with mobile phones and wireless computers proliferating.

Not only has the volume of communication increased, we believe that the velocity of communication has also increased in Japan and elsewhere in the Internet-using world. Although e-mail is asynchronous and does not necessitate instantaneous response, in practice many people respond quickly. Moreover, distant network members who did not have much contact when limited to face-to-face, telephone, or postal communication are now in frequent Internet contact. Telephone costs are plummeting, and mobile phones mean availability is increasing even when people are away from home or workplace. Internet communication essentially ignores distance and minimizes per message costs. The further the distance between network members, the more important is the Internet in maintaining ties (Quan-Haase et al. 2002; Chen, Boase, and Wellman 2002).

Thus, the impact of computer-mediated communication will be that people have larger-scale social networks: more people, more communication, and more rapid communication. Witness how those Yamanashi residents who use both webphones and PCs have larger social networks and more frequent communication with these networks. Large social networks with high frequencies and velocities of communication allow information—and perhaps knowledge—to diffuse more rapidly.

It is not clear if the high use of computer-mediated communication will foster more densely knit communities—good for conserving resources—or more sparsely knit communities—good for obtaining new information and other resources. On the one hand, some characteristics of the Internet foster denser networks: the ability of Internet users to communicate simultaneously with multiple others, and the ease of copying and forwarding messages to others. In such cases, it is more likely for the friend of my friend to become my friend. On the other hand, as social networks become larger, it is often more difficult for them to maintain their density. As the size of the network increases arithmetically, the number of ties must increase geometrically to maintain the same level of density.

The Yamanashi results suggest that webphone use may foster a complex digital divide. Social scientists have been discussing the digital divide—the gap between users and nonusers of the Internet—at least for a decade. More recently, they have highlighted the gap between those who merely have marginal access to the Internet and those who are active, informed users: what Castells (2000) calls "the interactors" and "the interacted" (see also Chen and Wellman 2004). Yamanashi shows us that even the "interactors" may themselves have limited use of the power of the Internet when they only use webphones. Limited screen size and access speed restrict the use of websites, keyboard limitations constrain the length and complexity of messages, and in Yamanashi at least, a more limited range of people is contacted. Moreover, webphone messages are overwhelmingly segregated exchanges between two persons, while PC-based e-mail involves bringing multiple others into conversations (Geser 2004). The result is a mixture of segregated, bilateral Web conversations integrated with group-based chats with physically present friends (see chapter 13).

The turn toward networked individualism before and during the age of the Internet suggests more people are maneuvering through multiple communities of choice, where kinship and neighboring contacts become more of a choice than a requirement (Wellman 1999; Greer 1962). This phenomenon started in Japan before the advent of the Internet (Nozawa 1996; Otani 1999), but webphones and PCs are probably accelerating it. Webphone users have the possibility of contacting whomever they want, whenever they want, and wherever they are located. This suggests a fragmentation of community, with people increasingly operating in a number of specialized communities that rarely grab their entire impassioned or sustained attention. The multiplicity of communities should reduce informal social control and increase autonomy. It is easier for people to leave unpleasantly controlling communities and increase their involvement in other, more accepting ones.

Networked individualism should have profound effects on social cohesion. Rather than people's being a part of a hierarchy of encompassing polities like nesting Russian dolls, they belong to multiple partial communities and polities. Some communities may be widely dispersed, such as those found in electronic diasporas linking far-flung members of emigrant ethnic groups (Mitra 2003). Some may be traditional, local groups of neighbors with connectivity enhanced by listservs and other forms of computer-mediated communication (Hampton 2001). In a "glocalized" world, local involvements fit together with far-flung communities (Wellman 2003) because the McLuhanesque "global village" (McLuhan 1962) complements traditional communities rather than replaces them. This is especially true today when almost all computers are physically wired into the Internet, rooting people into their desk chairs. Even as the world goes wireless, the persistence of tangible interests, such as neighborly get-togethers or local intruders, keep the local important (Hampton and Wellman 2003). Local and long-distance—webphone and PC—it is all one fluid and complex social network.

Notes

Research for this study was supported by grants from the Japan Society for the Promotion of Science, KAKENHI15330137; the Matsushita International Foundation; the Social Science and Humanities Research Council of Canada; and the U.S. National Science Foundation. The Centre for Urban and Community Studies, University of Toronto, hosted Kakuko Miyata during her research leave, 2002–2003. We thank Mitsuhiro Ura, Hiroshi Hirano, Tetsuro Kobayashi, and Kaichiro Furutani for their collaboration in the design of the survey; Bonnie Erickson, Bernie Hogan, Mimi Ito, Robert Ramsay, Irina Shklovski, and Rachel Yould for their advice about our research; and Vicky Boase, Monica Prijatelj, Uyen Quach, and Phuoc Tran for their assistance with this chapter.

1. We include SMS (short message service, sometimes known as texting) as well as regular e-mail in our analyses. Accounts of webphone use in other East Asian societies include Yan (2003) (China) and Chae and Kim (2003) (South Korea).

2. The World Internet Project Japan (2002) found a similar rate (91 percent) among those in their twenties.

3. Self-perceived technical confidence was measured on a scale ranging from 7 to 21. The scale was compiled from seven questions that asked respondents to rate their ability to do certain technical tasks: sending a fax, recording television programs using a VCR, sending e-mails by computer or webphone, typing with a keyboard, using a search engine from a computer or webphone, downloading a file from a computer or webphone, and installing a computer program.

4. Mizuko Ito, personal communication, November 5, 2003.

5. The per day estimates of frequencies of contact were translated from respondent-reported frequency codes. The original values were: 0—no e-mail sent or received yesterday; 1—one to five e-mails yesterday; 2—six to ten e-mails yesterday; 3—eleven to twenty-five e-mails yesterday; 4—twenty-six to fifty e-mails yesterday; and 5—more than 51 e-mails yesterday.

6. The rather low explained variance (R^2) is common in Japanese surveys because many respondents do not want to divulge their personal income, and wealthier individuals tend to have more supportive ties.

7. Organizational involvement was measured on a three-point scale (nonmember, member, active member) for each of ten formal organizations, such as labor unions or charitable organizations, and three informal groups. Scores were summed to indicate a measure of group association.

8. Sending New Year's cards in Japan is analogous to sending Christmas cards in North America.

9. The greater Japanese reticence in disclosing personal information suggests that the number of Japanese relationships cannot be easily compared with the number of North American relationships.

Accelerating Reflexivity

Ichiyo Habuchi

Much discussion concerning youth is set within a discourse on social problems. It is apparently difficult to understand some of the crimes of youth through commonsense. Although these crimes are infrequent, they cause ongoing anxiety because there is no easy explanation or settlement. As a result, a popular sentiment exists that new media are the source of social problems, driving young people to crime. Discourse on youth mobile phone use is a particular site of moral panic (Cohen 1972). For example, one well-known Japanese psychologist has written, "The reason men, especially youth, of today cannot distinguish between pseudospace and real space, is the *keitai* Internet" (Okonogi 2000). Furthermore, according to one social psychologist, youth using *keitai* become juvenile delinquents at higher ratios than nonusers (Nakamura 1996c).

This chapter focuses on the first step in the construction of intimate relations, or *deai* (encounters) using simple access to the *keitai* Internet. Much of the criticism of *keitai* as fostering criminality is directed at *deai-kei* (online encounter-type) sites, where young people look for new friends and lovers over the Internet. Encounter sites accessed through any kinds of media devices have been seen as a social problem, but the easy and casual access by *keitai* has aroused a higher degree of suspicion. In 2003, Japan introduced legislation designed to stop practices that enticed children onto online dating sites. Part of this was legislation restricting access by young people under the age of eighteen. Behind this legal effort was a fear, emerging in the 1990s, that these online sites were being used for the practice of *enjo kousai* (teenage girls dating and prostituting with older men). In public discourses on this problem, reports emerged that *keitai* encounter sites facilitated *enjo kousai*, which in turn was said to drive the popularity of these sites.[1] Further, the media reported several cases where *enjo kousai* led to rape, theft, and even murder.[2]

Because of the correlation between youth and mobile Web use, as well as the link between youth and crimes related to encounter sites, the public perception is that anonymous dating sites are a youth problem. However, the focus on these youth problems creates a skewed picture of the diffusion of *keitai*. Media communications do not have a single effect on social life but are contextualized in and grow out of a wide range of social contexts. This chapter draws from empirical data to create a more nuanced and precise view of youth culture and *keitai* use. Before we fix on a particularly lurid image of *keitai* in society, it is crucial that we establish a more balanced understanding.

Deai in Context

Since the 1990s, Japanese *deai* culture has been perceived as relying primarily on the use of media communications. People rarely reflect on what encounters meant before this period because these encounters did not rely on media. It was only after media took a central role that *deai* with anonymous partners emerged as a social problem. The impetus for this shift was the emergence of *telekura* (telephone dating clubs) in the 1980s. The magazine *Bessatsu Takarajima* (1980), known for its reporting on social customs, quotes one married couple as saying, "There is no way that couples who met through *telekura* would go on to get married." Tetsuya Shibui (2003) suggests that the sexual information boom of that period was an important backdrop to the public focus on *deai* as media culture. The association with the sex industry continued in the period after *telekura,* with the advent of voice mail services such as Dial Q2 and *dengon dial* (see chapter 9). Coupled with the association with youth, *deai* culture through media continued to have a particularly shady and disreputable image.

Research on anonymous *deai* reminds us of theories of security and insecurity. The issues of trust and of security have been a major theme in theories of modernization. Issues of interaction with strangers have been discussed since the early twentieth century (see Simmel 1902; 1903), and trust continues to be a key research topic grounded in a large body of empirical study. For example, in contemporary social psychology there is the theory that "compared to people who live in the countryside, people who live in the city have more trust toward strangers" (Yamagishi 1998) as well as the opposite theory (Ishiguro 2002). Accordingly, the contrast of "strangers versus acquaintances" continues to be discussed as an aspect of urbanization that runs parallel to modernization.

However, *keitai* cannot be understood solely in terms of urbanization. While it is easy to assume that increased media use is driven by urbanization and the concomitant increase in the numbers of people one can encounter (Matsumoto 1991), urbanization cannot fully explain many of the dynamics that I have found in youth relationships (Habuchi 2003). In the case of *keitai*, use is interpersonal, and there are multiple frames through which human relationships can be promoted, for example: *deai*, maintenance of relationships, and selection of relationships.

In the *deai* scene it is easy to imagine that the ease with which people can give out their numbers to strangers creates more opportunities for encounters (Okada, Matsuda, and Habuchi 2000; Ling and Yttri 2002). Growth in the social capability of the *keitai* Internet (including e-mail) could also be a factor, driving an increase in new encounters. Such assumptions could be supported by urbanization theory. However, in contrast to these types of encounters, the other two frames of *keitai* use (maintenance of relationships and selection of relationships) strengthen existing intimacies or non-anonymous collective intimacies (Habuchi 2003), and thus there are many social phenomena that cannot be explained solely by urbanization trends.

In other words, *keitai* is a medium with contradictory connotations that reflects the characteristics of particular users. The increase in encounters with strangers through *keitai* corresponds to the increase in the population one encounters, a factor related to urbanization. On the other hand, *keitai* can serve as a means of maintaining existing relationships when it is used to strengthen ongoing collective and social bonds. *Keitai* do not allow the entry of strangers into such collective cocoons.

Therefore, it is problematic to assume that *keitai* use unilaterally increases the number of people we encounter on a daily basis and that *keitai* use is free of the constraints of prior and limiting social relationships. The availability of opportunities and strangers is not sufficient to create encounters. In other words, even if a more advanced multimedia *keitai* Internet leads to increases in matchmaking *keitai* Internet sites, encounters are avoided or created through the proactive choice of the user. *Keitai* is an effective tool for the maintenance of relationships in which the encounter has already happened (Habuchi 2002a) as well as for selection of persons whom the user has already encountered (Matsuda 2000a; Ling and Yttri 2002). There is a zone of intimacy in which people can continuously maintain their relationships with others who they have already encountered without being restricted by geography and time; I call this a telecocoon.

Modernization theory examined insecurity in interactions with strangers due to urbanization. For our continued theorizing of urbanization, we should take into account a growing awareness of security and intimacy grounded in the condition I am calling the telecocoon. *Keitai* technology provides an opportunity to review theories of modernization that associate technological progress and the advance of urbanization.

The discussion is based on the following hypotheses:

- Youth *deai* culture creates an increase in the possible selection of intimate others because of the increase in opportunities for new encounters.
- Insecurity is generated through the reflexive awareness of the increasing substitutability of the position of the self in the opportunities for the other.
- Obsession with relationships is an expression of these insecurities.

Method and Data

I draw from two interview studies and a national survey as a basis for my analysis of *keitai* use and reflexive awareness in the position of the self.

Qualitative Data

The qualitative materials for this chapter are based on two interview studies. These intensive interviews were instrumental in forming my hypothesis and helping me to understand different cultural contexts. In addition, the interviews provided microlevel contextual detail that could not be addressed by the quantitative data.

A public interview survey was conducted in 1998 in Shibuya and Harajuku districts in Tokyo and in the Minami district in Osaka. In Tokyo twenty men and thirty-seven women were interviewed, and in Osaka, fourteen men and seventeen women. We chose these locations because they are large consumer centers that define youth cultures in the respective regions, and they are focal points where young people gather. The interviews were conducted near the end of the period when youth personal media were transitioning from pagers to *keitai*, an ideal time to study the reasons for the transition and initial impressions of *keitai* use. Moreover, it was also a time when Japan began to see more and more people using *keitai* on the streets, making public interviews ideal as a research method.

The second set of data was collected in 2002 from an intensive interview survey intended to search for deeper meanings and connotations of the use of *keitai*, which had by that time become a very popular medium. The survey was conducted in Aomori prefecture, one of the rural areas in Japan on the periphery of the main island of Japan. Population is sparse, at 1,460,000, and the average annual salary is one of the lowest compared to other prefectures. The information infrastructure is also relatively underdeveloped. Two to three two-hour interviews were conducted with eleven high school girls.

Quantitative Data

The quantitative data were obtained from research on the "expansion of *keitai* uses and its effects" by the Mobile Communication Research Group (directed by Hiroaki Yoshii), conducted in November and December of 2001 with men and women aged 12 to 69 throughout Japan. The research method was based on a two-stage stratified sampling procedure (200 points all over Japan) in which a questionnaire was distributed to 3,000 households. A total of 1,878 samples were collected with a collection ratio of 62.6 percent. Some of the significant findings are summarized as follows.

The ratio of *keitai* users in this survey was 64.6 percent. There was a large variance in the rate of users by age and generation; over 80 percent of people in their twenties used *keitai*, but the percentage decreased as age increased. Variance between men and women was small (70.1 percent: 59.3 percent) (figure 8.1). In the breakdown by occupation, college students showed the highest percentage, at 97.8 percent (figure 8.2).

Turning to usage charges, the survey showed average *keitai* charges at ¥7,100 per month. Men reported paying more than women (¥7,700:¥6,300). Asked about frequency of use of *keitai* telephony, the most common answer was "once or twice a

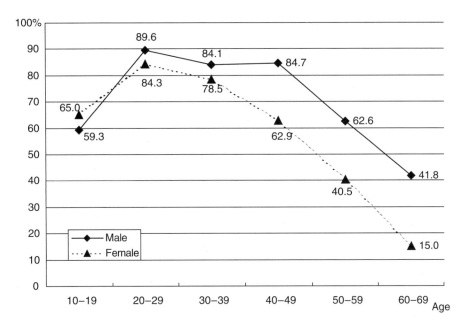

Figure 8.1
Keitai adoption rates, by gender and age. From Mobile Communication Research Group (2002).

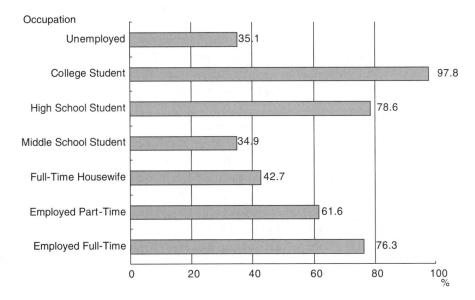

Figure 8.2
Keitai adoption rates, by occupation. From Mobile Communication Research Group (2002).

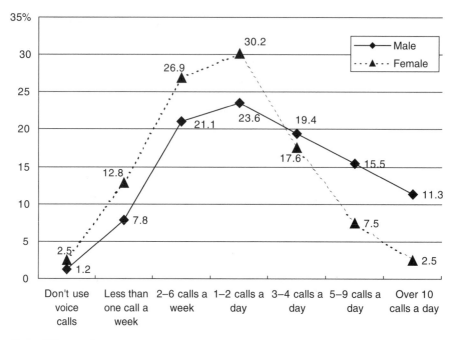

Figure 8.3
Keitai voice call frequency, by gender. From Mobile Communication Research Group (2002).

day" (26.7 percent), with men tending to use *keitai* as a telephone more often than women (figure 8.3). The percentage of *keitai* e-mail users was 57.7 percent of *keitai* telephone users (*n* = 1,213). Men had not used e-mail as often as women (48.6 percent: 69.5 percent). *Keitai* users were on average approximately ten years younger than nonusers (45 years: 32 years).

Seventeen questions were formed based on the qualitative research in 1998 (Okada and Habuchi 1999). These provide insight into some of the social and psychological consequences of *keitai* use. Subjects were asked to answer "often," "sometimes," "rarely," or "never" to the responses listed in figure 8.4. The top responses indicate the salience of issues of security and insecurity surrounding *keitai* use: "sense of security from being able to reach the other person at all times," "less irritation from not being able to reach the other person," "family is less worried," and "feel insecure going out without *keitai* or PHS" (Mobile Communication Research Group 2002).

Transformation of *Deai* Culture

The pager had a strong influence in Japanese youth subcultures that use mobile communications. Pagers fostered a culture of intimacy called *bell-tomo* (pager friends). The

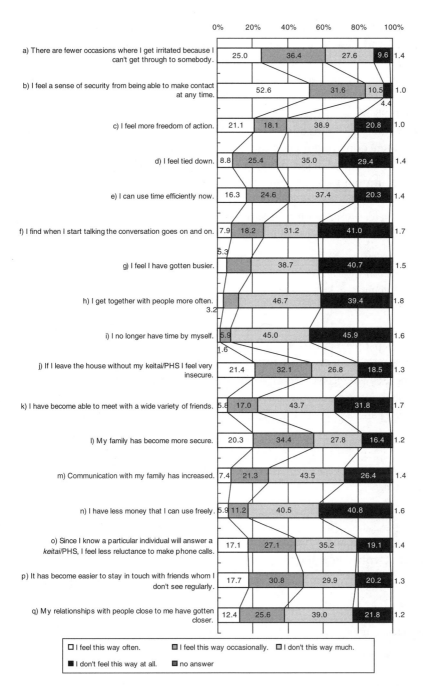

Figure 8.4

Effects of *keitai* use. From Mobile Communication Research Group (2002).

bell-tomo was a practice in which pager users would send messages to unknown people[3] in an attempt to find new friends. For example, someone might suddenly receive a message saying, "Do you want to be my *bell-tomo*? From Shota." If the receiver wanted a *bell-tomo*, he or she would send a reply. This pager *bell-tomo* culture was the forerunner of *keitai mail-tomo* culture.

Even in our 1998 research there were still traces of the *bell-tomo* culture, reflected in the following interview quotations. The *bell-tomo* culture at that time was acknowledged as a way of encountering friends and acquaintances. Consequently, when a user found an exclusive boyfriend or girlfriend, or when the person whom a user had met as *bell-tomo* did not turn out to be to their liking, there was already the tendency for the user to withdraw from the exchange at that point.

I spoke to a nineteen-year-old girl in Sankaku Park in Osaka's Minami district, and she described how she expanded her circle of friends through *bell-tomo*:

I made a lot of *bell-tomo*.
Oh, you have a lot? Around how many?
At first, there were around ten. I went out with two of them.
Girls?
Guys.

Of course, young people who have *bell-tomo* like this are not normative, and they represent a fad among those in a particular youth subculture. This response is not typical of the responses we received from young people we spoke to on the street. One nineteen-year-old college student whom we met at Sankaku Park had both a personal handyphone (PHS) and a pager but described *bell-tomo* in the following way:

Do you have any bell-tomo*?*
No.
Have you ever got any messages from strangers?
Yes, I get them often.
Have you ever replied?
No. I felt it very troublesome.

A seventeen-year-old girl and a twenty-year-old woman whom we spoke to in Tokyo's Shibuya district described how they lost interest in *bell-tomo*:

I had a *bell-tomo*, but ever since I changed to PHS, we started talking on the PHS, and we met in person once.
How was it?
He was totally different from what I imagined. So I haven't done it since then.

I used to have a *bell-tomo*.
And now?
I have a boyfriend now, so not at all. [She completely cut ties with her *bell-tomo*.]

With the weakening of the influence of the pager, the exchange of pager messages eventually made a transition to the exchange of messages through *keitai* e-mail, and the term for friendships based on the exchange of messages shifted from *bell-tomo* to *mail-tomo*. *Mail-tomo*, unlike *bell-tomo*, does not initiate contact through an unsolicited message, hence a subtle difference in the connotation of *mail-tomo* versus *bell-tomo*. For example, one seventeen-year-old high school student described how she asked a friend to find a *mail-tomo* for her as a first step in finding a boyfriend:

I want a boyfriend, so I asked my friends to find a *mail-tomo*.
Is this the usual way high school students find a boyfriend?
That's how it usually happens. If you think you can get along with the *mail-tomo*, you meet in person and decide whether or not you want to pursue the relationship.

Pager addresses were relatively simple because of the fixed range and number of characters in the addresses. By contrast, mobile e-mail addresses can contain a wider range of numbers and characters in much more complicated configurations. With *mail-tomo* it is much less likely that one can find friends or lovers, much less generate valid addresses, through randomly addressing messages. The process of meeting a new person thus shifted from the *bell-tomo* reliance on random and serendipitous connections toward a culture of a more intimate cast, which relied on connections with existing acquaintances.

In discussing romance and relationships with the opposite sex, youth often use the expression *"deai ga nai"* ("I have no encounters" or "I never meet anybody new"). Since Japanese youth cultures are partly segregated by gender, young people often have a sense of being "deprived of encounters" with the opposite sex. Japan has only a short history of several decades of male-female relationships in the Western mold. Until 1967 *miai kekkon* (arranged marriages) were more popular than *ren-ai kekkon* (romance or love marriages) (Yuzawa 1995). Marriage was usually forged for economic or class reasons; it was not formed for the interests of individuals but strategically, for the sake of the primary social group of the *ie* (house). As such, encounters with partners occurred through relatives in the form of *miai*. In recent years, marriage by *miai* has increasingly been seen as an artifact of feudal Japan and has steadily lost popularity, now comprising only 7.4 percent of contemporary marriages (National Institute of Population and Social Security Research 2003). Even today, however, certain aspects of *miai* culture still exist, and people can feel assured of a good encounter if the intermediary is a trusted acquaintance.[4]

Contemporary youth culture has an encounter process called *shokai* (introduction). A youth looking for an exclusive relationship is introduced to a person of the opposite sex by a person of the youth's own sex. This introduction is similar to *miai*. Just as marriage is a goal of *miai*, dating is a goal of *shokai*. This practice existed before mobile messaging technology. Nonetheless, with the advance of media technology, meeting a person through a face-to-face introduction has in some cases been replaced by the more convenient *mail-tomo* introduction.

Deai-kei businesses capitalizing on *mail-tomo* introductions have become a social problem in modern Japan. Most high school girls we interviewed said they did not use such *deai-kei* sites, but they completely let down their guard when a *mail-tomo* introduction is made through an acquaintance. Is there really such a social gulf between a *mail-tomo* introduced through a *deai-kei* site and a *mail-tomo* introduced through an existing relationship? This trust toward the intermediary is grounded in the fact that youth already had a culture of encounters through introduction.

According to a study by the Cabinet Office (2003), the percentages of fifteen-to-nineteen-year-old men and women who have accessed a *deai-kei* site are 16.2 and 29.6, respectively. The Mobile Communication Research Group (2002) study found the percentages of fifteen-to-nineteen-year-old men and women who reported using these sites at 12.6 and 7.4, respectively. The percentage of thirteen-to-nineteen-year-old youths who chose "*deai-kei*" as a type of site they frequently access was 7.2.[5] Currently, these *deai-kei* sites are criticized as systems that aid and abet delinquency. Setting aside this bias, I would like to consider young people's reasons for accessing these sites.

I met my boyfriend through an *deai-kei* site, but I don't use those sites anymore now that I have a boyfriend. [17-year-old girl, Aomori]

At first, I used the *deai-kei* sites because I wanted a *mail-tomo*, but gradually I started looking for someone who would treat me to karaoke. [17-year-old girl, Aomori]

I have a boyfriend.
Where is he?
Okinawa.
What?
I found him through a *deai-kei* site.
Have you ever met him?
No.
[17-year-old girl, Aomori]

The initial objective of high school girls when they use a *deai-kei* site is to look for a *mail-tomo* or a boyfriend. When they finally do find a boyfriend, many of them stop using the sites. On the other hand, when they are not able to find a boyfriend and continue using the sites, the encounters do not turn out to be the sweet fairy tales they had envisioned. The objective of many of the users is not the type of romantic relationship hoped for by high school girls but sexual in nature, which seems to gradually change the girls' motives for using the *deai-kei* sites.

Increase in Possible Choices of Others and Ontological Insecurity

I have described how new practices have been established on the terrain of *deai* culture. Why is it that in Japan we need technology-mediated *deai* despite the prevalence of introduction practices similar to *miai*? I would like to address this question through analysis of the national survey data. Here, I examine the relationships of people of all ages that are fostered through *keitai* e-mail by dividing them into three categories and comparing differing senses of security about self and interpersonal relationships.

In the first category are relationships between people who exchange *keitai* e-mail but have never met face-to-face. Second is the category of relationships initiated through *keitai* e-mail that have continued after the parties have met. The last category is old relationships that are maintained through *keitai* e-mail, with the correspondents rarely meeting in person.

People who have a relationship with someone they have never met but with whom they exchange *keitai* e-mail comprise 7.9 percent of all *keitai* users. A characteristic of such people is that they are unable to find suitable relationships from *chi-en* (community relations) and *ketsu-en* (blood relations). To the statement, "My interests and way of thinking are different from those of the people in my surroundings," 54.9 percent of this group answered yes, and 29.2 percent answered otherwise (T test: $p < .001$). Like the high school girls who look for a boyfriend through an introduction by a *mail-tomo*, it appears as though these people cannot find their desired intimacy from close-at-hand relationships. If we break out those between the ages of ten and twenty, 66.9 percent answered affirmatively and 27.3 percent answered negatively (T test: $p < .001$).

What are the characteristics of people who have met in person and whose relationships were initiated by *keitai* e-mail exchanges? Answering affirmatively to "my interests and way of thinking are different from those of the people in my surroundings" were 45.6 percent, compared to 30.1 percent of respondents who were not in such relationships (T test: $p < .05$). Again, if we break out those between ten and twenty years old, 45.6 percent also answered affirmatively and 30.1 percent answered negatively (T test: $p < .005$).

Keitai has expanded the arenas through which an individual can look for desirable interpersonal relationships. It is evident that users do in fact search for such alternative forms of encounters and that they want relationships that they find most suitable. Nonetheless, only very few people, fewer than 10 percent, have actually met in person after forming a relationship through the exchange of *keitai* e-mail.

I now analyze the ontological insecurity created by the increase in choices made available by the uses of new matchmaking technologies and practices. We measured general identity insecurity with the question "I really don't know what kind of person

I am." In the group having never met in person but exchanging *keitai* e-mail, 49 percent responded they didn't know "what kind of person I am," compared to 31.7 percent in the group that was not in such a relationship (*T* test: $p < .05$). Among those between ten and twenty years old, there was little difference: 53.4 percent answered the insecure-identity question affirmatively, and 45.4 percent answered negatively (*T* test: p not significant).

When people feel vague identity insecurity, their awareness of their circle of friends seems to become a more important factor in self-actualization. We then asked the question "I care about what my friends think of me." The correlation coefficient between the two variables of self-insecurity and awareness of friends was 0.372 (*T* test: $p = .001$). For this question regarding the awareness of friends, in the group exchanging *keitai* e-mail but never having met in person, 72.5 percent answered affirmatively, compared to 52.5 percent in the group not having such a relationship (*T* test: $p < .05$). There was no significant variance between the two groups (80 percent : 69.7 percent) in a data set of the youth group. These analyses suggest that identity insecurity and the sense of being in the minority with respect to interests and ways of thinking correlate with the exchange of *keitai* e-mail with strangers.

I have explored some of the differences in sense of security about identity and peer perception in order to understand the practices of building relationships through new media. The following conclusions can be drawn. Those responding that they have *mail-tomo* also tend to respond "my identity is different from those around me," and they are self-conscious of how they are perceived by their peers. In addition, they have a generalized insecurity about their identities. On the other hand, for young people, having *mail-tomo* correlated with the self-assessment that "they are different from others around them" but did not correlate with identity insecurity and self-consciousness. We could interpret this as evidence of the fact that young people overall experience a life stage where they are concerned about how their peers view them and have a degree of identity insecurity. This is indicated by the correlation between age and the variation between these two different factors (Pearson variation analysis for both: $p < .001$). I would conjecture that the practice of engaging with *mail-tomo* relates to features of certain young people, such as their seeking approval from peers and identity insecurity.

Increase in Possible Alternatives

As was evident in the quantitative data, people who seek intimate relationships through new media technologies feel that there are not many others who share their interests in the geographically local relationships represented by *chi-en* and *ketsu-en*. However, the proportions of people who seek these relationships are very low, and this kind of encounter culture has not gained legitimacy.

In our interviews about *deai*, youth described a sense that the pool of people with whom they could form relationships had increased. Identity insecurity is suggested by quantitative data. Based on these two findings, I would like to examine the hypothesis that insecurity creates a dependence on existing interpersonal relationships.

The increase in *deai-kei* sites has led to a changing perspective on relationships. "It's not cool to get emotional" was characteristic of relationships in the 1980s and early 1990s (Tomita and Fujimura 1999), but that attitude no longer characterizes male-female romantic involvement. Looking back, this earlier attitude exhibits a sense of confidence about relationships based in a stronger sense of self and identity that young people do not currently have. With the expansion of opportunities to select boyfriends or girlfriends and close friends, the possible number of people who can fill a particular relationship has also increased. What is important here is not the actual range of choices but the impression of an increase in options. These eighteen-year-old boys, while describing how it has become easier to make friends, also subjectively shut out the opposite sex. The perception that one does not have to be the person who has been selected as a boyfriend or girlfriend fosters the insecurity that he or she will in fact be replaced. This insecurity is represented in the following patterns of romantic relationships.

[Two 18-year-old boys, Osaka, Minami district]
Is there anything that changed after you started carrying a pager or a PHS? your lifestyle, relationships with friends, activities?
A: It's easier now to make new friends.
Are they people you meet through the pager? Bell-tomo? *Do you send your own messages?*
A: No, not *bell-tomo*. People I meet in town.
You exchange pager numbers?
A: Yeah.
You don't have to worry as much about giving out pager numbers. How about the PHS? Do you give your number only to close friends?
B: I don't give it out to girls.
Why not?
B: Because I have a girlfriend.
She's the only one who knows it.
B: That's right.

[17-year-old girl, Aomori]
Do you have male friends? Do you exchange e-mail?
No, no. Exchanging e-mail is cheating. Talking to a male friend is cheating. I won't forgive my boyfriend if he even talks to another girl.

[17-year-old girl, Aomori]
Don't you want a boyfriend?
Someone asked me to go out with him yesterday.
Who was it?
Oh, a *mail-tomo*. But I'm going to say no. It's too much trouble. I hate being tied down anyway.

There are some cases now where young people consider any contact with the opposite sex other than a boyfriend or girlfriend a betrayal. The flip side of this phenomenon is that some high school girls do not want an exclusive boyfriend because they do not like being tied down. One is the antithesis of the other, but both of these phenomena are manifestations of the insecurity generated by the increase in the possible replacements for the self in a romantic relationship. When a person becomes fixated on a boyfriend or girlfriend, it is because of this insecurity; when a person avoids a relationship, it is because he or she expects to be tied down because of the same insecurity and chooses to avoid the trouble and to circumvent the relationship itself. These phenomena should not be interpreted as a peculiar form of relating romantically but as the beginning of a broader transformation in modern intimacy. In other words, the interaction of intimacy and technology affects decisions in everyday life regardless of the awareness of the users of the technology.

Telecocoons

Are these changes in relational security and insecurity particular to the romantic love of young people? Here I draw again from the quantitative data to describe how this tendency in young people's relationships is a facet of a broader-based set of emergent meanings in how social relationships are being chosen. As described, only 7.9 percent of the study population has *mail-tomo* who have never been met face-to-face. By contrast, the conditions that exist in the telecocoon are tied to a more foundational aspect of intimacy. For example, 65.8 percent of *keitai* users responded, "There is someone whom I now rarely see in person but keep in good contact with via *keitai* e-mail." This illustrates how *keitai* is used as a device for maintaining existing familiar relationships in the absence of face-to-face meetings. Among these users, 45.6 percent responded, "This person is irreplaceable and very important to me." *Keitai* e-mail in Japan is generally exchanged between partners who are relatively close geographically and who see each other regularly face-to-face (Mobile Communication Research Group 2002). Based on these results, we can surmise that the people in this 45.6 percent are actively working to maintain a networked relationship. In other words, approximately half of those who work to maintain a relationship in the face of physical distance are inhabiting a telecocoon. What are the particular characteristics of these people?

People who form telecocoons do not feel that their interests and ways of thinking differ from people around them, and they are not particularly insecure about their identities. This suggests that telecocoon relationships are different from the relationships formed and maintained exclusively online. On the other hand, self-consciousness about how one is perceived by peers was high among people who formed telecocoons (60.3 percent), varying significantly from those who weren't telecocooning (49.3 percent) (T test: $p < .005$). The data also show a correlation between telecocooning and age (Pearson: $p < .001$), with young people exhibiting this behavior more frequently. I described earlier the increasing effort that young people put into maintaining romantic relationships. This is tied not only to a growing sense that the self in a relationship is replaceable but also to the social demand to make an active and committed effort to choose relationships (see chapter 6).

Discussion

I now discuss the role *keitai* technology plays in the formation of the self, based on my research on Japanese *keitai* use. E-mail and short conversations using *keitai* play a major role in interpersonal relationships. For maintaining interpersonal relationships, a significant factor in identity formation, there has never been a technology like *keitai* that has so completely achieved a one-on-one connection between the user and the device. This apparatus for sending and receiving communication has transformed how relationships are formed with others and can also transform the very essence of the self, although in minute and subtle ways.

Keitai functions as a part of the body of its users because it is in a completely one-on-one, intimate relation with them. Based on this hypothesis, we can identify the use of *keitai* as one of the actions that symbolizes reflexive modernization on a microlevel. Furthermore, I propose that it not only symbolizes but also accelerates the reflexive cycle: transformation of the self → acknowledgement of others → transformation of the group to which the self belongs → transformation of the self.

According to Castells (2000), self-identification is a process that adopts cultural characteristics in order to understand the self and its structural significance. It follows that the new forms of communication enabled by *keitai* influence the acknowledgment of the self. It is therefore important to first review some key issues in understanding modern subjectivity.

Keywords are *security* and *insecurity*. Beck, Giddens, and Lash (1994), advocating reflexivity as a characteristic of the modern age, describe modern society as follows: Modern society, with its inherent dynamism, undermines the sustenance of class and hierarchy, occupation, gender roles, nuclear family, industrial facilities, and corporate activities as well as the natural development of technological advances and conditions for economic growth. Reflexive modernization could possibly turn progress into self-destruction. Through this self-destruction, one modernization undermines another modernization and introduces a new phase of its transformation.

Such major trends of modernization have started incurring a heavy burden on the individual. The individual has been separated from the cocoon provided by the family or the village, shouldering the new burden of having to select tradition by individual choice rather than being forced to accept it by circumstances of birth. The individual has been removed from the security of given relationships and is obliged to live by his or her own choices and responsibilities. This is generally expressed as personalization.

With the development of an industrial society, personalization signifies the increase in the objective importance of the internal world and the loss of importance of social backgrounds. Anthony Giddens (1992) points out that an others-oriented lifestyle leads to a growing need for acceptance from a stable group of others. An individual who has lost ontological security has to struggle fiercely in order to achieve a conscious awareness of the self. Giddens (1992, 202) has termed this condition a reflexive project of self.

Giddens (1990) prescribes trust and "expert systems" of scientific and technical knowledge as mechanisms that disembed individuals from social relations, and this lends a useful dimension to the discussion of *keitai*. By viewing *keitai* as symbolic of a type of expert system, we can form the following hypothesis. *Keitai* drives society toward emptying the existing structures of time and space. Then these devices become associated with individuals and become a subject of trust, thus becoming a basis for creating self-identity. My analysis of youth relationships can be considered one concrete case of this broader hypothesis. My studies have much in common with previous research. For example, Ling and Yttri (2002) have described how Norwegian youth facilitate the development of self-identity by mobile phone communication with groups of friends.

Needless to say, the reflexive cycle between society and individual reflects an ethos of modernity that predated *keitai*. However, the important point here is that mobility, which is a prominent feature of *keitai*, has accelerated this cycle. The acceleration of the reflexivity of the self (Habuchi 2002a) can be considered a temporary liberation from insecurity.

On the other hand, youth are involved in a fierce struggle for a conscious awareness of the self (Habuchi 2002a). Consequently, their *keitai* use is different from the way other generations use it. The analysis of the qualitative research data in this article has focused on young people who are in critical stages of self-formation, and I have described the acceleration of reflexivity in the reflexive project of their selves in the new media environment. This kind of project of accelerating reflexivity in self-awareness may be leading to individualized existential security through an online social group. If so, this would be a true merger of the real and virtual through media communication.

Conclusion

This chapter has examined identity insecurity and *keitai* use by analyzing quantitative data and interviews with youth about their relationships.

Keitai have two types of expressive functions. The first is that it provides opportunities for new interpersonal relationships. Among modern Japanese youth, there is an electronic *mail-tomo* culture of introducing people, which draws from various other *deai* cultures. This is the practice of finding a *mail-tomo* of the opposite sex who is a friend of a friend and starting to date if they get along after exchanging messages. In other words, *keitai* provides an electronic version of pen pals.

The people who are involved in this *mail-tomo* culture lack confidence about themselves because they feel that their interests are different from others around them. They have established this culture to search for electronic relationships that offer a wider range of possible encounters, friendships they cannot find in their existing *chi-en* and *ketsu-en* relationships.

I have hypothesized that the increase in opportunities for *deai* increases the possible choices of partners in intimate relationships, which in turn increases a sense of risk that the self is replaceable. The feeling that one is involved in a possibly fleeting interpersonal relationship could be increasing relational insecurity and creating an obsessive need for exclusive relationships.

Another expressive function of the *keitai* is its capacity to maintain familiar and existing relationships. For example, a *keitai* provides one with the opportunity to sustain a bond with an old school friend with whom one hasn't met in person for a long time. As manifested in the quantitative data, more people use *keitai* for such purposes than those who use it to look for new encounters. Since using *keitai* to maintain familiar relationships strengthens collective social ties, this use serves to maintain the cocoon of the existing community. I have termed this sphere of intimacy that is free of geographical and temporal restraints a telecocoon. The people who form telecocoons can generally be characterized as being constantly attentive to their group of friends, attentive to the retention of the group's association.

The ceaseless cycle of interpersonal relationship building and *keitai* communication is evidence of the acceleration of reflexivity in microinteractions. The underbelly of the protective security of the *keitai*-enabled telecocoon is the risk and fear associated with not having a *keitai* phone. Although *keitai* technology should have freed us from constraints, it has also produced a new set of constraints—reliance on *keitai* ownership. What kind of new technology will we see in our society in the future to respond to this situation? This is yet another case of a cat-and-mouse chase between human desire and technological innovation.

Notes

1. Based on a 2004 report by the National Police Agency, crimes associated with *deai-kei* sites and involving *keitai* have been on the rise in recent years. In 2003, 95 percent of arrests related to *deai-kei* sites involved *keitai* use.

2. In 2000 a 26-year-old man was arrested in Kyoto for murdering and abandoning two women (one in her teens, the other in her twenties) whom he had met on a *deai-kei* site accessed by *keitai*. This incident was widely reported in the newspapers as "the *meru-tomo deai-kei* site serial murders."

3. Since pager numbers had predetermined prefixes, it was possible to send messages to unknown pager owners by pushing random numbers after the prefixes.

4. Marriage agencies and other businesses have been established for introducing couples for the traditional *miai*. Such marriage businesses hardly ever become a social issue like the online encounter sites because they thoroughly investigate the backgrounds of their registered members and provide a sense of security regarding the encounters. It is for this sense of security that the members pay a high membership fee.

5. Based on a 2003 national survey conducted by the Mobile Communication Research Group in South Korea, among those aged 12–19, 2.5 percent had accessed a *deai-kei* site via *keitai*. Of course, one reason for this is the fact that in South Korea *keitai* Internet use is extremely low compared to that in Japan. Access rates for PC-based *deai-kei* sites were at 10 percent. This latter percentage is much higher than the comparable rate in Japan.

Hidenori Tomita

Arashi no Yoruni (*One Stormy Night*) (1994), by Yuichi Kimura, is an award-winning children's picture book. It is about a goat and a wolf that accidentally find shelter in the same shack one stormy night. The pitch-black darkness of the shack conceals them from one another. Without knowing each other's identities, the goat and the wolf talk throughout the night and become friends. At dawn, they promise to meet again before parting ways. In a setting where partners remain faceless, friendships are born from two hearts coming together, even if one is a dreaded wolf and the other is a delicious goat.

New relationships born through new media can resemble the relationship between the goat and the wolf in *Arashi no Yoruni*. One medium that supports such relationships is *keitai*. However, people generally do not try to truly understand these new kinds of relationships because they do not fit into conventional social structures and frames for interpersonal relationships. We are only just beginning to understand the power of the social worlds called forth by *keitai* as they continue their spread across the world. This small, handheld device may have the power to transform many fundamentals of twenty-first-century society.

This chapter describes new kinds of interpersonal relationships created by *keitai* communication, specifically, encounters with anonymous others. First I introduce the concept of the intimate stranger, a new category of social relationship where anonymity is the condition for intimacy. I then trace the historical development of this social category, beginning with fixed-line telephone services that emerged in the early 1980s, and discuss current practices on the PC Internet and *keitai*. Until now, new kinds of romantic relationships fostered by the PC Internet were a focus of public attention. In the case of new relationships fostered by *keitai* use, they have been associated with *enjo kousai* (teenage girls dating and prostituting with older men) and juvenile delinquency, and have been subject to regulation. Despite this, the new interpersonal relationships fostered by the PC Internet are steadily making inroads to the *keitai* Internet, and I discuss current trends in both. In contrast to the PC Internet, which has fostered a range of online communities, *keitai* and the *keitai* Internet are oriented toward personal networks. I conclude with a discussion of the implications of these cases for theories of anonymity and the construction of online identities.

The Intimate Stranger

New styles of telephone communication emerged in the latter half of the 1980s with *telekura* (telephone clubs); *dengon dial* (dial-up telephone messaging), introduced by NTT in 1986; and party line and two-shot services (similar to 900 numbers in the United States), introduced in 1989. These services heralded new forms of mediated communication at the crossroads of anonymity and intimacy. The two-shot service was canceled after becoming a source of major social problems, with people becoming obsessed with the service and incurring astronomical fees. Contact with intimate strangers later reappeared in the form of pager *bell-tomo*, which peaked in popularity in 1996. Although these media differ as to whether they are text- or voice-based, and synchronous or asynchronous, they all enable users to make intimate contact with strangers. I call people with whom intimate, anonymous contact is made in cyberspace intimate strangers (Tomita et al. 1997; Tomita 1999).

The concept of the intimate stranger can be easily understood if we cross the two axes of anonymity and intimacy (figure 9.1). An anonymous other who is not intimate is a complete stranger. A person who is not anonymous but also not intimate is an acquaintance. A person who is not anonymous and who is intimate is a friend or lover. We generally live our lives surrounded by these three categories of people. Now, however, we have the intimate stranger, a person who is anonymous but also intimate, a type of relationship that has been made possible by new media.

The unique characteristics of communication with intimate strangers can be understood in comparison to earlier communication patterns. Originally, the telephone was used for conversations with friends or to relay messages. When we talked with people we did not know, it was for business or for making an inquiry to customer service—in short, the purpose was purely instrumental. When traditional telephone communication was positioned on a grid of acquaintance versus stranger and instrumental (serving as a means to an end) versus consummate (existing as the end itself) (Yoshimi et al. 1989), one quadrant remained empty: the one representing stranger and consummate. This quadrant was filled by the new telephone communication between intimate strangers (figure 9.2).

Until recently, new forms of telephone communication with intimate strangers were seen as a social pathology because of their association with the sex and pornography industries. However, developing intimacy with anonymous others in cyberspace is becoming more commonplace with the widespread adoption of the Internet. Friendly conversations have always taken place between strangers, for instance, between people on vacation or between a bartender and a regular customer, but these are not the same

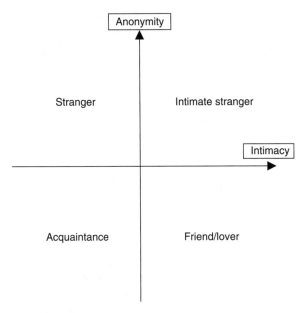

Figure 9.1
Axes of intimacy and anonymity.

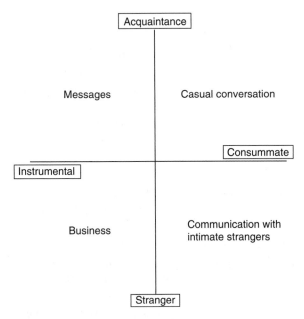

Figure 9.2
Communication with intimate strangers.

as the communication between intimate strangers being discussed here. Strangers who meet during vacation do not necessarily have a high degree of intimacy, and business considerations come into play in the conversation between a bartender and a customer (Milgram 1977). The biggest difference is the contrast between face-to-face interaction and the facelessness of cyberspace. The anonymity guaranteed by cyberspace protects us from the hidden dangers of modern society. Cities are also spaces of anonymity; nonetheless, we are always vulnerable to dangers that could be triggered from physical contact with others. On the other hand, the anonymity provided by cyberspace enables us to disappear in an instant and to disengage from online relationships at any time. With a relationship maintained under the protective wing of anonymity, there is a dramatic acceleration of the deepening of intimacy. At the same time, this same safety in anonymity also causes problems, such as online slander and defamation or duplicity about identity.

For communication between intimate strangers, one medium has been the audio communication of *dengon dial*, party line, and two-shot; another is the text communication of computer networks and Internet *mail-tomo*. Somewhere in the middle lies the pager *bell-tomo*, which is text communication but does not require a computer and utilizes the more accessible telephone. Also, by being a mobile medium, the pager has generated a culture of its own. And now the birth of the *keitai* Internet integrates these disparate communication trends.

Voice Services and the Origins of the Intimate Stranger

Deai (encounter) through Internet and *keitai* has been a focus of public attention and concern because of the potential for people to meet and date others who they otherwise would not have. These *deai* practices have precedents, however, in the prior technologies of print media and the fixed-line telephone. In particular, certain services in fixed-line telephony that began the 1980s constructed the social category of the intimate stranger.

Telekura and the New Pickup Scene

Telekura first emerged in the mid-1980s. At a rate of about ¥3,000 per hour, men waited in private rooms for calls from women. Women made calls to a toll-free *telekura* number. When a woman made a call, the phones in all the rooms rang at the same time, signaled by a light turning on. The first client to pick up the phone got the call. This system was eventually changed to clients' getting calls based on the order that they had checked in. Most women made the calls after getting a flyer or a tissue pack with the *telekura* number printed on it handed out to them on the street. For times when there were few calls, shills were on hand to play the role of anonymous general callers.

Men go to *telekura* sites to pick up women, but women's reasons for participating are more diverse. Some are looking for a date, others for sex, and others just for somebody to talk to. Even a simple chat can lead to another conversation or a date if the encounter is enjoyable. In the case of *telekura*, after talking on the phone, the man and women often decide on a time and place to meet, so the woman who is calling needs to be close to the *telekura* site. Although picking up a date is the goal of this activity, finding a time and place to meet is not the only outcome. If the phone conversation does not go well, the date does not happen. Further, some women call *tekekura* just to pass the time, with no intention of actually getting together with a partner. Even if a time and place are decided upon, both people don't always show up. In short, regardless of the ultimate motive, an enjoyable phone conversation is important in all cases.

The Allure of *Dengon Dial*

NTT's *dengon dial* service was introduced in 1986. The service was designed to function like a noticeboard at a train station, where messages could be left by telephone. Users called into the *dengon dial* center, and by keying in a call-back number of up to six digits and a four-digit personal identification number they could record and listen to messages. In the time before *keitai* and pagers, this was a useful service that could be accessed anywhere across the country, like a centralized version of today's voice mail. The service, however, began to be used as a way of exchanging messages with strangers through numeric combinations in the opening dial-in—like 123456—that anybody could think up and access. People began to use relay dial systems where one could leave some thoughts as a message and strangers would respond in sequence. Pickup dial systems where people could leave a voice personal ad also emerged with the *dengon dial* service. The *dengon dial* system eventually was used for *telekura* and Dial Q^2 as well (see also Okada 1993).

Dial Q^2 as a Social Problem

NTT began the Dial Q^2 service in 1989, a service that collected information fees like the 900 numbers already launched in the United States. For a fee, Dial Q^2 enabled anybody to gain telephone access to voice-recorded information. NTT collected the associated information fee as part of phone billing, making it an easy payment system for the user and a simple collection system for the information provider. A wide range of programs emerged to take advantage of the Dial Q^2 system. Among these, the most popular were the two-shot and party line services. Two-shot was a program where a man and woman could go on a blind phone date. Party line was one where several men and women could talk freely over the phone. With the two-shot service, men would call a Dial Q^2 number starting with 0990, and women would call a toll-free number starting with 0120. The men and women were then connected for a phone date; the name two-shot refers to a couple. *Telekura* sites were restricted by law to certain zones, and most were in red-light districts. By contrast, Dial Q^2 was not restricted by location, and the service could be accessed from anywhere. Because both male and female numbers could be dialed easily from home, these services exploded in popularity. As a result, use spread to all corners of Japan.

In the loneliest hours of the night, when friends were sleeping, two-shot on Dial Q^2 provided a "late-night counselor" who would kindly ask, "What's wrong?" There were also cases where elderly people living alone would use the service to mitigate their loneliness. Soon, services were also established that played pornographic recordings or solicited *telekura* clients through Dial Q^2.

However, the information fee that was added to the phone toll for the two-shot service was high, generally ¥10 for 4–6 seconds connected to the service. Some extreme cases were reported, such as ones where people killed themselves after not being able to pay the fees they had incurred, and one where a father killed his son who was obsessed with Dial Q^2. There was also a case where a man beat up a high school girl whom he got to know through a party line service, and a case where a dating club operator was using the two-shot service. As a result, NTT stopped taking new subscriptions for the two-shot service in June 1991, and it reduced the maximum fee for the party line service to one fifth of the prior limit. This reduced the financial incentives for content providers. As a result, both two-shot and party line services eventually disappeared, as did the *dengon dial* system based on Dial Q^2, in February 1994 (see also Tomita 1994).

The Second *Telekura* Wave

Telekura services, which had lost popularity after NTT launched Dial Q^2, began reviving after the two-shot, party line, and *dengon dial* systems became subject to restrictions. In the earlier period, *telekura* sites had relied on storefronts for their operations, but the second wave included *telekura* with no physical location. The two-shot services so popular with Dial Q^2 were called home *telekura* at the time. The second *telekura* wave involved a kind of no-storefront *telekura*, and payment was not made through Dial Q^2 but through bank transfers or purchases of fee cards in vending machines. In contrast to Dial Q^2, which could be easily accessed from home, these new *telekura* services meant that users had to go out of their way to make payments. Almost all *telekura* uses thus became oriented toward pickups and dating.

Cases where minors used *telekura* for prostitution (figure 9.3) became recognized as a social problem and were given the label *enjo kousai*. As a result, all prefectures around the country implemented regulations aimed at restricting the use of *telekura* by minors.

New *Bell-Tomo* Friendships

"Pocket bell" pager services began in 1986. At that time, the pager was a device designed for businessmen. As the devices evolved, in 1996, into pagers that could receive,

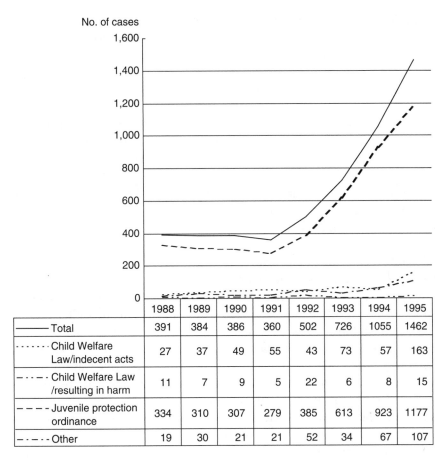

No. of cases

	1988	1989	1990	1991	1992	1993	1994	1995
——— Total	391	384	386	360	502	726	1055	1462
······ Child Welfare Law/indecent acts	27	37	49	55	43	73	57	163
– · – · Child Welfare Law /resulting in harm	11	7	9	5	22	6	8	15
– – – – Juvenile protection ordinance	334	310	307	279	385	613	923	1177
– · – · · Other	19	30	21	21	52	34	67	107

Figure 9.3
Arrests related to abuses of *telekura*, 1988–1995. Courtesy of National Police Agency (1996).

convert, and display digits as letters, they became a huge fad among high school girls (see chapter 2). Using this one-way pager communication, a new way of exchanging messages with strangers was born, the *bell-tomo* (pager friend).

In this case, there was no intermediary as with Dial Q^2 and *telekura*. The practice was simply to exchange messages through the tiny pager. Because the pager is portable, messages could be received at any time and place. When a message arrived from a *bell-tomo*, the recipient could go to a pay phone and enter a reply through the keypad. Then another message could be returned right away. Even though the messages might be simple—"Good morning," "How are you doing?" "What are you up to?"—by continuing these exchanges, friendships might be born, and in some cases, even romance. These encounters with strangers, with no involvement of the sex and dating industries, could be considered a world created by the communicators themselves. Further, the intimacy and pervasiveness of the pager gave it a more direct and unmediated feel compared to the earlier encounter practices.

Deai Online

One characteristic of media communication is that it enables people to maintain a certain level of anonymity. This is the case for both the telephone and the Internet. In an Internet chat room, participants can maintain a friendly conversation without knowing each other's identities. By repeatedly encountering each other in such chats, people can develop friendships and romantic relationships. In other words, this is a world where intimacy arises from anonymity. These types of relationships existed before the advent of the Internet with early PC bulletin boards.

PC Communication and Online Marriages

At the beginning of the 1980s, computer-mediated communication became popular among a small sector of Japanese society. With the launching of NEC's PC-VAN online service in 1986 and Nifty Serve the following year, online communication became more mainstream, and in the 1990s it really expanded. In the case of Nifty Serve, by the end of May 1993 members numbered 500,000; by the end of April 1995, over 1 million; and by January 1996, 1.5 million. At that point, Nifty Serve was the largest online service in Japan, with over 570 online forums, over 190 online shops, and approximately 1,350 national and international databases.

Bulletin boards and chat rooms were the most popular among the variety of ways of communicating online. People who were otherwise strangers would talk about their interests and hobbies, exchange views, and chat. Some cases emerged where people met online and got married. At the time, these marriages were called *paso-kon*, a pun on *pasokon* (personal computer), using the *kon* character from *kekkon* (marriage). In this way, relationships created through computer-mediated communication also became a site for intimate strangers. Homes with personal computers were still relatively few at that time, however, so only a limited number of people had access to these online interactions.

Internet Romance

Commercial Internet access in Japan began in 1993. In 1995, Microsoft released Internet Explorer as part of Windows 95, and the Japanese Internet population began to grow. It only took five years from the start of commercial Internet access for the percentage of households with access to reach over 10 percent (figure 9.4). ICQ, a standard for instant messaging, appeared in 1996, and AOL, one of the top three portal sites, provided Instant Messenger for its subscribers in 1997. Soon after, other portal

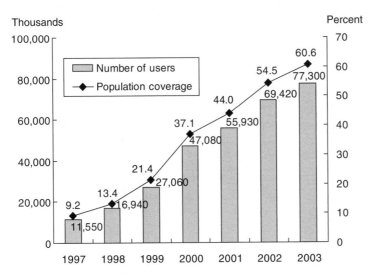

Figure 9.4
Internet user population and adoption rates, 1997–2003. From Ministry of Public Management (2003). Reproduced with permission.

sites followed suit, launching MSN Instant Messanger and Yahoo Instant Messenger. Each portal site also had chat rooms, and many people, regardless of age and interest, started chatting online.

The Internet provided new opportunities for encounters between men and women. A U.S. movie, *You've Got Mail* (1998) depicted a romance mediated by e-mail. The female lead, using the screen name Shop Girl, found the love of her life via e-mail. She fell in love with her e-mail partner, about whom she knew little. They were able to communicate honestly with each other only because they did not know much about each other.

In Japan there was a popular TV drama, starting April 14, 1998, called *With Love—Chikazuku Hodo ni Kimi ga Toku Naru* (*The Closer You Get, the Farther You Become*), about a romance mediated by e-mail. The male lead was a 27-year-old ex-leader of a popular band, now a popular television commercial songwriter, and his girlfriend was a 23-year-old bank employee. They lied to each other about everything—their occupations, where they lived, their daily lives, and so on. The lies were told, however, so that they could continue communicating their genuine feelings.

The common themes of these two productions were Internet romance and the characters' realization that some things are more important than people's social status. They suggest that the "real self" is not the individual defined by the social positions and roles of the everyday material world but is the individual found online who is

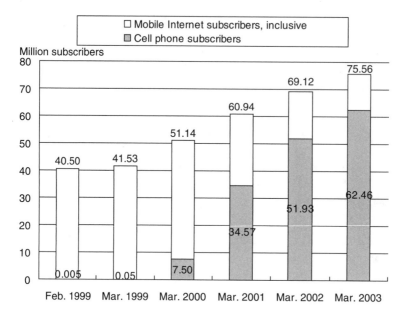

Figure 9.5
Subscribers to *keitai* and the *keitai* Internet, 1999–2003. From *White Paper: Information and Communications in Japan* (2003). Reproduced with permission.

free of such definitions. As a result of living in an era that allows the presentation of a pluralistic, flexible self in cyberspace, modern people are seeking encounters between the "real me" and the "real you," consensual concoctions.

Today, in this information society where all sorts of information is digitized and copied, *originality* has become a sensitive concept. Cyberspace enables the creation of a "real me" that is different from an "original me." The stripped-down nature of e-mail enhances this tendency. Now this kind of online communication and presentation of self is accessible through *keitai*.

Keitai Deai-kei Sites

NTT DoCoMo's *keitai* Internet service, i-mode, was launched in 1999. Soon after, the other *keitai* providers began offering Internet service from their handsets (figure 9.5), and a variety of digital content was produced for *keitai*. Among this was the *deai-kei* (encounter/dating) site.

These sites differ from chat rooms for conversation among strangers in that they are more specifically romantic matchmaking services. Internet *deai-kei* sites generally involve providing information on age, sex, and address, and searching for compatible others. Portal sites offer free matching services. *Keitai deai-kei* sites rely on the same

Table 9.1

Estimated Number of *Deai-kei* Sites, 2001 and 2002

	September 2001	September 2002
Deai-kei sites accessible from PCs	884	2,038
Deai-kei sites accessible from mobile phones	2,569	3,401

Source: Kioka (2003).

Table 9.2

Arrests Related to *Deai-kei* Site Incidents, 2000–2002

	No. of Cases Total (%)	No. of Cases Tied to *Ketai* (%)	No. of Cases Tied to PCs (%)
2000	104 (100)	59 (57)	45 (43)
2001	888 (100)	714 (80)	174 (20)
2002	1,731 (100)	1,672 (97)	59 (3)

Sources: National Police Agency archive, February 6, 2002; Kioka (2003).

basic structure. After providing background information, users can see messages posted by others and can respond to the ones that appeal to them. Subscription-based *keitai deai-kei* sites charge fees of approximately ¥300 a month.

The young people who have used these sites have gotten public attention after becoming embroiled in certain incidents and crimes. The National Police Agency puts the number of *deai-kei* sites in 2002 at approximately three thousand (table 9.1), but this appears to be a conservative estimate. Among these, about two thirds charge a fee and the rest are free sites (Kioka 2003).

According to National Police Agency archives, arrests related to *deai-kei* sites have been growing yearly, from 104 in 2000, to 888 in 2001, to 1,731 in 2002 (table 9.2). In particular, the number of arrests related to "child prostitution/child pornography crimes" was the highest in 2002, at 813 cases, followed by "violation of youth welfare and guardianship" at 435 cases (figure 9.6). Cases that involved *keitai* use represented a growing percentage every year (see table 9.2). Use of *deai-kei* sites to seek *enjo kousai* with girls who are still minors became a subject of widespread concern. As a result, the government passed a new legislative initiative September 30, 2003, called "Legal Plan to Address Entrapment of Children through Internet Dating Industries," or *Deai-kei* Sites Legal Plan for short. This law was unique in that it enabled the authorities to arrest the minors who were engaged in *enjo kousai*.

Society recognized *deai* between couples who met through the PC Internet as a new form of romance. By contrast, with the *keitai* Internet *deai* became a social problem and the subject of new legal restrictions. Both *deai* over the PC Internet and the *keitai*

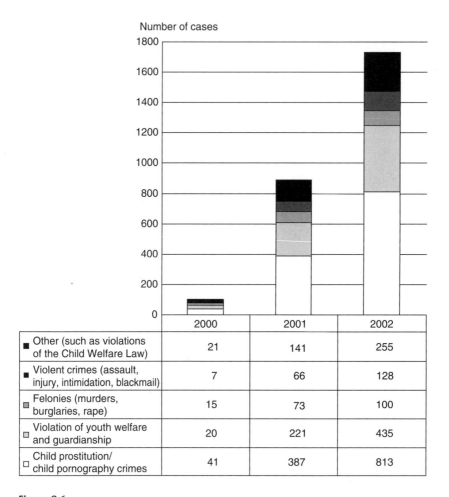

Number of cases	2000	2001	2002
■ Other (such as violations of the Child Welfare Law)	21	141	255
■ Violent crimes (assault, injury, intimidation, blackmail)	7	66	128
▫ Felonies (murders, burglaries, rape)	15	73	100
▫ Violation of youth welfare and guardianship	20	221	435
▫ Child prostitution/ child pornography crimes	41	387	813

Figure 9.6

Arrests related to *deai-kei* site incidents, 2000–2002. From National Police Agency archive, February 6, 2002; Kioka (2003).

Internet, however, foster more flexible identities than had previously been possible. The *keitai* Internet has an even more significant potential. Why are the social reactions to the new relationships fostered by the *keitai* Internet so different from the reactions to the those on the PC Internet? I address this question by focusing on the differences between *keitai* networks and PC networks.

Keitai Networks: From Online Community to Personal Networks

There are important differences between accessing the Internet with *keitai* and accessing with PCs. In contrast to how the PC Internet supports the construction of online communities, the *keitai* Internet is geared toward the construction of personal networks.

Online Communities and the PC Internet

The Internet is home to numerous special-interest groups of shared hobbies and interests. These online communities provide new spaces for social exchange, enabling people to talk about their interests, share news, and post questions. They also enable people to find friends with whom they can converse on any topic and who will freely share information about themselves. For example, Yahoo Japan's bulletin boards have over one thousand categories, supporting enormous communities of users who visit every day. Chat provides opportunities for real-time communication, and bulletin boards allow for longer, more reflective discussions to unfold over time. More recently, audio and video input has also been integrated into chat rooms. Most of these online communities on the PC Internet are composed of people who do not know each other in other contexts.

Keitai Personal Networks

People most commonly give their *keitai* numbers and e-mail addresses to friends, acquaintances, and family members. As a result, they can be reached whenever those people want to contact them. As I have already described, PC Internet connectivity has become more widespread in Japan. In the case of the PC, however, the user must sit in front of the computer, in contrast to *keitai*, which is carried with the user at all times and which will receive a message as long as it is turned on. In this way, a *keitai* can be considered a device of continuous connectivity. Because of *keitai*, people can maintain a sense of being continuously connected to friends, families, and lovers (see chapter 13).

When heavy *keitai* users lose their *keitai*, they often lose contact with their friends. This is particularly true among young people; it is difficult for them to even conceive of peer relations now without *keitai*. Initially, *keitai* was a device for augmenting existing relationships. Now, as *keitai* have become more pervasive, it is becoming difficult for people to have relationships without *keitai*. In other words, where the PC Internet supported the formation of online communities, the *keitai* Internet supports the formation and maintenance of personal networks.

Instant Messenger as the Internet's *Keitai*

After becoming close to somebody through chat or a bulletin board, users often go to instant messenger (IM) to have one-on-one conversations. IM is a free service where people can see which of their buddies are online and exchange messages and begin chatting immediately. Before the advent of IM, there was no easy way of seeing whether others were online. Just because a buddy is online, however, does not mean that one necessarily sends a message right away. Rather, IM constructs a space of ongoing peer connectivity and thus is similar to how *keitai* functions. IM constructs a personal network on the PC Internet, creating stronger relational ties for online communities.

Keitai Online Communities

Online communities are increasingly mobilizing *keitai* communications, particularly Internet businesses that are integrating *keitai* into their online community offerings. For these businesses, online communities, starting with chat and bulletin boards, have been seen as a powerful marketing tool to drive traffic to their sites. As an extension of these strategies, *keitai* online access has also been receiving attention. The major portal sites have created online sites specifically for *keitai*, with chat and bulletin boards just like the PC-centered sites. In other words, it is becoming increasingly possible to construct online communities through *keitai*.

Telephone Online Communities

Communities constructed through the medium of the telephone existed before the Internet era in the form of party line and *dengon dial* utilizing Dial Q^2. The early text-based PC Internet and the voice communications of party lines and *dengon dial* are coming together with the advent of the broadband Internet. Now it is important to consider Internet online communications, fixed-line telephones, and *keitai* in an integrated way. It is wrong to think that online communities cannot exist via telephone.

The current move to bring *keitai* into the online community space is grounded in *keitai* Internet uses, and voice communication has been largely ignored. While Internet chat rooms can support groups of people communicating in text, voice, and images, it is strange that *keitai* online communities do not make use of voice communication. With the advent of mobile messaging (the *keitai* version of IM), *keitai* users can participate in PC Internet chat rooms, linking the PC and *keitai* Internets. In other words, both the PC Internet and the *keitai* Internet support online communities and personal networks as well as one-to-many and one-to-one communication.

Given this situation, what distinguishes the PC and *keitai* Internets? The issue of mobility is central (Feather 2000). Although laptops are also mobile technologies, they differ from *keitai* in that they are not always turned on while in transit. The difference between *keitai*, which is always on and always carried along, and the PC, which tends to be tied to a particular spot, defines the difference between the PC Internet and the *keitai* Internet. *Keitai*'s continuous connectivity enables a sense of directness, of being able to contact an individual directly, grounded in the one-on-one modality of *keitai* communication. Another distinctive feature of *keitai* is the compatibility with action in physical space, how it functions to coordinate action in real space by enabling users to keep in touch and coordinate action up to the moment when they meet up. This feature of the *keitai* Internet is supporting distinctive forms of use within urban space, fostering new connections between intimate strangers.

New Communication Tools for Urban Space

1998 saw the release of Lovegety (distributed by Erfolg Co.), a small device designed to enable boys and girls to hook up while out in town. The devices were manufactured with different hardware models sold separately to boys and to girls. When within five meters of one of the opposite sex, they would notify the user with light and sound. Users could also select between three different modes: "I want to talk," "karaoke" (meaning, let's go play), and "get2" (meaning, I want to get a boyfriend or girlfriend). When a Lovegety found another user with the right mode, it lit up. In the same year Astel Co. released a PHS called Coofy that had much the same functionality as Lovegety. The Coofy had a transceiver to receive "angel wave mode" signals, and when two receptive Coofys were within 150 meters of one another, they would signal with vibrations and sound. If the other person was registered in the user's PHS phone book, the number would also appear on the PHS display, signaling that a friend was nearby. In 1999, ImaHima released the free service ImaHima, integrating locational information, mobile community, and IM. A user could notify IM buddies information about his status—where, what, "how I am feeling"—and send short text messages and e-mail. ImaHima became an NTT DoCoMo official site in 2001. In 2002, Frepar Networks and Atlas Co. started selling Navigety, a device similar to the Lovegety that users plugged into their *keitai* phones. The Navigety signaled if other users with the same intent were within 2.5 kilometers, and it supported the sending of messages.

These kinds of new mobile communication services, designed to facilitate encounters within urban space, continue to appear. At the same time, wi-fi hotspots are also becoming more prevalent in urban Japan. Users can access the Internet from a laptop computer if they are located close enough to the wireless hub. In this way, the PC Internet and the *keitai* Internet are competing in the urban environment.

Discussion: Anonymity and Online Personas

As *deai* practices and the category of the intimate stranger develop and evolve through a changing palette of communications technologies, we discover new dimensions and challenges to our theories of anonymity and identity construction. Drawing from the material I have presented about Japanese telephone and online *deai*, I now turn to the relation between anonymity and online personas.

Sherry Turkle (1995; 1999) focuses on the opportunities provided in cyberspace for individual play with identity and experimentation with new identity. Erik H. Erikson's (1968) work on identity describes adolescence as a period of moratorium. Moratorium was a crucial concept in the 1950s and the 1960s, and scholars described school life as a period of moratorium (Okonogi 1978). However, Turkle points out that moratorium today is not provided by school life but by cyberspace through the creation of virtual personas. Through the act of switching "windows," we create multiple online personas. The anonymity guaranteed on the computer screen gives people the opportunity to express undeveloped aspects of their identities and to simultaneously foster plurality. Turkle adds that life on the screen plays out and materializes a series of cultural trends that approach the concept of identity from the viewpoint of plurality and flexibility.

In our communication with intimate strangers, we present our pluralistic and flexible selves. It is a method for relaying our honest feelings. These multiple cyberspace identities are all the "real me." We used to sit in front of the computer connected to the Internet to enjoy our online personas in chats and e-mail. Now such online personas can be carried with us wherever we go with Internet-enabled *keitai*.

Keitai deai-kei sites are very popular and more convenient than conventional PC e-mail. Most important, these sites allow users to send messages at any time and place, thus sharply increasing the frequency of e-mail exchanges. Internet-enabled *keitai* put the world of Internet romance into our pockets. As a result, intimate strangers are transforming into presences that can be nearby at any time.

Gary Gumpert (1987) describes how the development of various media removed spatial restraints, making it no longer necessary for people to be in the same location to carry on a conversation. He also states that many city dwellers belong to an unmapped community that has no sense of neighborhood. Moreover, Sidney D. Aronson (1971) observes that the wide adoption of the telephone established psychological neighborhoods that are not confined to the physical boundaries of where one resides. In a similar vein, Joshua Meyrowitz (1985) points out that electronic media create a distinction between physical places and social places (see also Watanabe 1989; I. Nakamura 2000).

On the other hand, Chantal de Gournay (2002, 202) states that "the current configuration of social relations (excluding functional and organizational relations) can no longer be likened to the image of a network" and that "as soon as relations are partitioned into segments functioning as pairs, they cannot form a network because they are intransitive." It is easy to discuss the possibility of an establishment of Internet communities, but the most definitive representation of interpersonal relationships created by telephony is not a network but "proliferation communication." This point is critical in examining telephone and *keitai* communication: As de Gournay (2002, 202) writes, "I can be the friend of my friend's friend, but I cannot be the lover of my friend's lover."

If the telephone fixed to a location represents social place and social positions and roles in the traditional sense, *keitai* holds the potential for actualizing the individual liberated from these place-bound definitions. In contrast to the fixed-line telephone, *keitai* are characterized by the "obsolescence of address." Even if a telephone number is not listed in a phone book, the fixed-line telephone is still fixed to a physical and social location. In other words, the fixed-line telephone number is the same as the address, identified by a series of numbers that represent the physical and social location. In contrast, *keitai* is a telephone of an individual, not tied to a particular location, and its number can be perceived as the individual's number. From this standpoint, the relation between the fixed-line telephone and *keitai* mirrors the relation between the "original me" and the "real me." In other words, the traditional "original me" was restricted to social positions and roles like the fixed-line telephone. In contrast, the "real me" is the other self in cyberspace, the identity that has been freed from the restrictions, like *keitai*.

There is, however, always a gap between the "real me" and the "original me" as realized through online personas. The incongruity that many people feel when they first meet a *mail-tomo* in person originates from the discord between the "real me," which they had envisioned from the online persona of the *mail-tomo*, and the physical reality they are facing. When the face in cyberspace differs from the face in the real world, we have a tendency to deny the real-world face and to believe in the cyberspace face. We try to maintain the consistency of the image within us, even if it means rejecting reality ("This is not him/her!"). Meanwhile, we are afraid that the "real me" in cyberspace will be dismissed because of everyday reality ("original me") ("So this is what he/she is really like. How disappointing."). Such potential disconnects exist in *keitai*, which we always carry around with us.

Conclusion

Beginning in the 1980s, Japan has seen the emergence of new forms of telephone communications that have enabled people to form intimate relationships with anonymous others—intimate strangers. Intimate strangers have continued to exist with changing media communications, including *keitai*.

Keitai have unique characteristics that differ from the fixed-line telephone, including the recent integration with the Internet. The new kinds of romantic relationships developed on the PC Internet continue to be the object of public attention. Encounters between men and women through *keitai*, however, have been subject to heavy regulation because of associations with juvenile prostitution and delinquency. Despite this, the new kinds of relationships developed on the PC Internet are steadily being exported to the *keitai* Internet, leading to even more flexible identity formations. The PC Internet and the *keitai* Internet are not identical, however. While the PC Internet excelled at creating online communities, the *keitai* and the *keitai* Internet are keyed toward the creation of personal networks. Currently, the PC Internet and the *keitai* Internet are competing for their place in the urban environment.

Georg Simmel (1903) analyzed human behavior in the modern city (see also Goffman 1959, 1963, 1967; Hall 1966; Milgram 1970). He characterized the urbanite mental attitude as "cold," suggesting not only indifference but an attitude that could turn to hatred and aggression when strangers come into close contact with one another. On the other hand, a big city provides freedom to individuals. City dwellers are free of scrutiny and prejudice found in small towns and villages, and they can act as they please without being self-conscious. Simmel identified loneliness and desolation as the negative aspects of a big city and individual freedom as a positive one (Nagai 1986; Ishikawa 1988).

Keitai allows us to move around freely in urban space. We can enjoy conversations with friends without being tied down to a specific location. Such uses of *keitai* can create discord in urban space. Another characteristic of *keitai* is its effectiveness in times of emergency. *Keitai* adoption has correlated with a sharp increase in the number of calls to 110 (police) and 119 (emergency). We feel a sense of security just by carrying a *keitai*. This sense of security motivates parents to give their children *keitai* when they move away from home for the first time or start living in an unfamiliar city. *Keitai* alleviates the loneliness and anxiety of urban space. In other words, *keitai* can be a medium to prevent getting lost in the urban desert. In the city, with its high degree of anonymity, *keitai* responds to both the freedom and loneliness that Simmel identified.

The intimate stranger is a natural outgrowth of this urge to prevent loneliness in new urban spaces. We travel and reconfigure ourselves, and we need new and unfamiliar others with whom to test our identities. *Keitai* accelerates our chances to meet and have shared experiences with intimate strangers. The pace of this change has met with early resistance from social, legislative, and corporate governors. But we should see the continued evolution of this new kind of social relation, as the curiosity of users and the shape of the technology continue to promote identity liberation. People familiar with the company of intimate strangers are developing new sets of social mo-res and customs to protect themselves and promote the best possible outcome from these unusual new partnerships; future studies of this phenomenon can use the history I've described here to understand the evolution of a new social form native to the *keitai* internet.

The competition between the PC Internet and the *keitai* Internet in urban space is reflected by a similar competition in the home. Recent years have seen a growth of *keitai* use in the home at the expense of fixed-line telephone use. It is not surprising that *keitai*, as a technology that can be used at any time and place, would be used in the home as well. Both the PC Internet and the *keitai* Internet are being accessed in the home (see chapter 11).

We can expect to see even more flexibility in where and when to use the PC and the *keitai* Internets, leading to a more pervasive presence of the intimate stranger in more settings of everyday life. People can now be connected to intimate strangers at any time and place, and we can expect them to become an indispensable way of relating to others.

IV Practice and Place

Daisuke Okabe and Mizuko Ito

Despite an often oppressive crush of humanity, trains and subways in Japan are remarkably quiet. Although many passengers type into *keitai* keypads or scroll through pages on tiny screens, nobody talks on *keitai*. Even the sounds leaking from a young person's Walkman are considered a violation of this norm of silence. Pervasive announcements and signage prod commuters toward behavior that minimizes their audible presence in this shared space, but the subtle interactions between passengers are the most effective mechanisms for maintaining this social order. Suppose that a ring tone breaks this silence, or somebody sitting in a subway car starts a *keitai* conversation. Most likely, people nearby will glance quickly at the source of the noise. If the offender speaks particularly loudly, she may get a glare or an expression of disapproval (even if there are people chatting more loudly in the next seat). This kind of scene is a familiar one in everyday life in urban Japan.

As Matsuda has described in chapter 1, *keitai* have suffered from a bad reputation, particularly regarding their use in public places. Although *keitai* use is restricted in other public places, trains and buses are the sites of the most intensive efforts at regulating use. In the initial adoption period in the early 1990s, *keitai* were more often the object of envy than public regulation. However, after *keitai* were widely adopted by youth from 1996 on, public transportation facilities stepped up their efforts to dictate limits to *keitai* use. The current social consensus is that "voice calls should be avoided but *keitai* e-mail on public transport is okay." These norms for *keitai* use on trains and buses are the result of a decade-long process of developing social standards and regulations.

This chapter describes *keitai* users' experience of this social order and how it is achieved and maintained at the level of everyday interaction as well as at the level of ongoing public regulation. After first outlining our conceptual framework and research, we provide a description of the nature of the social order on trains and user perceptions, drawing from interviews. Next, we present cases of moments when the social order was disrupted by *keitai* calls. The final section describes the history of ten years of public regulation efforts leading up to current developments in this area.

Research Framework

Even outside Japan mobile phone use in public spaces was one of the first social issues surrounding the mobile phone to arouse public and research consciousness. The most obvious manifestation of the social and cultural consequences of mobile phone use was the colonization of public spaces with private conversations that had no relation to the local setting. In Japan, as elsewhere, mobile phones have been considered a social blight on the urban landscape, disrupting the normative cultural logics of shared public places. Studies have documented people's perception of manners (Ling 1998), the relation to experiences of public urban space (Kopomaa 2000), and interactional details of how people manage the disjunctures between physically co-present encounters and mobile phone conversations (Ling 2002; Murtagh 2002; Plant 2002; Weilenmann and Larsson 2001).

Building from this base of interaction-based studies, our work describes how some of these disjunctures are managed and regulated in the specific setting of trains in Japan. Like much of the groundbreaking work that has preceded ours, we draw on Erving Goffman's (1959; 1963) studies of behavior in public places for inspiration. We depart, however, from a strictly interactionist perspective by also describing how the social situation under question has changed historically, in relation to a shifting network of technologies and people. To draw from the terms of science and technology studies, this is a process of describing the social construction of a technological system (Pinch and Bijker 1993) by tracing the contours of a shifting actor network (Callon 1986; Latour 1987). Our overall aim is to describe how the current social order in Japan surrounding *keitai* use in trains has been achieved through the sedimentation of both day-to-day interactional achievements and larger-scale institutionalized forms of regulation. We contend that both microinteractional and structural-historical factors need to be analyzed in order to understand how new cyborg (Haraway 1991) sociotechnical practices are stabilized as social norms.

The chapter draws from a range of ethnographic observations and interviews that we conducted in 2002 and 2003. The interview material is drawn from the diary-based studies described in more detail in chapter 13. Central to the descriptions in this chapter are observations that our research group conducted on trains in the greater Tokyo and greater Osaka regions between July 2002 and February 2003. Okabe and two student observers participated as passengers on trains, generally in pairs, observing *keitai* use at different times of day, on different days of the week, and on different train lines.

These observations had two components. The first was detailed documentation of cases of *keitai* use that occurred in the immediate vicinity of the observer riding the train. The observer would note the details of the person using *keitai* as well as the reactions of those in the vicinity. The second component was a raw count of the observed frequency of *keitai* use (voice or e-mail/Internet) in a single train car, segmented according to the time between train stops. Between these stops, the observer would also count the number of people in the train car so we could have an indicator of how crowded the space was.

Keitai in Japanese Trains Today

In our observations of the frequency of different forms of *keitai* use on trains, we found that the general social norm of "no voice, e-mail okay" was borne out in the actual practice of passengers. For example, one 41-minute observation on a busy train line represented the highest volume of use that we recorded. During the period of observation, there were thirty-seven instances of observable *keitai* e-mail use (both receiving and sending e-mail), and four instances of voice calls. In a 30-minute observation with the lowest volume of use, there were ten instances of *keitai* e-mail use and one voice call. The overall average of *keitai* voice calls in any given 30-minute span was one to two calls.

In our twenty-four interviews with *keitai* users, almost all responded that they would freely engage in e-mail exchanges but were hesitant to make and receive voice calls. For example, respondents described how they might decide not to answer a voice call if the train was crowded, or they might move to a less crowded location to take a call, or they might take the call but cut it right away. Most also responded that they were annoyed when somebody took a voice call on a train and talked in a loud voice. These responses were consistent across all age groups. Here is a typical response:

[18-year-old boy, high school student, Kanagawa prefecture]
When I hear somebody's *keitai* go off on a train, it bothers me. I think, "I'll always keep mine in silent mode."
How about e-mail in trains?
I do e-mail a lot, to kill time. I think e-mail is probably okay. If I get a call, I do usually answer it, but I keep my voice low. I do feel bad about it and don't talk loud.

Although the social consensus for *keitai* use seems well established, public transportation agencies continue their ongoing public regulation efforts through posters, signage, and announcements. Currently, on trains and buses across the country, the following type of announcement is played after each stop: "We make this request to our passengers. Please turn off your *keitai* in the vicinity of priority seating [for the elderly and disabled]. In other parts of the train, please keep your *keitai* on silent mode and refrain from voice calls. Thank you for your cooperation." In the past, there was no standardization of announcements across the different train lines. As of September 2003, however, the seventeen companies, including both the formerly public Japan Railways and the various private train and subway lines, standardized on this announcement.

Keitai manners in trains in Japan are part of a broad palette of behaviors that are policed explicitly and persistently by public transportation institutions. Most similar to the issue of *keitai* is the problem of noise pollution through Walkmans, which has been the subject of controversy and regulation since the 1980s (du Gay et al. 1997). In addition, in stations and on trains, posters illustrating appropriate and inappropriate behaviors are pervasive, as are announcements cautioning, directing, and instructing passengers toward certain behaviors. Posters illustrate and warn against such transgressions as leaning a wet umbrella on another passenger's leg, eating or applying makeup, groping female passengers, getting fingers pinched by train doors, taking up too much space on seating, or leaving a backpack on rather than holding it at a more unobtrusive level. Announcements warning against running through closing train doors are repeated as each train is about to leave the platform. As passengers get off the train at each station, announcements remind them not to leave anything on the train. Even more than buses, trains in urban centers are characterized by precise technical and social regulation and very low rates of disorder, whether it be poor manners, a late train, graffiti, or litter.

Keitai Involvements

In addition to the formal efforts of public transportation institutions to keep trains and passengers running in good order, passengers engage in ongoing acts of mutual surveillance, regulation, and sanctioning that keep other passengers in line. Even before the advent of *keitai*, passengers on Japanese public transportation regulated behavior through mutual surveillance, so it is not surprising that these practices have extended to *keitai* use as well. In Goffman's (1963) terms, the space of the train is a well-defined social situation, with specific expectations of mutual "involvement" or participation in the space. Deviance from these expectations is noticed and acted upon by other passengers, often through nonverbal displays.

We follow Goffman's lead by presenting cases of transgressions of the social order as indicators of the underlying social order. *Keitai* voice calls are a prime example of practices in which people engage with the understanding that they are transgressing the social situation of the train, and we have observed some instances of this transgression and the subtle social surveillance and sanctioning that accompanies these acts.

February 5, 2003, 12:50 p.m., Osaka. The train is fairly crowded. A woman in her twenties, dressed casually, is standing by herself near one of the doors. Her phone vibrates, indicating a voice call, and she takes her *keitai* out of her purse. As she answers the phone, she covers her mouth with a magazine and her hand, and starts to speak. Her gaze is directed downward and to one side, and is still. She speaks very softly, and I can't make out what she is saying. At the moment that the woman answers the phone, two people sitting nearby glance quickly at her. She speaks for about three minutes, ends the call, and gazes out the window.

Most instances of *keitai* voice calls resemble this kind of scene. The phone is in silent mode, and the receiver decides to take the call but conveys through an introverted gaze and posture and a low voice that she is trying to minimize disruption to the social situation of the public space. She also makes use of what Goffman (1963, 38–42) might call a "portable involvement shield," a prop—in this case, a magazine— "behind which an individual can safely do the kind of things that ordinarily result in negative sanctions." In this case, the *keitai* user is clearly demonstrating her understanding of behavioral expectations even while transgressing them, so she is subject to only mild sanctioning by those in the vicinity. In other cases, a passenger might display respect for the social situation by moving to a less crowded area while taking the call.

The social situation of riding a train is constructed and maintained through these kinds of ongoing interactions and displays. When *keitai* users "unavoidably" have to take a call in a train, the situation demands that they continue to display involvement and respect toward the shared setting by performing their taking of the call as a secondary involvement. If they don't display the appropriate level of involvement and consideration to the social situation, then they are subject to more visible forms of sanctioning by other passengers. Drawing from Goffman (1959), Ling (2002, 4) describes public mobile phone use as "interacting on a double front stage," where the user must manage accountabilities to both the online conversation and the local setting. He suggests that "the verbal and the gaze/gestural effects can be used in opposite ways for various publics."

Ged Murtagh (2002) describes some of the nonverbal ways in which other passengers indicate what they feel to be appropriate or inappropriate in other passengers' behavior with mobile phones. He points to gaze and posture (changing direction of the face or upper body) as subtle negotiations that construct shared context and the implicit boundaries of what constitutes a public nuisance. The most common form of nonverbal sanctioning behavior is the gaze. Glancing or glaring at a *keitai* user is a way for other passengers to engage in public regulation of behavior. The following interview excerpt gives some sense of behavioral expectations and the effectiveness of the gaze as sanctioning behavior:

[20-year-old woman, college student, Osaka]
What to you think are the situations and places in which keitai *should not be used?*
On the train, I try not to make voice calls.
What if you get a call?
I do answer it.
When that happens, do you, like, try to appear apologetic? Maybe cover your mouth with your hand?
Yes.
If people around you started glancing at you, would that bother you?
Hm. Probably if somebody looked at me, I would stop.

If another passenger is making a voice call, what do you think?
Hm. If they talk just for a short time in an apologetic way, I think maybe that's okay? But if there is, like, a high school girl talking in a way loud voice, then it pisses me off. I think they should cut it out.

If civil inattention is the interactional norm for strangers in public transportation, then glancing at a fellow passenger is an indication of disruption of this social situation. Just as Goffman looked to the social transgression of mental ward patients to describe the social orders that we take for granted, even small breaks in the normative order can indicate the boundaries and contours of that order. In some instances, when the passenger taking a voice call is not displaying sufficient respect to or involvement in the shared situation, other passengers might use the glare as a stronger form of sanctioning.

December 27, 2002, 8:55 p.m., Osaka. The train is fairly crowded. One man in his forties, wearing a suit and apparently on his way home from work, is standing in the space between two doors. Suddenly, his *keitai* rings, and he answers it immediately. Smiling, he says, "What's up? Hm. Hm. I'm in the train now so I'll call you back later, okay?" Two suited men in their fifties standing near the door a short distance away glare quickly at him when his *keitai* rings. They glance at him a few more times, but the offender seems not to notice.

In this case, the other two passengers interpreted the *keitai* user's actions as deviating from the norms of the social situation in a way that was observable by the researcher. On the other hand, we observed instances of taking a *keitai* voice call on a train where a failure to defer to the shared situation was *not* subject to sanctioning by other passengers.

The social sanctioning of voice calls by other passengers is not a behavioral universal on trains. Although the general rule is to refrain from voice calls while on a train, this rule is subject to situationally sensitive interpretation, leaving some margin of variability in interpretation. In this sense, the prohibition against *keitai* voice calls in trains is a somewhat less fixed social standard than more established and universal standards such as making space for others to sit. Although Japan is characterized by a high degree of consensus about appropriate behavior on public transportation, *keitai* use still has some interpretive flexibility surrounding it. Actions such as a glance or a glare, the announcements that try to specify behavioral norms, and passenger perceptions that vary based on region, age, and time of day, all construct a relatively defined but somewhat flexible social situation that is open to interpretation. The following case illustrates some of the variability we observed:

January 25, 2003, 7:15 p.m., Tokyo. It is a Friday night, and because the train is headed for the city center of Shinjuku, it is very crowded. The train is also very loud, with many people talking at a regular volume with no apparent concern for the other passengers. A casually dressed man in his thirties is sitting in the second seat from the edge of a long seven-seat bench. His *keitai* rings. After quickly looking at the screen (probably to check who the call was from), he answers immediately. His voice is at normal conversational volume, and he seems unconcerned about the others around him. The other passengers take no notice of his actions.

In contrast to the prior examples, in this case the other passengers do not display any sanctioning behavior despite the fact that the *keitai* user's phone has rung and he has answered it with no display of situational deference or involvement shielding. Even the announcements on the train seemed to have been drowned out by the unusually high volume of passenger chatter. The *keitai* user keyed his behavior to the specific context of a late night train ride full of people headed for a night out in a raucous city center.

Even as *keitai* have been adopted by Japanese of all ages and social stripes, there is still a gap between different social actors, such as youth, business users, train companies, and the elderly, in their sense of what is appropriate *keitai* use. For example, in the Kansai region of western Japan, the announcements on trains about *keitai* use were not as strict as those in the Kanto region, and they were implemented later. In our interviews and observations, we also saw that passengers in the Kansai region were more forgiving about *keitai* use on trains. According to Shunji Mikami (2001), age also colors how people view *keitai* use. In the case of a crowded bus or train, 50 percent of 16–19-year-olds and 58 percent of 20–24-year-olds felt that "you should never use a *keitai* and should turn it off" in contrast to 81 percent of those aged 50–54 and 88 percent of those aged 55–59. In other words, older people had a stricter sense of how *keitai* should be used.

It is difficult to posit a uniform standard for how *keitai* should be used in trains, and actual *keitai* use is keyed to the tendencies of particular users situated in specific contexts. Perception has also changed along with the evolving landscape of *keitai* adoption. We turn now to examine the state of social understanding of appropriate *keitai* use in trains over the past decade.

The Historical Development of *Keitai* Manners for Trains

The current set of norms (including their variability across regions, age groups, and specific settings) is a result of a series of negotiations between different social actors spanning a decade. Newspaper coverage on the topic of *keitai* manners in public transportation emerged widely around 1996 and reached a peak in the last half of 2000 through 2001. At the present time the topic is no longer one of interest in the mainstream press.

We begin with a review of articles appearing between 1991 and 2001, and then analyze how *keitai* manners were socially and culturally constructed into the form that appears natural today. We draw broadly from approaches in the social construction of technological systems (see Bijker and Law 1992; Bijker, Hughes, and Pinch 1993) to describe the gradual shift from interpretive flexibility to closure that went hand in hand with the changing identity of the *keitai* user and the mobilization and enlistment of varied social actors.

In their discussion of the development of the bicycle, Trevor Pinch and Wiebe Bijker (1993) describe how different social actors ascribed different meanings to the device, and it was through the negotiation between these different actors that the meaning and form of the bicycle was gradually stabilized. Initially, bicycles were considered the domain of young athletic men. Women and elderly men gradually came to be considered "relevant" as social actors in relation to the bicycle, raising safety concerns that resulted in a series of controversies over appropriate bicycle design. Eventually, rhetorical and design closure moved to a model of a bicycle with an air tire and a smaller front wheel that solved both the safety needs of everyday riders and the speed needs of athletic riders.

We can apply a similar model of stabilization to the development of the meaning attached to *keitai* use in public transportation, as the relevant social groups shifted from business users to youth to the general public, resulting in periods of transition and controversy and gradually arriving at a social consensus.

Business Users as Relevant Social Actors

As chapters 1 and 2 have described, in the early 1990s *keitai* was an expensive status symbol identified with executive business users. Isao Nakamura (2001b) analyzes the changing image of *keitai* by dividing its evolution into three stages with particular user demographics and characteristics. According to his model, the period up to 1995 was characterized by the image of business use because 90 percent of *keitai* users were businessmen. Even with the economic downturn and downsizing that characterized the period from 1991, *keitai* retained the image as a tool for the adult man; women office workers and teenage girls were associated with the pager rather than *keitai*.

As Matsuda describes in chapter 1, the meanings associated with *keitai* at that time included the sense of "the successful businessman" as well as an image of the "self-important businessman" displaying his status as a wealthy and in-demand person. Japan Railways began announcements regarding *keitai* use in its bullet trains in 1991 to restrict use. A newspaper article in that year (*Yomiuri Shinbun* 1991) indicates the emergent social controversies surrounding *keitai* use:

You have probably seen the figure of a man walking briskly through town while talking on a *keitai*.... [As one person puts it,] "The original location for phones, even when in a restaurant, is in an unobtrusive place that would not bother other patrons with the sight and sound of the call. Is it really necessary to go out of one's way to make a call in a busy place full of people? Shouldn't one choose a more appropriate time and place? It may be fashionable, but I wonder about walking around with one in your hand." ... This viewpoint is apparently common, and Japan Railways has started an announcement on its Tokai bullet train: "*Keitai* calls while seated are bothersome to other passengers, so please make the extra effort to make your calls in the deck area [between cars]." Indeed, in a first-class car, it must be unpleasant to have to listen to a loud conversation.

At this time, adoption rates were still low, and people strutting through town with *keitai* were a distinct minority. Even at this time, however, we began to see the first glimmerings of efforts to regulate *keitai* use in public spaces.

An Emergent Controversy

Keitai use in trains emerged as a topic of widespread social concern in 1995 and 1996, along with a sudden rise in adoption rates, reaching 15 percent of the population in 1996 (see chapter 2). In 1995 *keitai* subscriptions surpassed pager subscriptions, and the user base for *keitai* was broadening. Unique styles of *keitai* use by high school girls became a focus of public attention. The following newspaper excerpts (*Asahi Shinbun* 1995; *Yomiuri Shinbun* 1995) provide some hint of the social consciousness at that time:

With the advent of the PHS, *keitai* use has suddenly skyrocketed. In commuter trains, I am starting to notice scenes of people hanging on a strap with one hand and talking on a *keitai* held in the other. Although *keitai* are convenient, complaints are also frequent: "Their voices are too loud." "It's annoying to have to listen to somebody else's private conversation." Although train companies are calling for manners to prevail against voice calls in trains, this is still at the level of "just a request."

On the train ride home from work after 8 at night, I suddenly hear the sound of a *keitai* ringing. Since it was still during rush hour and the train was crowded, I couldn't see her, but I could hear the voice of a woman starting to talk on the phone. The previously quiet space inside the train became unsettling because of the phone call. A conversation I had no desire to hear reached my ears through no choice of my own. The conversation was not about anything urgent that needed to be attended to while on the train. She continued this trivial conversation at length.

NTT DoCoMo first released a handset with a silent mode vibration, the Digital Mover F101 Hyper, in December 1995, so the preceding articles refer to a period when phones had only ring tones to notify users of incoming calls. Because of this, people expressed their annoyance at the sound of the *keitai*'s ringing as well as at having to listen to phone conversations. From this time on, the public called for "*keitai* manners" on trains, and *keitai* use joined the list of actions considered "bothersome to others" on trains.

Youth as Newly Relevant Social Actors: From Controversy to Stabilizing Consensus

From around 1997 train companies responded to ongoing passenger complaints by trying to regulate *keitai* use more aggressively. For example, Japan Railways started broadcasting a stricter announcement in its regular (nonbullet) trains: "We ask for your cooperation in refraining from using *keitai* while on the train." Prior to this, announcements tended to say, "Please be considerate of others when using your *keitai*." This was the period when *keitai* had been adopted by about 22 percent of the overall population.

Also in 1997 all carriers began short message services, and after the launch of i-mode in 1999, text messaging became even more popular, particularly among young people. According to Nakamura's (2001b) three-stage model, in the period from 1995 to 1999 the core user group shifted to the youth demographic. He links the growing preponderance of young users to the discourse of *keitai* as "social nuisances." Along with this, public discourse also began to tie young people's *keitai* use to bad manners in public spaces like the train:

On a train on my way home from work, a young man, probably a student, is talking loudly and at length on his *keitai*. I reprimand him. To my surprise, he says, into the phone, "Some weird old guy just said something to me" and continues his conversation.... He was still talking as he got off the train at the next station.

Young people who are inconsiderate to others are everywhere.... In this incident with the *keitai*, other passengers were glancing at the young man, thinking "I wish somebody would tell him to stop." (*Yomiuri Shinbun* 1997)

A national poll by the *Yomiuri Shinbun* reveals what those over thirty think of young people.... "Use of *keitai* with no consideration of time and place" was ranked first, at 48 percent, among young people's activities that are annoying, followed by "failing to do proper greetings," at 41 percent.... These kinds of people are most noticeable in the larger cities. *Keitai*, which have been a focus of public outcry because of use in trains and buses, is used by 41 percent of the population in the provinces and 58 percent in the big cities. (*Yomiuri Shinbun* 1998)

As the *keitai* shifted from an exclusive association with businessmen to greater association with youth, there was a matching increase in efforts to regulate *keitai* use in public spaces. From around 1999, during the morning and evening rush hours, Japan Railways started an announcement, "Please refrain from using your *keitai*," and put up stickers in the trains to same effect. These efforts also coincided with the growth of text messaging, a timely and appropriate way to circumvent the social stigma against talking on trains.

What was the general perception of *keitai* use in trains at the time? The Telecommunications Carriers Association conducted a survey in 1998 in nine different locations across the country about perceptions of *keitai* use in buses and trains (1,300 respondents aged 16–59). According to this research, approximately 80 percent of respondents answered positively to the following statements: "It doesn't bother me if people check messages on their voice mail without talking or engage in non-voice mobile communications" and "It doesn't bother me if people's *keitai* are on vibrate mode and there is no ring tone." In other words, a new norm was taking hold: *keitai* use on trains is not bothersome if it doesn't entail voice communication.

Together with the rising tide of regulatory announcements on trains, consensus was beginning to form regarding manners for *keitai* use on public transportation. Young people's bad *keitai* manners on trains were taken up widely in newspapers and magazines and on TV; accordingly, a new social order was constructed around prohibition of *keitai* use in trains. At this time, the overall tendency was to reject *keitai* use in trains, and the announcements reflected this. However, rather than being a natural outcome of proper *keitai* use, this norm emerged from a historically specific process of enlisting and disciplining a new social actor, youth, in the meanings and practice of *keitai*. This contingent social order was achieved through a complex interaction between various social actors—public transportation organizations, *keitai*-adopting youth, and adults in positions of power in relation to those of youth—signage, announcements in public transportation, discourse in mass media, and changing *keitai* technology. As Matsuda describes in chapter 1, the meanings surrounding *keitai* shifted dramatically toward issues of regulation and control as youth became dominant actors in this space, and the case of public transportation is one arena where this shift played out.

The Pacemaker as a Newly Relevant Social Actor

As of March 2002 *keitai* adoption rates stood at approximately 60 percent of the Japanese population. The social consensus at that time was that voice calls are not appropriate on trains but e-mail is no problem. Although users try to avoid ring tones and voice conversations disrupting the space of the train car, they do not actually turn their phones off but rely on silent mode settings to keep their *keitai* quiet.

Within this relatively stabilized social order, however, a newly relevant social actor enters the mix, the pacemaker,[1] or perhaps more accurately, the cyborg entity of the "passenger with a pacemaker." Up to this point, the primary social actors on the passenger side were youth *keitai* users, businessmen *keitai* users, and passengers disturbed by *keitai* noise. Public discussion of the relation between pacemakers and *keitai* electromagnetic waves emerged as early as 1996.[2] At that time, however, coverage by the popular press was still spotty, and the pacemaker was not immediately incorporated as a relevant and influential actor in determining *keitai* use patterns on trains. Unlike the more visible and pervasive technology of *keitai*, the pacemaker is an invisible and more uncommon social actor that required more explicit public advocacy to be noticed. It was only relatively recently that pacemakers and pacemaker users became relevant social actors in these negotiations.[3] Here is a newspaper excerpt from *Yomiuri Shinbun* (2002):

Many people simply put their [*keitai*] in silent mode without turning off the power. Along with the spread of mobile phones, more people feel they cannot turn off their phones because of work-related calls. Although it is important for train lines to repeatedly broadcast announcements to "turn off the power while on the train," I don't think this alone will result in better manners.

We have known of the effects of a *keitai*'s electromagnetic waves on pacemakers for some time now. Some of the large train lines in central Tokyo have been calling for passengers' cooperation by setting a new rule that *keitai* need to be turned off in train cars with even numbers but can be left on in silent mode in odd-numbered cars, though voice calls are still prohibited. I feel that train companies should be more innovative in working to make trains an environment where people with pacemakers can ride with peace of mind.

The new visibility of the passenger with a pacemaker as a relevant social actor in defining the meaning of *keitai* in public places results in a new series of controversies about appropriate social behavior. *Keitai* is again subject to interpretive flexibility as people debate a new set of considerations that apply to a smaller sociotechnical group of pacemaker users but with more serious and life-threatening ramifications than "just manners." Although the period from about 1997 to 2001 was characterized by rhetorical closure and practical consensus about appropriate use of *keitai* on trains, we are in the midst of another transitional period where the nature of institutional regulation, public perception, and everyday practice is changing because of the influence of a new social actor.

The issue of pacemakers and *keitai* is still unstable, in part because of the instability of the scientific facts surrounding the interaction between the two technologies. The outcome of this current negotiation between *keitai*-using passengers, pacemaker-using passengers, the public, and transit facilities is still uncertain. This case, as well as the historical development of appropriate behavior on trains, indicates how social order is built through ongoing social and cultural construction and maintenance work at the level of everyday interaction, institutional policies, and public discourse. As Joan Fujimura (1996, 14) has said about the construction of "packages" of facts, theory, and practice in the sciences, "Resolutions can be short-lived or long-term, but they are rarely, if ever, permanent. Even consensus requires maintenance."

Conclusion

In many ways, the negotiations between *keitai* users and pacemaker users can be located in a much broader set of social and cultural negotiations between different "sociotechnical entities" (see chapter 11) about what constitute appropriate cyborg couplings and cyborg behavior. Public places like trains are stages upon which these negotiations between a changing set of cyborg entities is played out face-to-face. These stages are also the object of a political process through which different social actors enlist other technologies (e.g., *keitai*, pacemakers) as well as other institutional entities (e.g., the press, transit agencies) and scientific facts (e.g., the relation between electromagnetic waves and pacemakers, empirical studies of youth *keitai* use) to their cause.

In chapter 4, Fujimoto describes a broad set of historical and cultural trends that have led up to the current paradigm clash between teenage girls and middle-aged men, played out in places like trains. In his scenario, girls have enlisted the technology of *keitai* and middle-aged men have enlisted the technology of the newspaper to carve out territories in public space. Teenage girls' talking on *keitai* and middle-aged men's rustling newspapers represent two poles of the negotiations we have described surrounding *keitai* use in trains.

We have described a complex picture of both the micro and macro politics regulating the use of a new technology as part of an ongoing historical and everyday process. As a result of these political maneuverings, all parties to the networked negotiations are in some respects transformed. *Keitai* acquires new meanings and features that become naturalized through time; pacemaker users become newly visible and consequential; public transit institutions become accountable to a new set of health and social welfare issues; and intergenerational tensions are reinvigorated with the addition of a youth-identified technology. The web of negotiations and interactions that characterize even the particular setting of public transportation is indicative of the complexity of the process through which new technologies become established in an evolving social and cultural ecology.

Notes

This research was conducted with the help of Kunikazu Amagasa, Hiroshi Chihara, and Joko Taniguchi, who were graduate students at our lab at the time we were conducting the field work for this project. This work is indebted to their careful fieldwork as well as to the ongoing discussions we had as a research team. This work was supported by the DoCoMo House design cottage at Keio University Shonan Fujisawa Campus. We thank lab director Kenji Kohiyama for his ongoing encouragement and support.

1. In line with approaches in actor-network theory, we treat both human and nonhuman entities as potential actors within sociotechnical networks and negotiations (Callon 1986; Latour 1987).

2. Research by a task force on investigating the effects of electromagnetic waves, put together by the Ministry of Posts and Telecommunications, the Ministry of Transport, and various industry groups, concluded that having *keitai* within 22 centimeters of a pacemaker could result in the latter's malfunctioning. Japan has approximately 300,000 pacemaker users.

3. In response to concerns by pacemaker users, electronic manufacturers developed a device designed to block the electromagnetic waves of *keitai*. The Telecut, sold for ¥6,000 in specialty electronic centers like Akihabara, became a hit product, with approximately 10,000 units sold in a year.

11 The Gendered Use of *Keitai* in Domestic Contexts

Shingo Dobashi

Youth have been the driving force behind widespread *keitai* adoption in Japan despite the fact that the device was first introduced in companies as a business tool (see chapter 2). Now, although *keitai* have been adopted by a wider range of generations, young people, particularly those in their twenties, still have higher rates of use (Mobile Communication Research Group 2002). As Matsuda describes in chapter 1, most of the public and research discourse since the mid-1990s has focused on youth, stressing changes and the novelty of these changes that *keitai* has brought about in the lives of young people. There has been extensive debate over how *keitai* has changed interpersonal relationships among youth in Japan, how it has changed their buying behavior, and how it has affected delinquency and uses of *deai-kei* (encounter) sites. Whether the discussions take on a positive or a negative note, the interest is in how *keitai*, or young people who use the *keitai*, have become a unique entity, often in opposition to traditional social relationships, customs, and culture. Without doubt, Japanese youth have adopted *keitai* in an extremely brief span of time, and it is also an unmistakable fact that there has been some sort of change in their communication patterns.

However, I would like to note that this tendency to transform and deflect existing conditions does not affect all *keitai* users. All technologies, including *keitai*, do not exert the same influence on all users. Users are affected in different ways depending on their social positions and the specific contexts of use for each technology. More accurately, the conditions created through *keitai* use have not actually been brought about by *keitai* alone in a technologically determinist sense. They are rather a *specific consequence* of a synergy between the technology of *keitai* and its use by individuals in specific social positions. In short, the characteristics of *keitai* use observed among youth are a specific consequence of its use by young users and not inherent in the technology itself. It follows that *keitai* takes on different characteristics for users in different social positions.

Based on this argument, I posit that the influence of *keitai* can be understood not only as a transformation of existing social order, customs, and culture but also as a simultaneous process of maintenance and reinforcement, depending on the social position of the user. This chapter identifies these latter dimensions of *keitai* by examining its use by housewives in the domestic context. Needless to say, I do not suggest generalizing the characteristics of *keitai* use through my case of housewives because these characteristics are specific consequences of context, just as they are in the case of young people. Nonetheless, in order to study the more intricate details of the social influences of *keitai*, it is of utmost importance to understand varied use by people of different social positions. My focus on housewives is significant because these users have not been the subject of many discussions.

This chapter is based on an interview study of the use of *keitai* and other domestic technologies, and it describes how *keitai* reinforce housewives' social roles and reproduces the domestic gender order. It also examines how housewives define and position the technology of *keitai* in their activities at home and in the social relationships of the family.

The Home, the Housewife, and *Keitai*

It may seem odd to study the use of a cellular phone, a mobile device intended for outside use, within the context of the home. However, as seen in figure 11.1, survey results from the Mobile Communication Research Group (2002) show that the home is the predominant place where *keitai* are used, both for voice communications and text messaging. Part of this is simply because people generally spend more time in their homes than anywhere else. Nonetheless, the survey results indicate high *keitai* use in the home, and thus the importance of studying the social influences of *keitai* use there.

A number of issues arise when studying *keitai* use within the home. First, what kind of place is the home? A home is much more than a configuration of physical space; it is also a social and cultural space incorporating various values, norms, and gender and generational power struggles. Moreover, the home is not a static entity but rather is continually reborn through these contests over values, norms, and power. In other words, several dynamic layers within the physical environment work to continually construct and uphold the notion of home (Morley 2000).

Prior research indicates that these dynamics also have a strong influence on the adaptation of new technologies for home situations. For instance, Roger Silverstone, Eric Hirsch, and David Morley (1992) identify this situation as the domestication of technology. Their argument is that technologies that have been produced and given meaning in the public domain are not accepted into the home as is. Instead, as they are brought into the domestic setting, such technologies are redefined to fit into each home and the home's values, norms, and conventions. Put differently, while people's

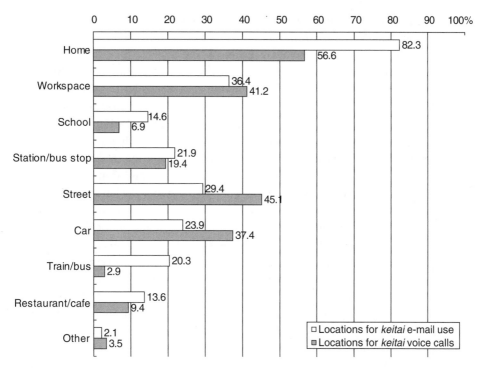

Figure 11.1
Locations for *keitai* e-mail use and voice calls. From Mobile Communication Research Group (2002).

lifestyles within the home are significantly affected by new technologies, the technologies are incorporated as a part of creating and preserving the home. This is precisely where the focus should be placed in studying *keitai* in the home: How are *keitai* used and perceived within the home? How, if at all, does this domesticated technology contribute to the creation and preservation of the home? How are *keitai* finding their place in the context of the home? Such questions naturally arise when thinking of the home as more than a physical space, recognizing the social and cultural forces at work as well.

Why focus on housewives? First, while this is a gender-biased view, housewives are generally expected to play a greater role in the creation and preservation of the home than any other family member. Things like household chores and childrearing are disproportionately distributed to the housewife through domestic gender roles prevalent in Japanese society. Thus, examining the use of *keitai* by housewives, who have these expectations placed upon them, becomes important in studying what *keitai* mean to a household. Most likely, *keitai* can either strengthen or destroy a family's unity and bonding, just as other media in the home can. In that case, looking at how housewives deal with the dualities of *keitai* and give them new meaning becomes a key part of studying how *keitai* are positioned in the home.

Second, women have historically played a large role in integrating new technologies into everyday life. This holds true for informational media as well as for home appliances. A model example is the fixed-line telephone, a predecessor of *keitai*. In the United States, as Claude Fischer (1992) has explored in detail, telephones were viewed by the telephone companies as a communications device primarily for business. Therefore, the home took second place to offices or retail outlets in the telecommunications industry. What's more, using the telephone for chatting, an activity engaged in mainly by housewives, was seen as incorrect use of the telephone, deviating from its true purpose (Rakow 1988). However, the social use of telephones by women prevailed and became a key force behind the commoditization of the telephone (Fischer 1992). In fact, the U.S. telecommunications industry began engaging in a major rollout of advertising campaigns that focused on the social aspects of telephone use after the 1920s.

Matsuda (1996a, 199) explicates what we can read from this history of the telephone: "Women telephone users changed the 'nature' of the telephone through their everyday practice." The public meaning of telephones, as defined by telecommunications industries, was altered by women's use, and in the process they redefined the meaning of the telephone to fit the daily routines of the domestic environment. The same implications can be witnessed in the history of the television or radio or other home electronic appliances.[1] Such accounts describe how housewives played a pivotal role in the domestication of new technologies.

The current reality is that housewives' adoption of the *keitai* is still lower than that of other groups. A Mobile Communication Research Group (2002) survey indicated unusually high percentages of adoption by high school and college students (78.6 and 97.8, respectively). Full-time workers were at 76.3 percent, and part-time workers at 61.1 percent, in contrast to full-time housewives, who were only at 42.7 percent. We have yet to see whether housewives will drive a similar trajectory of *keitai* domestication as they did with telephones, radio, and television. However, if the primary *keitai* user has shifted, first, from businessmen to younger generations, and continues to shift in the future, housewives must be included as one possible demographic group that will take on that role.

With these theoretical and historical issues as a backdrop, the following sections look at how housewives use *keitai* in the home, based on findings by my research group.[2] We conducted a series of interviews on how information media devices were received in the home, focusing on recent advances in digital informatization, including personal computers, the Internet, and *keitai*. This ongoing study commenced in April 2002 and involved approximately twenty households.[3] In the following sections,

I use the PC as a point of reference and contrast PC Internet access to Internet access through *keitai*. The PC and *keitai* are often brought into the home around the same time, and studying different dimensions of how the two entities evolve clarifies what *keitai* means within the home.[4] The significance lies in the relationship between the *keitai*, a technical entity, and the housewife, a social entity. The following sections focus on the time and spatial structure of housewives' activities, their management of family relationships, and their views on owning *keitai*.

Mobility of Domestic Activities and Mobility of *Keitai*

Key to analyzing the uses of *keitai* by housewives in the home is an understanding that their activities are often multilinear and mobile. When we think of the tasks involved in housework and child care, it is easy to imagine multiple activities requiring simultaneous attention. For example, NHK's national time use surveys conducted every five years break down housework into four categories, "cooking/cleaning/laundering," "shopping," "child care," and "miscellaneous housework," as well as eleven more specific items, "meal preparation/clean-up," "cleaning," "laundering/ironing," "shopping for food/clothing/housewares," "nursing baby," "child care and play," "supervising homework," "picking up and dropping off children," "cleaning/organizing," "going to the bank or government office," and "taking care of the ill or elderly."

These activities are clearly crucial to maintaining the life of the household, largely mandatory, and repeated on a daily basis. Further, these housework and child care responsibilities are responsive to the daily rhythms of other household members. Even if a housewife can claim some amount of discretionary time, it does not necessarily mean that she can use this time as she pleases. In comparison to many other countries, contemporary Japan is characterized by a much more distinct division of labor by gender, with women shouldering the bulk of household responsibilities (Yano 1995, 153–159). Housewives are often expected to be always available at the center of the household.

Consequently, it is difficult for many housewives to concentrate for long on one task in one place for a long period of time. Housework and child care require them to handle various tasks concurrently, which greatly restricts their media use at home. Moreover, housewives often recognize *keitai* as a medium that is highly adaptable to the rhythm of their activities. This point is consciously acknowledged by a woman in her thirties, who speaks of using e-mail while taking care of her child:

I haven't used the computer since my child was born. After I quit my job, I thought I wouldn't be able to use the computer at home, so I gave out [my *keitai* e-mail address].[5] [With a child] I have to run around the house, and he always wants to play with [the computer] if it's on. That's why I use *keitai* for e-mail. I could use the computer at night after I put my child to sleep, but I hardly do because I'm exhausted by then and he still cries a lot at night.... So, in order to be able to e-mail anytime, that's why.... If I'm on the computer, he starts moving around again. Without *keitai*, I wouldn't be able to do e-mail or stay in touch with other people.

Housewives said the same thing about housework in general, not just about looking after children. In our interview survey, there were hardly any housewives who preferred the computer over *keitai* for their day-to-day e-mail use.[6] As reasons for this preference, they cited being able to use *keitai* while doing housework or looking after children. Although there is room for debate, it is evident that *keitai* is assisting housewives' tasks and communications and is being used as a tool for empowerment. More important, though this is speculative, "being mobile" seems to perform a critical role even in the house, a fixed place that is not generally considered a site of mobility. For housewives, outdoor use is not the only relevant potential created by *keitai*; the significance of *keitai* is that it can be used simultaneously with other activities. This potential may seem extremely trivial, but it is a decisive factor in the choice of media for many housewives.

This point can be clarified by the fact that housewives are often required to make particular spatial arrangements for using a stationary computer in the house. A woman in her sixties explained,

Since we coincidentally bought a new computer when we rebuilt the house, we decided to put it here from the start. It's set up in the kitchen. I check my e-mail or something while cooking dinner.... So I use it pretty regularly. Before, we had it in the bedroom.... In that sense, it's more convenient now. When I have something else to do, it's better to have it here than in the other room. That's why we set it up here from the beginning.

When a housewife is among the computer users in the family, we often observed relocation of the computer from the bedroom to the living room.[7] It is a simple fact that to take advantage of computers and all other technologies in our daily lives, they must be integrated into our daily activities. Planning where to set up a device is the most straightforward method. Needless to say, *keitai* contrast with computers in the sense that such considerations are unnecessary, and it is therefore an extremely suitable technology for many housewives whose responsibilities are multilinear and require mobility.

If these issues regarding the flow of action and placement are spatial factors, *keitai* can also enable flexibility in time. For example, the following statement is made by a woman in her thirties, who indicates advantages in accessing the Internet with *keitai* rather than with the computer, and who expresses her awareness that *keitai* is a medium that enables more flexibility with time.

I use i-mode quite often to kill time or just as something to do. Like, I browse the ring tones and wallpapers and order books. I also look at sports news. I guess it's to kill time, a casual thing to do when I'm free. Killing time on the computer is a serious way to kill time, but on *keitai*, it's more laid back. I can turn it off in a second when someone is here. It's convenient for 2 or 3 minutes whenever I feel like it.... I use the computer around twice a week at most and sometimes only once a month.

When describing the "inconvenience" of using computers at home, an informant pointed out that her time is *komagire* (in little bits) because of her various chores. The ease with which *keitai* can be integrated into a *komagire* schedule is an important benefit for housewives. Depending on what she wants to do online, she needs a certain amount of time in order to make it worth starting up the computer to access the Internet. By contrast, Internet access using i-mode and other *keitai* Internet modes can be meaningful even in a *komagire* schedule.[8] As this informant indicates, they are both leisurely media, but when leisure can only be obtained in piecemeal ways, the *keitai* Internet is the more meaningful medium that fills short gaps in the day with some degree of satisfaction.

There is also a difference in rhythm between *keitai* e-mail and computer e-mail; in many cases, *keitai* e-mail is more suitable for housewives. The following quotations, from women in their thirties, describe how *keitai* is used for sending short messages that require relatively immediate response; a common example of how communication with *keitai* e-mail is clearly more useful than the computer. When engaging in activities like housework and child care that usually continue throughout the day, housewives often need to engage in such interactive communication. This is also indicative of their facility in integrating *keitai* into their lives.

I usually send messages that are short and don't necessarily have to be sent on the computer, like "Are you eating at home tonight?" or "What time are you coming home?" ... Or, if I'm meeting some people in the neighborhood, rather than calling them one by one, it's easier to send one short message to everybody at once, like "I'll meet you at the park at whatever time." I wouldn't use the computer to send messages like that.

I hardly ever use the computer because I wouldn't get an immediate response. My husband sits in front of the computer [at work], and I use my *keitai*. We don't talk about anything really important. I do sometimes use the computer, but I have to go through the trouble of starting it up. But with my *keitai*, I know right away when I get e-mail because it rings, and I can write back immediately. It's just more convenient.

In light of these advantages of *keitai* for housewives, it appears to be an empowering medium that supports their activities. However, there is a subtle undertone in these remarks; on the one hand, *keitai* assists housewives' activities, but on the other hand, it reinforces and reproduces their roles. In other words, *keitai* enable housewives to use e-mail and the Internet while taking care of their existing responsibilities. From a different vantage point, *keitai*'s ease of integration into the domestic schedule masks the disadvantages of housewives (such as a *komagire* schedule), which would otherwise be noticeable, for example, when trying to use a computer. I would further suggest that by effectively supporting the activities of housework and child care, *keitai* helps the housewives' *komagire* schedule remain *komagire*, which in turn preserves and strengthens the traditional gender division of labor. This is directly manifested by housewives who see *keitai* as the most suitable medium for themselves and who at the same time forgo using the computer at home.

These processes clearly reflect proactive decisions by the participants. Nonetheless, these proactive decisions are made within the frame of social roles that have been inequitably predetermined. Many housewives in the survey pointed to their spouses and children as persons with whom they exchange *keitai* communication. This indicates that *keitai* for them is a tool to maintain existing family ties rather than to develop new social relationships. This also shows that the uses of *keitai* by housewives are largely defined by their traditional social roles. There has been much discussion regarding how *keitai* has transformed existing social conditions. The case of housewives' use does not necessarily conform to that description and is by contrast more deeply associated with the gender order that has customarily been rooted in domestic space.

Managing Family Relationships

These issues can be explained more clearly by examining *keitai* use in family communication, particularly parent-child communication. This section describes the importance many housewives in the survey placed on communication with their children. However, it is important to note that such communication rarely involves the father. This fact is a reproduction and accentuation of certain particularities of Japanese family and parent-child relationships through new technology, particularly the gender imbalance characterized by the "absent father" and the "ever-present mother."

The absent father and the ever-present mother in the Japanese household require some explanation. Manifestations of the absent father and the ever-present mother exist in the gender-based difference between the mother and child, and the father and child, in the volume of communications. Even today, the norm of "men outside, women inside" with its corollary of "corporate warrior + housewife" leads to a large disparity between the amount of time that children spend with their mother and their father.[9] As a result, the mother-child relationship is maintained as the primary relationship at the expense of the spousal relationship and the father-child relationship.

Historically speaking, while the modern spousal relationship in the West is based on the idea of romantic love, the modern Japanese family was originally founded on the idea not of spousal love but of different gender roles in the household. Of course, contemporary Japanese marriages are generally based on mutual affection rather than being solely arranged by families, but this historical context is still the cultural ground that facilitates "mother-child bonding," leading to women's developing a stronger affective bond with their children than with their husbands (Setiyama 1996, 228–229). In fact, even today, "the magic word of 'the birthing sex' gets mobilized to make the social role of the mother appear to be something rooted in fundamental biological nature" (231). It follows from this line of thinking that a mother's love is the most important element of a child's healthy development and that it is the mother who should shoulder the burden of child rearing.

Of course, this kind of general understanding is not sufficient for grasping the particularities of parent-child communication that I am describing here. At the same time, in my current research, there is little doubt that these gender norms are operating in the manifestation of "absent father" and "ever-present mother" in *keitai* communication. For example, in the following interview with a woman in her fifties and her child, the lack of involvement of the father in parent-child communication is more or less taken for granted as something that does not need an explanation.

Mother: It's convenient to communicate by [*keitai* e-mail]. So even if he [the child] leaves in the morning without saying anything, we write to each other here and there during the day. He may keep to himself at home, but he writes back wherever he is, although his messages may be short. In fact, we communicate more often.

Is your husband involved in the e-mail interaction?

Child: No, somehow I don't with him. He uses a cellular phone. [Everybody except the father uses PHS.]

Mother: He [the husband] says that one person in the family should have a cellular because there are places that you can't connect with PHS. That's why he carries a cellular. I don't think we can e-mail him.

When you contact each other, is it by voice or text?

Child: We use both. In either case, it is simple communication, like about buying some bread on the way home or what time I will be getting back.

Mother: [*Keitai*] is more than adequate, so I don't think I need a PC or anything like that. We use [*keitai*] pretty much every day.

Child: Yes, pretty much every day. In my record of calls, it is all my mother.

Background issues also affect the involvement of the household members in family communication. These could include practical or emotional issues. Regardless, such differences between the husband and wife are linked to the differences in their respective positioning and defining of *keitai*. Matsuda (2001b) has made a relevant observation in the quantitative analysis of a questionnaire survey conducted in 2000. She indicates that there are distinct differences between genders in the development of social networks that are formed and maintained with *keitai*. Married men most commonly engage in *keitai* communication with business-related people even if the topic at hand is nonbusiness; on the other hand, married women, including those with an outside job, mainly use *keitai* with family. As Matsuda characterized *keitai* for such women as a private phone, it is evident that *keitai* for women, compared to men, is directed more toward the private realm.

Consequently, the differences between husband and wife concerning communication with children, as mentioned above, can be perceived as an integral characteristic of the gender differences observed in Matsuda's analysis. In other words, for men, *keitai* use contributes to expanding their networks mostly in the public world on grounds of business, and for women, *keitai* use is more directed to a private world of communication with children. Matsuda cautions that such differences between the genders should be regarded as reflective of "positions in social structure" rather than as independent choices by individuals. The results here similarly reinforce the role of *keitai* in reproducing the gender-biased role of the wife or mother.

A different observation made in the following interview with a couple in their fifties also confirms this tendency. The differences between husband and wife in parent-child communication are related to the differences in their criteria for evaluating *keitai*.

How do you decide whether to call someone or send e-mail?
Wife: It depends on the topic, but there are things that can't be said but can be written in e-mail.
Husband: I don't think so. You have to push the buttons a lot, and it's troublesome.
Wife: That's true, but it's what you want to say and the emotions.
Husband: I disagree. It's so much trouble.
Wife: It doesn't matter if it's troublesome. That's your opinion. There are things in this world that can only be said in e-mail. Like when I get into a fight with the children, some things are difficult to say on the phone. So I write, "I'm sorry, I didn't mean what I said." Then they write back and say, "I'm sorry, too. I also went too far." On the phone, they would say, "Stop nagging."

This dialogue represents differences in the way *keitai* technology is evaluated in the case of e-mail communication with children. The husband evaluates the user-friendliness of *keitai* as an appliance, whereas the wife praises the effects of *keitai* from the emotional dimension of communication with her child. Sonia Livingstone (1992), in her analysis of an interview study of various domestic technologies, observed a significant difference between husbands and wives in the assessment of each technology. The most prominent difference she points out, which also applies to the preceding example, is that men evaluate technology by its functions and performance, and women evaluate technology vis-à-vis specific contexts in daily life.[10] Of course, we should not glibly generalize from these findings, and we should not assume that the relations between men and functional dimensions, and women and emotional dimensions, are based in some essential gender difference. Instead, use and assessment of *keitai* are structured by certain historically and socially constructed gender norms and expectations for parent-child relations.

The key point here is that the housewives' or mothers' opinion of *keitai* is largely based on its role in maintaining family ties, especially the mother-child relationship. This is also true for other technologies. Although represented very differently, technology is evaluated based on its fostering of the mother-child relationship. In the following interviews with women in their fifties, the mother-child relationship is the framework for the informants' use and assessment of the computer and the Internet:

When we built the house, we designed it so that the children have to go through the living room to get to their rooms.... So we put the biggest TV in the living room so that the children would spend more time there.... But they still went out and bought big TVs for their own bedrooms.... In any case, the computer is there [in the living room], and they have to be there to use it, so I guess the computer is what keeps us together now.

Child: My mother asks me to look things up for her on the Web when she's planning a trip.
Is that convenient for you [the mother]?
Mother: I might have trouble when they're gone [after my two sons move out], so I do want to learn before they leave. But I ask them to do it because it becomes a part of our communication.... Maybe I can just save my Internet questions for when they come back home. If they feel undernourished, I think they will come back home for a visit.

What is conveyed in these examples is the significance of setting up the computer in common space shared by the family and the significance of deliberately leaving Internet expertise with the children. As articulated in the second example, the significance is the maintenance of mother-child communication. As repeatedly indicated by researchers, *keitai* and the computer are basically personal media, in contrast to family media like the TV. Public discourse has frequently addressed the domestic adoption of personal media, often positing a cause-and-effect relation between the personalization of media and the individualization of the family. The typical scenario has been of children who lock themselves up in their rooms with personal media. However, the uses of technology by mothers sometimes involve their efforts to integrate such personal media into the family as a whole and into the mother-child relationship. This practice is evidence that housewives have played an important role in the implementation of new domestic technologies.

I would like to stress that the informants' comments reveal that their practices cannot be reduced to the more obvious acts of adopting and using technology. These women incorporate technologies into the home and family contexts through a wide range of activities, including the construction of meaning and emotional nuance. With a conscious decision "not to use the Internet," the mother who nurtures communication with her sons by "asking them to use the Internet" positions that action in a mother-child relationship on an emotional level.[11] Such subtle behavior cannot be identified by studying media use behavior alone; in fact, users are often not even aware of their own behavior unless questioned about it in the context of a discussion of family relationships. Although subtle and often submerged, these remarks point to the gender-biased inequality I have been describing, defining the meaning and use of new technologies in each household and leading to a broader influence on the public definition of such technologies.

Sense of Ownership of *Keitai*

Based on this analysis of the relationship between housewives and *keitai*, I would suggest that among the different types of information technology, *keitai* could very well become a particularly intimate object for housewives. As mentioned, the adoption rate of *keitai* among housewives is not high compared to other groups. However, when single women, who are very active users of *keitai*, go on to get married and have children, we can expect that the rate of *keitai* use among housewives will increase. More important, our research indicates that *keitai* undeniably supports housewives' activities and role awareness, and is an important part of their social relationships. In this sense, I would speculate that housewives' relationship with *keitai* will become more intimate than their relationships with other technologies.

To corroborate this speculation, I introduce information regarding housewives' sense of ownership toward the computer and toward *keitai*. Not surprisingly, most (though not all) informants felt a stronger sense of ownership toward the *keitai* and felt emotionally closer to it than toward the computer. The following view from a woman in her fifties is a typical example:

It didn't feel strange to have this [*keitai*] even right after I bought it. I was just happy to have it. But the computer is so big compared to *keitai*, and I wouldn't know what to do if it breaks down. I'm not very good with machines. My husband would be able to do something about it, but not me. I don't know anything about the system. It's better to leave it alone. Yes, there is a big distance between the computer and me.

As stated by this informant, the most explicit factors that create the difference in perception of *keitai* and computer are physical factors, such as the size of the machine and the complexity of the system. In that physical sense, the computer is relatively intimidating and is perceived as something scary. Naturally, there is little basis for the common stereotype of women being weak with machines. Nonetheless, in specific domestic contexts, as seen in the preceding example, many housewives do not feel an affinity to the computer, stating that they are "not a computer user" or have "no experience using them at work or school." In contrast, *keitai* is regarded as "small and easy" and a medium toward which it is easy to feel a sense of ownership. It does not feel "strange."

The key issue here is that affinity for a technology is simultaneously associated with gender and with generational positioning within the family. In the preceding example, the self-awareness of the housewife that she is "not good with machines" stems from the comparison with her husband, who is "good with machines," which manifests a traditional gender-biased norm that technical things are the man's domain. There are also numerous other informants who feel that the computer is not theirs and belongs in the husband's or the children's domain, even though they are actually reasonably active computer users. Reinforced through comparison with other family members, this mentality can at times intensify the belief that *keitai* is the suitable medium for women. In other words, the affinity between *keitai* and housewives arises not only from their use of *keitai* but also in reference to other technologies and other members of the family.

Another interview in our records, with a woman in her fifties, illustrates this situation in a different light. In this family there is one computer in the home, which is shared by the husband, wife, and children—a common scenario in the modern Japanese household.

For computers, fear is what I feel the strongest, and [in the beginning] I could only do what I already knew. I didn't want my family to get mad at me for pushing or deleting something I wasn't supposed to.
Who would get mad at you?
Everyone in the family.
Do you still feel the same fear?
Yes, I feel that I shouldn't do more than what's necessary. It probably wouldn't be the same if I had my own computer. I prefer *keitai*. I think I feel this way because I'm better with it. Compared to the computer, that is. I haven't gotten that far with the computer.

This informant is conscious not only of the differences in the devices' functions but also of the difference between the computer as a family-owned device and *keitai* as a personally owned device. Consequently, her fear of the computer originates more from her concern of causing problems in the shared device rather than from her self-awareness of being "not good with machines." In many households, the primary users of the computer are the father or the children. Despite the shared ownership, unequal access is often an unspoken understanding. In contrast to the father's and the children's justified business or academic uses of the computer, women who do not have an outside job do not have such explicit reasons for use. As a result, uses of the computer by housewives are positioned as less important compared to uses by other members of the family. It is thus not easy for the housewives in this situation to feel a sense of ownership toward the computer.

On the other hand, *keitai* is usually completely owned by an individual,[12] and it is easy to operate; hence it follows that it is easy for the user to feel a sense of ownership toward *keitai*. *Keitai* is compatible in various ways with the self-identity of the housewife/mother, who consequently feels more intimate emotions toward it. Naturally, that does not completely deter housewives from using the computer. Nonetheless, the process through which the emotional definitions of the two new technologies are differentiated is affected by the gender-biased, generational conflict of power in the family, which decisively structures the sense of distance between users and technologies.

It is quite possible that such domestic micropolitics are intricately intertwined with the digital divide observed between men and women at the macrolevel. It is true that the general gap between men and women is shrinking with respect to their use of the computer and the Internet. However, this trend is grasped from a general indicator of rates of use, and on a micro level there is still considerable disparity between the genders regarding felt distance between user and technology. The two housewives quoted in the preceding excerpts are registered in our statistics as "computer user" and "Internet user," but there is a qualitative gap in how they use these technologies in everyday life. Further, they usually do not perceive this disparity as something into which they have been coerced but accept it as a matter of course in connection with their identity as housewives. As discussed, when housewives regard *keitai* to be more suitable for them than the computer, their judgment is proactive. This suggests that as the relation between *keitai* and housewives becomes more intimate, the digital divide between men and women could expand in ways that are difficult to measure.

The Housewife as a Sociotechnical Entity

This chapter has examined the domestic positioning and definition of *keitai* by focusing on housewives and by comparing *keitai* to the computer. Although *keitai* add new dimensions to housewives' communications, such as efficient integration with housework and more contact with children, the overall outcome tends to reproduce rather than transform the traditional order. As found in prior research on domestication, housewives in domestic contexts give their own unique meanings to *keitai*, and through this process *keitai* have become deeply incorporated into the lives of homes and families. This process of domestication did not result in mere interaction between two discrete entities—*keitai* as a technical entity and housewives as a social entity. Instead, I suggest, *keitai* as a technical entity and housewives as a social entity have become one indivisible entity.

Actor network theory in technoscience studies pioneered the perspective that society and technology are inseparable. This approach was a departure from prior theories that made a clear separation between technology and society, focusing analysis on the influence between the two (Callon 1987). According to actor network theory, "society" is not purely "social" and necessarily includes technological elements. It follows that the social and the technical cannot be theorized as separate entities. This perspective may appear strange from modern scientific viewpoints governed by dualism, but it is an appropriate lens for viewing communications technologies like *keitai* and their increasingly central role in mediating our relationships and activities. Society cannot exist without technology, and technology functions only upon being embedded in society (Callon and Law 1997). It follows that we should not be focusing on an analysis of how a new technology like *keitai* affects society but rather on an analysis of the process of constructing hybrid sociotechnical entities such as "a society with *keitai*" or "a person with *keitai*."

Housewives' use of *keitai* can only be properly understood through such a framework. The relationship between the identity of a housewife and *keitai* is not one of mutual influence but rather one of constructing and defining a shared identity. In other words, the reality of "being a housewife" is defined by *keitai* technology, and the technology we call *keitai*, which harbors various meanings, is molded into a specific definition as a part of the housewives' activities and self-identity. Therefore, the interaction between housewives and *keitai* observed in this research does not merely illustrate housewives as a social entity influenced by the technical entity of *keitai*, or *keitai* as a process developed by the social entity of housewives. It more accurately illustrates the simultaneous development of both processes into one undividable entity. In short, the development of their relationship is the process of sociotechnical development of the social and also the technical entity of "housewives with *keitai*." I might also call this process one of the cyborgification of housewives (Haraway 1991); "housewives with *keitai*" is a hybrid entity that is not entirely human or artificial. Through coupling with the technology of *keitai*, gender identity is rewritten (though in an extremely conservative way).

This is not a particularly new phenomenon; many domestic technologies were similarly involved in the formation of the housewife as a sociotechnical entity. The most common examples are the washing machine, vacuum cleaner, refrigerator, and other technologies directly associated with housework (see Cockburn and Furst-Dilic 1994; Lie and Sørensen 1996). If these technologies were taken away from modern housewives, many of them would not be able to fulfill their roles. In this sense, housewives never were simply social entities but were also mature technical entities. As I have discussed, "housewife" is basically a social concept. But housewives have been more than just social entities; they have also always been technical entities. Even media technologies, like the telephone, television, and radio, though not technologies used for housework per se, have also been directly and indirectly involved in the definition of "being a housewife" by their maintenance of social networks, engagement in social communication, and formation of social identities.

Seen from this viewpoint, *keitai* adds a new facet to the housewife as sociotechnical entity. Housewives and mothers who started using *keitai* in connection with their own social roles are using the technology to uphold at least a part of their existence as housewives and mothers. It is especially obvious in their incorporation of *keitai* in the context of family communication, and we can readily imagine that without *keitai* being a housewife and being a mother would take on different realities for women who frequently exchange e-mail with their husbands and children. *Keitai* technology is not affecting them from the outside but is being integrated as an inseparable internal component of their roles.

The development of housewives as sociotechnical entities consists not only of the functional aspects of *keitai* and the practical aspects of the homemaker and child-caring roles but also of the combined symbolic meaning of *keitai* and gender-biased self-image. The simultaneous establishment of such practical and symbolic aspects greatly enhances the affinity between *keitai* and housewives. This process of co-constitution, where housewives position *keitai* as an appropriate tool for them for both symbolic and practical reasons, illustrates the power of the dynamics of the household, a place that is conservative in nature. As housewives take on this sociotechnical identification, they are simultaneously acquiescing to inherently inequitable gender roles. This chapter has repeatedly raised issues regarding gender order to illustrate this point, and I stress once again that new technologies, regardless of their potential to transform the status quo, can have the opposite effect in day-to-day use, and that the distribution of power in the home is being remade in ways that are extremely difficult for the family members to identify.

The same domestic context, when observed from the standpoint of other family members, will naturally evoke different stories. As many researchers have found, seen through the eyes of children *keitai* is usually a personal communication tool that bypasses parents and symbolizes their liberation from parental control. It is undeniable that the definition and positioning of *keitai* differs depending on one's position in the family, which raises an interesting topic for future examination. We must seek to understand how *keitai* and other technologies restructure users' social positions and the social contexts of its uses, into which they are simultaneously integrated as an inseparable component. This could become a critical call to study in greater detail those processes often glossed as domestic informatization.

Notes

1. Lynn Spigel (1992, 26–29) studied the process of radio's being accepted into the home. The radio, which was first regarded as a technical appliance in the male domain, gradually shifted into the female domain and hence the home. In other words, early radio was positioned as a "device to catch faraway signals," and its image was one of a "rugged technical appliance," which delivered

an unmistakable masculine connotation to an object that did not belong in the home. Subsequently, through changes in design and program content, radio became an object for "feminine leisure" and was welcomed into the living room. As such, the transformation of the social positioning and definition of radio used women or housewives as one pivot before radio became major "domestic entertainment."

2. The research was conducted by a group of five researchers under my initiative.

3. The survey was conducted in central Tokyo and the northern areas of Yokohama, located near Tokyo. In Tokyo the informants were people who had lived in their neighborhoods for a relatively long time. In north Yokohama the informants were residents of typical suburban housing complexes.

4. Widespread adoption of personal computers in Japan started around 1995. Before then, there was a certain degree of computer use in the home by hobbyists, but it was around 1995, during the simultaneous surge of popularity of the Internet and Windows 95, that it was adopted by a wider range of users. The number of *keitai* users, moderate until then, also started to skyrocket. In 1999, with the introduction of the i-mode service, Internet access functions were built into *keitai* terminals.

5. The comments in brackets are supplementary information I have provided regarding the quotations from the interviews.

6. This applies only to everyday personal e-mail. Computer-specific functions are at times necessary for the exchange of business e-mail, and computers are also often used for personal e-mail when the messages are long or being sent between people who rarely see each other. Such e-mail interaction involves the act of writing letters and making the time especially to do so, but the day-to-day e-mail discussed here is the exchange of short messages between people who see each other on a regular basis.

7. Such relocation of the computer occurs when the family buys a new machine or when there is a new user within the family. When a computer is introduced into the home on condition that it be shared by the members, it is often set up from the start in the living room or other shared space.

8. It is important to note, however, that at present information obtainable from the *keitai* Internet is very limited compared to that available from the PC Internet. The most popular *keitai* Internet service in Japan is ring tone downloads, followed by wallpaper downloads. Other uses of the *keitai* Internet are not very common (Mobile Communication Research Group 2002).

9. Although these data are somewhat old, Amano, Ito, and Mori (1994) note some clear trends in their lifestyle study, conducted at Tama New Town (a typical new urban housing development). Although the gap was not large for weekends, on weekdays, in families with a mother who stayed at home and a father who was employed full-time, the father spent on average 53 minutes a day with the children and the mother spent 245 minutes, attesting to a large gender gap. In families where the mother worked part-time and the father worked full-time, these numbers were 60

minutes and 228 minutes, respectively. Even among families where both parents worked full-time, the gender gap remained large, at 72 minutes and 178 minutes, respectively.

10. Livingstone bases her observations on interviews concerning the TV, telephone, video, and audiovisual devices.

11. To further explain the background of this family: The two sons have their own computers in their own bedrooms. They are both very heavy users. When asked whether the sons spent less time in the living room with the parents because of having computers in their rooms, the mother said no; but when the sons were asked the same question at a different time, they felt that they did. These differences in perception are also a manifestation of the relatively strong awareness of the mother regarding the adverse effect technology has on family and parent-child relationships.

12. In the early stages of adoption, *keitai* was sometimes shared among the family. *Keitai* in such cases was positioned as an outdoor communications tool and was carried outside by the family members as necessary. In our study, one family used *keitai* in this manner.

12 | Design of *Keitai* Technology and Its Use among Service Engineers

Eriko Tamaru and Naoki Ueno

Keitai is evolving from a telephone to an information terminal. In the workplace *keitai* has shifted from being used simply to make contact with others to being used as a key tool to support teamwork between distributed workers. Since it connects people, *keitai* functions as a social tool. Although *keitai* are widely used in the modern Japanese workplace, there is yet little social scientific research on workplace uses of *keitai*. This chapter addresses this gap through a practice-based study and analysis of *keitai* technology design and use among mobile workers.

Approach and Framework

Many approaches in sociology and technology studies have relied on a model either of technological determinism or social constructivism. Technological determinism works to understand the influence of technology on society. Social constructivism works to understand the influence of society on technology. What both of these apparently contrasting approaches share, however, is the underlying tendency to view technology and society as fundamentally separate and to query the relation between the two entities.

In contrast to these approaches, current work in technoscience studies has posited that technology and society are inseparable, and that technology design is social systems design (Hughes 1979; Latour 1987; Callon 1986). For example, Michel Foucault (1975) describes how the panopticon involved much more than the design of the watchtower itself. The structure where guards are able to watch the prisoners but the prisoners are not able to see the guards is more than a design of space; the watchtower also designs the social relationship and power structure of the prisoners and the guards. Moreover, the politics between the guards and the prisoners cannot be designed without this artifact of the panopticon. As such, the designing of artifacts and the designing of social systems are mutually constitutive. In other words, designing artifacts is in fact designing social systems.

These sorts of issues were also evident in workplace studies in the 1990s. In much human interface research, usability research, and design practice, communication and coordination tools were being analyzed and designed without taking into account the activities and networks in which they would be utilized. However, since the 1990s with the growth of practice-based workplace research, a variety of cooperative activities were studied, and tools and information design were positioned in a more socially contextualized way (for example, Engeström and Middleton 1996; Goodwin and Ueno 2000). Studies of this sort have demonstrated that tools are never used in isolation but are integrated into cooperative activity and social networks. Through use and association with other tools and resources, a tool that has been designed for a single purpose is often used in ways that go beyond its intended design. Accordingly, the design of a tool is substantively influenced by its uses in cooperative activities and networks.

In addition to providing these insights on tool use, workplace studies also described the ways in which the design of tools was part of the construction of sociotechnical networks through cooperative activity. For example, Susan Newman (1998), in her ethnographic studies of middleware designers, describes how development means creating software that satisfies given specifications but also involves developing a sociotechnical network. In other words, developing middleware involves building a network of affiliated companies, clients, competitors, industry standards, technological resources within the company, and computer engineers. When project leaders, system engineers, managers, and consultants engage in this development, they make objectives in the software visible on various levels and negotiate, account for, and construct acceptable compromises between conflicting interests and inconsistencies in the sociotechnical network of the middleware. To rephrase the words of Callon and Law (1997), the specifications of software reflect economic and technological constraints as well as complex processes of interaction among various actors.

By viewing technology design as the construction of sociotechnical networks, we arrive at a view of technology and social structures as co-constituitive. Tools only affect social networks if the social structure is reconstructed in relation to the tool's implementation. It follows that in approaching *keitai*, it is crucial to keep in view how it is embedded in practice, social networks, and organizations. *Keitai* are generally considered to be a personal technology, and most applications focus on individualized and private uses. This tendency is evident in business uses as well. For example, *keitai* were originally imagined as individualistic tools for the executive (see chapter 2). Our research, however, describes how *keitai* is used in the workplace in ways that differ radically from these prior images of use.

The copier service technicians we studied use *keitai* not as a private technology but as a device for linking to their organizational network and for describing, coordinating, and visualizing their group activities. The members of a service technician team, distributed across their service area, monitor on *keitai* a continuously updated list of service calls and a list of which calls team members are attending to. By referencing these lists, the technicians are able to maintain, moment by moment, a grasp of the overall state of their team's activity and to coordinate their work in relation to this knowledge. In this way, a particular work team is made visible and organized through *keitai* technology. At the same time, through its uptake by a team of this sort, the *keitai* is constituted as a technology for collaborative work. The design and refinement of this *keitai* list system involved the service technicians themselves in a participatory design process.

In addition to functioning as a device for organizing and visualizing the team, the *keitai* system also enables technicians to make their entire service area visible as needed. Through the online *keitai* system, the technicians are able to make decisions about how and when to move around their service area. In other words, service technicians are able to reconstitute time and space through their use of *keitai*.

In short, *keitai* makes visible social organization and the service area, a reconstitution of time and space, as well as a visualization and engineering of collaborative work through participatory design. This case demonstrates how the meaning of a technology is not inherent in the object itself but is constituted through an interaction between the technology, practice, and social organization. We view the design of a *keitai* system not as the design of particular technical features but as the design of a social system. From this perspective, we take a detailed ethnographic look at the design of a *keitai* system and its interaction with the social network in which it is embedded.

Ethnography of *Keitai* Use in the Workplace

Our case studies are of mobile technology used in the cooperative activities and teamwork among technicians who maintain copiers, printers, and networks, otherwise known as service technicians. Julian Orr (1996) conducted an ethnographic study of how service technicians share skills and technical know-how within their community. However, their work has changed considerably both in Japan and in the United States since the days of Orr's research in the 1980s. Copiers are now digitized, systemized, and diversified. Also, in the United States, a wireless system has been introduced that enables team members in different areas to keep in constant touch with one another. They also have a knowledge management system that compiles repair tips supplied by the engineers themselves. Even in Japan, the adoption of mobile technology like *keitai* and laptop computers has changed the nature of teamwork.

This chapter is based on our fieldwork on the activities of Japanese service technicians working in the service division of a Japanese company manufacturing and selling copiers (Tamaru 2002; Tamaru and Ueno 2002; Kawatoko 2003; Ueno and Kawatoko 2003). This division is responsible for the service and maintenance of copiers in the field. In the service division there are three major organizational types. One is distributed throughout the country and includes the service technicians who actually service the machines and the back office that supports them. The second, of which there are four across the country, are the organizations that provide technical support for certain groups of local units. The third is the central office, which provides overall management of national service operations, supporting field technologies as well as providing a pipeline to the development divisions.

Our fieldwork cut across all three of these organizational types and was conducted between 2000 and 2002. In the field we shadowed the technicians and observed their repair work, their use of the dispatch and knowledge management systems, and their collaboration with team members. After returning to the office, we also conducted detailed interviews. We conducted this fieldwork at three different service areas with a range of characteristics. Our research with the regional and national technical support and management units was primarily in the form of interviews. We begin our description with a few specific examples.

Case 1: Giving Advice on Repairs

Suzuki, a color copier specialist, often receives inquiries from his colleagues regarding color copiers. He keeps a record of the repairs he has handled in his personal notebook, and in his laptop he carries electronic manuals and engineering documents on repairs issued by the administrative department. One day, when he was driving to a client's office, he received a call on his *keitai* from coworker Sato. Sato, working on a complicated malfunction of a color copier, could not find a way to fix it and needed help. As soon as he received the call, Suzuki pulled over to a safe spot to aid his colleague. "It's the DM1200; the XP code won't go away. What can I do to fix that?" Inside the car, Suzuki thought, "I've heard of this case somewhere before" as he referred to the memo inside his notebook. But he could not find the information he wanted. Next, he started up his laptop, browsed the DM1200 manual, and searched the engineering documents. Soon, he said, "I've found it," and called Sato via *keitai*, but the call did not go through. Then he called the back office. "Is there anybody having trouble with the DM1200?" The back office had also been informed of this difficult case. The staff in charge was working with another supporter whose *keitai* number was given to Suzuki. He called that person, and his call got through. He relayed the information he had found in the engineering documents. His colleagues had already been notified of the same information from another source. "OK, then try that. The parts are in the office. I'm glad it's solved," said Suzuki as he hung up and continued his drive to his client's office.

Case 2: Adjusting Schedules

Takeda, after finishing a repair at a client's office, accessed the self-dispatch system with his *keitai*'s network application in order to plan his next call. The self-dispatch system displays a list of calls from clients and a list of the current location of each team member. Taking into consideration various factors, such as the clients' call list, the condition of the machines under their watch, geographical distance, and difficulty of the repair, the service engineers decide which client to visit next. When Takeda viewed the system, there were seven pending calls. Among them was a request from company A in the building across the street. However, there was also a request from company B, which Takeda was in charge of. Just at that moment, he received a call on his *keitai* from Tanaka, who saw where Takeda was on the self-dispatch system and who wanted to coordinate their schedules. "Can you go to company B for me? And I will go to company A across from here," said Takeda. Takeda then returned to the self-dispatch system and entered into it that he was going to company A. He wrote down the name of the person in charge and the phone number of company A, which he called on his *keitai*. "Is this Saito-san? I will be there in about five minutes," he said to let the company know that he was on his way.

Case 3: Procuring Parts

When inspecting a machine at a client's office, Yamada found a problem that was not mentioned in the original request. He needed to exchange a part in one unit in order to complete the repair. Usually he carries around parts necessary for fixing the most common problems and readies in advance the ones he thinks might be required for specific requests. In this instance, however, he did not have the part he needed. Hoping to find a colleague nearby who had one, he viewed the self-dispatch system on his *keitai*. He then called a team member who was closest to where he was. "Do you have the part to the A unit of DX500?" No, he did not. He then called three other members, but none of them answered. He had momentarily given up trying to reach anyone and had gone back to work on the original problem when a colleague, who saw that he had missed a call from Yamada, returned his call. Unfortunately, he did not have the part either. Yamada got the client's machine up to basic function and then left to return to his office to get the part. Even as he returned to the office, he repeatedly checked the self-dispatch system and called team members who were nearby. Unfortunately, nobody had the part he wanted, and ultimately he had to return to the office to get it. Yamada went back to the client's office in approximately 30 minutes, exchanged the parts, and completed the repair.

As illustrated in these examples, the technicians we observed in our fieldwork used *keitai* for more than making contact with one another. The technology facilitates team members' collaboration and also supports the coordination of their activities and the exchange of information and know-how.

Keitai as Technology That Makes the Service Area Visible

Keitai technology has become an indispensable tool for service technicians in organizing and maintaining their teamwork. In this section, we focus on the technologies that visualize the ecosystem of the service area, examining how the tools were designed,

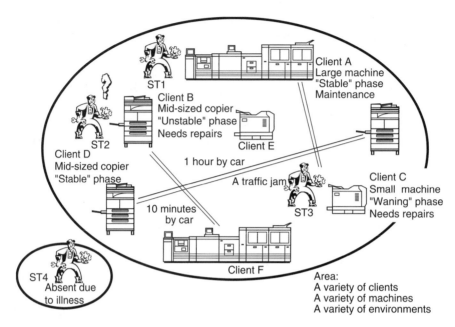

Figure 12.1
Ecology of the service area.

how they are being used by the service technicians, and the interaction between design and use.

Ecology of the Service Area

The teamwork and the tools of service engineers cannot be understood apart from the specifics of their work settings. The workplace of a service technicians is distinctive in that it is not a stationary location, like an office, but rather dispersed throughout a particular area. One team of five to ten technicians is in charge of a specific service area (figure 12.1). Clients and copiers are scattered throughout this service area, and the technicians go to these offices upon request to fix machines or conduct regular maintenance checkups.

Each service area, depending on the clientele found there, has unique traits and different types of problems that are likely to occur with a copier. For instance, in the service area in central Tokyo where many governmental offices are situated, there are many aging buildings and often an unhealthy environment for copiers. The humidity level in such buildings is high, inhibiting the function of the roller that feeds the paper. Many offices in this area also use recycled paper, which has a less-than-ideal degree of friction, occasionally causing the roller to stop functioning properly. Furthermore, the copiers in this area have a high degree of wear because one copier is used to produce excessively large numbers of copies. On the other hand, in a service area where there are many copy shops, printing companies, and design offices, service requests often concern the quality of the copies. These can be highly sophisticated requests; for example, a client might ask to have a barely noticeable smear removed or to have minute adjustments made in the color output.

The problems that occur in one service area, even with their idiosyncrasies, do not always follow the same pattern. The types of problems change over a span of a few months or even daily. For example, copier models are divided into separate categories depending on the length of time since they have been introduced to the market. The models that were released between six months to a year ago are in the "unstable phase," those that were released several years ago are in the "stable phase," and the those that were released before that are in the "waning phase." Copier problems are characterized in relation to these phases. For instance, ones in the unstable phase often have problems of unknown origin, including product defects. In these ways, the conditions of the service area change in daily cycles and in cycles of a few months in duration, and accurate diagnosis must be keyed to these temporal rhythms.

For the service engineers, the copier does not exist in isolation but is a part of the service area, an ever-changing social-instrumental environment with unique characteristics. As we have described, the ecology of the area is a critical resource in the diagnosis and repair of copier problems.

Making the Service Area Visible

Team members must cooperate with each other in various ways despite being dispersed throughout the service area. For example, in addition to individual history, copiers and their attendant troubles must be understood in relation to the condition of other copiers of the same model in the area. Further, the changing conditions of the area cannot be made visible by one service technician.

Even on any given day, the conditions of the area could change by the hour. The service technicians must respond to these changes as a team. For instance, in order to determine an individual's next service call, the team's schedule must be interactively adjusted based on the urgency of the problems, the location and status of each member, and traffic conditions. How are the conditions of the area that change by the hour made visible, and how are the tasks allocated to each team member? We discuss this process along with an examination of the design of technologies that make the area visible.

Making the Service Area Visible by the Call Center Using a Physical Board System In the repair department where we conducted our fieldwork, the employees used a dispatch system for dispatching service technicians to clients' offices to repair and maintain machines. The design of the dispatch system has changed in tandem with technological developments. The dispatch system of the 1980s was a large board, a grid with the names of the service technicians in charge of the areas and the time. When a client called the center to request a repair, the client card was placed in the appropriate position on the board. It was a call center–oriented design, and the conditions of the area as a whole were only visible to the staff at the office. After receiving a call from a client, a dispatcher would assign a service technician to a task by taking into consideration the urgency of the problem and the location and workload of each team member. As a kind of modern panopticon, the call center had a bird's-eye view of the conditions of the service area and controlled the movement of the service technicians.

A major irony was inherent in this system. The service technicians, who were well versed in the geography of the area, the clientele, the conditions and histories of the individual copiers, and traffic conditions, were not able to see the overall conditions of the area, the call list, or the status of team members at a specific moment, and so could not use their information to coordinate and allocate team tasks. Yet the dispatchers at the call center, who were unfamiliar with all these details, had a bird's-eye view of the entire region and the responsibility of coordinating the work.

It was difficult for such a call center system to accommodate the unstable conditions of the service area and to dispatch the service technicians with responsiveness and flexibility. Consequently, when the call center received a call from a client, it could not necessarily dispatch a technician whose experience and physical location were most appropriate for the task. The technicians partly compensated for these problems with unsystematic and informal practices. For instance, some of the technicians coordinated their schedules and shifted tasks among themselves by getting information about pending visits and the location of team members over the phone from the dispatchers. This kind of communication with the call center enabled some of the technicians to make the status of clients' calls and their colleagues partially visible, and they worked to be responsive to these conditions (figure 12.2).

Board System after the Implementation of Information Technology   With the introduction of information technology in the 1990s, information on clients and calls was computerized. This transformed the physical board system into terminals that connected to a host computer. The tools were new, but the system remained call center–oriented. Consequently, even during this period only the call center could see the entire area whereas the service technicians could not.

The computerization of call information had a secondary effect. At each service technician headquarters, a computer terminal was set up on which dispatch information could be viewed. While these computers were originally intended for a different purpose, the technicians taught themselves how to use the systems and began to monitor the calls and thus know where they were going to be dispatched next. They could only use the computer in the morning and the evening when they were in the office, but

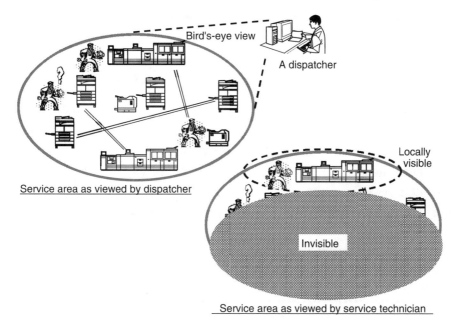

Figure 12.2
Visibility of the service area.

still this provided an opportunity for them to make the area visible. As a result, the experienced technicians started giving advice to the dispatchers on the allocation of tasks ("for this machine, you should send so-and-so"), and the unsystematic activities of the service technicians started taking on more significance.

Transition to the Self-Dispatch Method with the Implementation of Mobile Technology Around 1995 the introduction of mobile technology, such as laptop computers and *keitai*, replaced the old call center system with a new system called the self-dispatch system. In the self-dispatch system the service technicians were not controlled by the dispatchers at the call center but were able to allocate the tasks among themselves.

With the self-dispatch system, each service technician could access the server at the computer center from a laptop and view the call list, which had previously only been accessible at the call center. Figure 12.3 shows a sample display of the system. At the left of the screen, there is a list of clients who have called to request a repair. The technicians would consider the various conditions of the moment and choose the most appropriate job for themselves. The call would then move from the left to the right of the screen onto the list of calls already being handled.

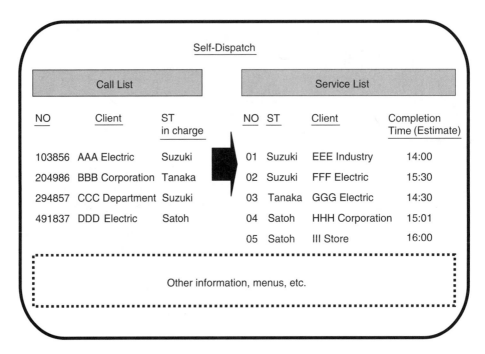

Figure 12.3
Sample screen of the laptop-based self-dispatch system.

Through their use of the self-dispatch system, the service technicians started inter-acting with each other in different ways. For example, when they responded to a call, they would enter the estimated time of completion of the task, say, 14:00. One team, however, made its own rule: after they arrived at a client's office and judged the prob-lem to be too complex to handle alone, or if the repair seemed like it would take a very long time, instead of entering the usual 14:00, they would add a 1 to the end and enter 14:01. This practice allowed the technicians to inform other team members working far away that they were facing a difficult problem. They would sometimes receive an offer of assistance via *keitai* ("Do you want me to go help you?") from someone who saw an entry such as 14:01. In other cases, they would receive repair advice via *keitai* from a team member who specialized in the particular model. With the self-dispatch system, since both the conditions of the clients in the entire area and the status of the team members are visible, the jobs can be allocated based on specific details, such as a copier's history and current condition, characteristics of the client, urgency of the request, and the traffic in the area at that particular time. In this way, the implementa-tion of the self-dispatch system together with *keitai* made the current conditions of the entire area visible to each team member, made it possible for each to choose the most pertinent task of the moment, and enabled team members to interactively support and cooperate with one another.

Self-Dispatch System Using *Keitai* Technology *Keitai* became increasingly popular after the introduction of various network applications like e-mail and the *keitai* Internet. The self-dispatch system was redesigned in response to these new technologies. The basic design concept was adapted from the system using laptops. The method of application, however, was completely different. Although mobile tools are in theory to be used anytime, anyplace, in reality the laptop system meant that the conditions of the area were only visible before or after a technician visited a client's office. It took some time for the laptop to boot up, and there were often times when the *keitai* signal was unstable in high-rise buildings. Consequently, the area was visible only a few times during the day, when the technicians were in the office, and before and after their visits to clients' offices. In contrast, with the self-dispatch system using the *keitai* Internet, service technicians can access a visualization of the area's conditions almost anytime, anyplace. Even when taking a short break during a repair at a client's office, they can view the self-dispatch screen on their *keitai* and confirm present conditions of the area. In other words, whereas during the laptop days the technicans had to explicitly look at the area, now the conditions are continually accessible and visible to them. It has given them a constant awareness of the area's ever-changing ecology, boosting efficiency in their maintenance activities and task coordination (figure 12.4).

Restructuring the Relation between the Call Center and the Field

A map is a medium for visualizing space that is geographically distant and out of sight. It does not simply represent geographic information. Representations differ depending on the map's purpose, for example, political governance or the distribution of goods. In turn, the map participates in the restructuring of social behavior and space (Black 1998). In the same way, because the dispatch system is a tool for visualizing a service area, it is related to the activities and social structure of the workplace. In fact, the design of the system's representations has changed through social interactions between the call center, the field, and the back office as well as various structures of control and social order. In this section, we describe changes in the design of the dispatch system and related changes in the structure of activities and organization.

Restructuring the Network of Activities

We have described how the transition from the call center system to the self-dispatch system restructured activities of the service technicians in the field, enabling them to make their own choices about which calls to respond to. There were also changes in

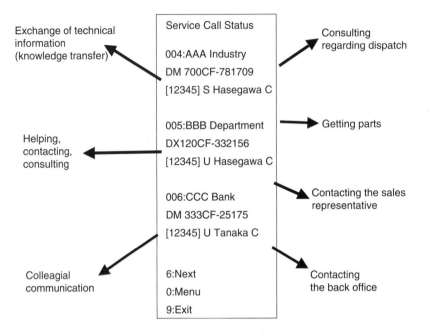

Figure 12.4
Wide-ranging communication mediated by the *keitai* self-dispatch system.

aspects other than the method of dispatch. The service technicians restructured repair methods depending on the state of the field.

Their repair activities are divided between fixing problems that cause a machine to malfunction and preventing the recurrence of problems. When there are a lot of pending calls, they now prioritize fixing machines and choose clients in need of such service. Activities to prevent problems are conducted when there is more time. In other cases, after checking the self-dispatch screen, they call a team member who seems to be taking a long time with a repair and lend support ("How is it going?" "Are you all right by yourself?" "Is there something I can do?"). In these ways, the engineers have become better able to take into account the entire area in determining appropriate actions for maintenance of the machines.

The call center system was a type of a modern panopticon where the center had all the control and access to a visualization of the area. The service technicians in the field could not make the area visible, and they felt like they were constantly under the control of the call center. By contrast, the self-dispatch system, by enabling each technician to visualize the conditions of the entire area as well as the status of each team member, acts as an "open tool" (Hutchins 1988). According to Hutchins, a map is an open tool when a team uses it for navigational purposes. All members of the team can see one member's markings on the map, and from this they can visualize how this member views their situation. As such, the map is an open tool that enables a team to share the "horizon of observation." Similarly, the self-dispatch system, designed as an open tool, has facilitated the responsive and cooperative activities of team members.

Restructuring the Rights of Dispatch

In the call center system the rights of dispatch were held by the call center rather than by the technicians, but the center's resources for visualization were actually insufficient for an effective allocation of tasks. Lacking relevant information on the ecology of the area, it sometimes inappropriately assigned jobs. For example, at times a service technician had to travel from one end of the area to the other. In other cases, a technician was assigned a task beyond his or her competence. Such situations caused tension between the controlling functions of the center and the technicians controlled by the center. The element that managed this tension and prevented the call center system from breaking down was the unsystematic, informal coordination of tasks by veteran service technicians.

With the transition to the self-dispatch system, the rights of dispatch were transferred from the call center to the technicians themselves. This represented a diffusion of control where technicians were given control over their own work. When we interviewed service technicians, they said, "Now it no longer feels as if we are being commanded, but it feels as if we move based on our own decisions."

On the whole, the self-dispatch system appears to be operating smoothly. However, during busy times, being able to visualize the whole area makes some technicians feel rushed or feel that "it's easier to be directed in times like this" (Kawatoko 2003). When there is a lot of work to be done, the self-dispatch system can create tension among members. They are "only human," and they do not always take the most suitable actions for the service area. For example, technicians have their strong and weak points in technical expertise and sometimes avoid repairs on copier models with which they have little experience. Also, since they know which models cause the most complicated problems, they tend to avoid those machines. Such dynamics can cause discord among field technicians. These are times when the supervisors intervene and give dispatch orders based on what is best for the service area as a whole.

As Kawatoko (2003) points out, in the call center system, tension existed between the center as controller and the technicians as the controlled, but in the self-dispatch system, the tension, along with the control, has been diverted to the field. This tension is relieved by a new kind of control from the back office, which supports the smooth operation of the self-dispatch system. In other words, the self-dispatch system has not removed all control from the call center but has changed it to a milder form of control exerted by the back office and among the service technicians themselves.

In this process, technology has not unilaterally restructured the relationship between the call center and the field and the relationships among the technicians. The social networks and practices of the technicians and the call center also constitute the technological system. The success of the self-dispatch system depends on the relationship among team members. If the members have a supportive relationship, the system will run well, but with an unsupportive relationship the dispatch system will disintegrate. In the field where we conducted our observations, the self-dispatch system was functioning effectively because a foundation of teamwork and voluntarism already existed in the team. According to the person responsible for the implementation of the system, there was much concern regarding the system's success. The informal coordination of dispatches by the veteran technicians during the days of the call center system was, however, a factor that suggested turning accountability over to individuals. This new voluntary system will not necessarily succeed in all communities of repair technicians; its success is largely dependent on the social relationships among the members of the community that use it.

Organizational Restructuring

The new dispatch system changed the organization responsible for repairs as well as the dispatch method of the service technicians. The most significant change occurred in the role of the dispatchers. In the physical board system, the service area that could be covered with one board was extremely restricted because of the board's physical limitation. A large number of dispatchers were in charge of several service areas all in close proximity to where they worked. With the implementation of information technology, the call centers were gradually integrated and consolidated, increasing the number of service areas handled by one dispatcher and decreasing the number of dispatchers. When the self-dispatch system was introduced, the dispatchers were left with only a very limited role, and their number was drastically reduced. In this way, the change in the design of the dispatch system involved a restructuring of the organization and roles involved in the repair work.

The restructuring of social networks by technology and the influence of social structure on technology do not occur unilaterally but are interactively constitutive. Designing the self-dispatch system means not only designing a tool but also designing the relationship between the center and the field, the rights of dispatch, relational tension, the role of the center, and organizational structure.

Design Process: Construction of the Design Network

We have examined how the self-dispatch system is not simply a method for allocating work but also a tool for making the service area visible and transforming previously individualized work into a cooperative activity. Elizabeth Churchill and Nina Wakeford (2002) suggest a new design framework and approach for creating new mobile technologies. They argue that in contrast to popular characterizations of mobile workers as individualized "road warriors" and "globe-trotters," the actual practice of mobile work is driven not by mobility per se but by collaborative relationships and timely access to others. In light of this framework, our work with service technicians is illustrative. In contrast to the traveling mobile executive or the office-hopping free-lancer that often comes to mind in discussions of mobile work, copier repair work involves close, ongoing collaboration that is constitutively tied to a spatially distributed work site. The central problem is not mobility and travel between different sites of work, but rather is about coordinating action across a dispersed work area, or, as Eric Laurier (2002) suggests, this requires "assembling" a spatially distributed shared workspace or region. John Sherry and Tony Salvador (2001, 116) note that research on collaboration through computers has tended to focus on "tight collaboration," such as collaborative authoring or shared meeting and work environments. They suggest that more attention should be paid to "passive awareness and coordination," where people adjust their activities based on easy access to the status of team members. The self-dispatch system we describe is an example of this kind of awareness and coordination.

This section describes how the integration of the new technology into coordinated activity was the result not only of the uptake of the technology but also of a process of participatory design. By focusing on the details of the design process for the *keitai* Internet platform, we examine how the dispatch system was redesigned to become a tool for visualizing and organizing the service technicians' workplace and social network. Further, we analyze the mobilization and restructuring of the design community involved in the redesign of this system.

Converting from a Centralized to a Self-Regulated System

We have described how informal fine-tuning by experienced service technicians helped overcome some of the problems in the earlier centralized dispatch system. This existing context was reflected in the design of the self-dispatch system because two out of three system developers had had actual field experience as service engineers, and they were aware of how veteran staff facilitated the call center's coordination of dispatches. They also conducted design observations of their colleagues while performing the work of service technicians. These observations and their own prior experience as service technicians led them to conclude that service calls should be planned by the service technicians themselves. In most cases, systems and tools are not designed by the people who will use them, and users' input about work needs and processes are rarely solicited before design. A design department may produce a detailed spec sheet without having recorded the tasks and activities of the users in detail. Then, because of lack of understanding of users' activities and the underlying social networks, systems or tools will be unusable or subject to frequent breakdowns. A typical example is cited by Graham Button and Richard Harper (1993) in the development of an order-receiving system at a factory. In that case, the implementation of a computerized system reified a linear work flow that did not reflect the more flexible mechanisms by which orders were anticipated and processed.

In the case of the self-dispatch system, the presence of former service technicians in the design team ensured that the system would reflect the informal practices of the technicians in managing and adjusting the centralized dispatch system. These designers' understanding of service technicians' practice led to the idea of a self-regulated dispatch system. Their thorough understanding of the activity networks also led to the construction of a redesigned system that would not break down. This kind of redesign would not have come about simply by implementing a new mobile technology. Mobile technology provides a platform for designing a distributed dispatch system. However, if this same system were implemented with a group that did not have a prior practice of adjusting service calls among team members, there is a risk that service technicians would simply answer calls in the order that they appeared in the system. The redesign was possible because of a crucial connection between mobile technology and team members who had a sense of teamwork and who performed ongoing adjustments to their work based on their relation to the overall area and their colleagues. In this way, design is not simply design of technology but design of the overall social system, including the users' activity networks.

User Participation in Deciding Specifications and Prototyping

The administrative department handled the development and implementation of the self-dispatch system, but several service technicians—the end users of the system—participated in the design process and trials. For the development project, a task team of some twenty members was formed; the members came from various divisions involved in the dispatch system, such as the administrative department (project leader), the field (site of trial), and the call center. The manager from the administrative side had field experience as a service technician and pointed out that a general-purpose system was often impractical despite the large cost of its development. Consequently, he intentionally approached the development of this system from a new angle and invited service technicians from the field to participate in the design process. Through the technicians' participation, the designers were able to concretely address the technicians' activity networks. And rather than proceed in a top-down fashion for the official launch, the technicians who participated in the trials worked as instructors, gradually disseminating the system nationwide.

The most direct method of linking with users is to have them participate in the design process. User participation in design is often difficult, however, for a variety of reasons. For example, the difficulty of users' understanding a software specification can become an obstacle. Bødker and Grønbæk (1995) have proposed a method for overcoming such difficulties in involving users in design, which they call participatory design. In their case study, a dummy HyperCard database was prepared as a prototype for designing a database for a local government. With input from users, developers revised the prototype on the spot using HyperCard to create links and buttons. As they repeated this process, the developers were able to draw out concrete design suggestions from users. Through the use of a concrete prototype rather than a software specification, the cooperative design of the system established a new network of designers and users.

As in Bødker and Grønbæk's case, the technical specification of the self-dispatch system was developed through a prototyping process that facilitated communication between the developers and the users. The developers first created a simple screen that was not meant to be the finished design but was created to draw opinions from the service technicians in the field. This prototype was revised through interaction with the task team members. Sitting in front of the prototype screen, they discussed the system's concept, functions, and operation method over several days.

This link with users was crucial in considering the design of the self-dispatch system as a social system. Users made many comments on the particulars of the system. For example, they had concrete proposals on how to best represent a service technicians' status (at a client's, at a meal, delivering parts, etc.). The designers refined the specification based on the design input from users and their understanding of what visualizations would enable them to grasp the activities of their colleagues and organize service calls appropriately. Creating a link with users is not simply a matter of designing a system that is easy to use. The participatory design process created a link between the design of the self-dispatch system and the activity network. This in turn led to a reconceptualization of the design as it shifted from the design of a dispatch system, to making the area and group activity visible, to the design of a collaborative tool to support teamwork.

Coordinating with Providers of the Network Infrastructure and *Keitai* Technology

Initially, the mobile self-dispatch system utilized a laptop computer. The laptop system had some limitations: the machine was heavy, it took a long time to boot up, and the network connection was not reliable. Because of these factors, some service technicians resisted the new system and continued to call the back office by phone to get their dispatch orders. The *keitai* system was proposed as a solution to these problems. *Keitai* were used as a voice communication tool in the laptop days, but with the advent of the *keitai* Internet, *keitai* could become a platform for the self-dispatch system itself.

The primary design challenge for the *keitai* system was how to represent the necessary information on a tiny screen. The system developers used only the bare minimum of information necessary for dispatch and eliminated much of the information that had been presented in the laptop-based system. However, the field technicians realized that the system was meant not only to manage dispatch, and they pushed for inclusion of as much of the information that had been on the laptop system as possible. In contrast to the designers, who wanted to prioritize usability by limiting the volume of information, the service technicians wanted to prioritize the visualization of information that would enable them to appropriately maintain coordination with their team and area. As a result, the *keitai* system prototype was redesigned based on user input.

The testing and implementation of the system was extremely thorough in that the designers prototyped with an actual working system rather than a mock-up tool like HyperCard. First, they selected a region comprising several service areas that was a microcosm of the regions in Japan as a whole. They ran a trial of the system for a few months in that region, and the results were fed back to the developers, who subsequently revised the design. However, the infrastructure of the *keitai* network was still very unstable and caused many problems during the trial. In order to resolve such problems, the system implementers needed to form an alliance with the companies that provided the network infrastructure (in this case, within the same corporation). The service technicians participating in the experiment documented the locations and the symptoms of network trouble on a map. Based on these data, the providers of the network infrastructure started working on solutions. Without this coordination, the implementation of the self-dispatch system would have failed even if the actual system had been successfully designed. For systems relying on new technologies, success hinges on coordinating with technology providers. The design network is continuously renewed through the forging of links with diverse stakeholders, including users and technology providers.

Reconsidering the Meaning of *Keitai* in the Workplace

This chapter has examined the case of a self-dispatch system as a tool for making the area of service technicians visible, and has described how the system was designed as a social system through interaction between the practices of use and forms of social organization. We now reconsider *keitai* technology in relation to this system.

Keitai is commonly thought to be a personal technology that ties together individuals. In our example, however, *keitai* became a group technology that connected people working together as a team and that gave shape to and reflected their social networks. The self-dispatch system makes visible the service area that is overseen by a team of service technicians, allowing them to organize appropriate responses to service calls. They are able to visualize where their team members are, what they are doing, and whether they are encountering problems. In this way, the technicians are able to maintain a close collaboration despite the fact that they are distributed across space. In other words, *keitai* functions as a group tool for the team of service technicians.

The laptop system also enabled visualization of the area and group coordination, but it was not sufficient for the demands of the continually changing circumstances. *Keitai* technology addressed these needs in important ways. The *keitai* system was always close at hand, enabled continuous visualization of the circumstances of the area, linking a distributed work team, and allowed a sense of group identity, awareness, and co-presence. By positing that *keitai* can be a group or a personal technology, we are not suggesting that *keitai* is inherently one or the other. Rather, these characteristics of the technology are constructed within the context of social practice and organization. Through mobilization in the self-dispatch system, *keitai* actualized its identity as a group technology. By the same token, the self-dispatch system, through the *keitai* platform, strengthened its identity as a collaboration tool that enabled the visualization of group activity. In this way, technology and practice co-constructed one another.

Keitai is thought to tie individuals together regardless of time and place. With the self-dispatch system, engineers were able to make dynamic decisions about where to move in their area based on a visualization of their work conditions. In this way, the technology made the service area visible, becoming a technology for reconstructing time and place. In contrast to the idea of anytime, anywhere communication, the self-dispatch system is tied to activity in and visualization of a very specific physical location. This process of making the service area visible is tied to a strong awareness of place for the service technicians, and fixes their attention and activity to this place. Further, making circumstances of the team visible in a particular area strengthens the bonds of the team and highlights the boundaries with other teams. These qualities are in sharp contrast to the image of *keitai* as a technology that dissolves the boundaries of time and place. In our case study, *keitai* amplified the characteristics of the social organization and relations of the service technicians, whose practice is tied to the maintenance of relations in a fixed location.

The case we have presented demands a reconsideration of the meaning of *keitai* technology in the workplace. *Keitai* is not inherently a personal technology or one that transcends the boundaries of place and time. Rather, the meaning of *keitai* is located in specific practice and use contexts. Ultimately, the design of *keitai* technology is part of the design of social systems that include the activity networks of users and their social networks.

13 Technosocial Situations: Emergent Structuring of Mobile E-mail Use

Mizuko Ito and Daisuke Okabe

The integration of mobile phones into social life is still in its infancy in most parts of the world, triggering a set of sociocultural convulsions as institutions, people, and places adapt to and regulate its use. As is typical with technologies that alter patterns of social life, the mobile phone has been subjected to an onslaught of criticism for the ways in which it disrupts existing norms of propriety and social boundaries (see Matsuda chapter 1). While celebrated as a technology that liberates users from the constraints of place and time, the mobile phone has equally been reviled as a technology that disrupts the integrity of places and face-to-face social encounters. Sadie Plant (2002, 30) writes that "even a silent mobile can make its presence felt as though it were an addition to a social group, and ... many people feel that just the knowledge that a call might intervene tends to divert attention from those present at the time."

The case of heavy *keitai* e-mail use in urban Japan provides one window into the new kinds of social situations (see Goffman 1963), or more precisely, technosocial situations, emerging with the advent of widespread *keitai* use. This chapter reports on an ethnographic study of *keitai* users in the greater Tokyo area, examining new social practices in *keitai* e-mail communication, and how they are constituting technosocial situations that alter definitions of co-presence and the experience of urban space. The central argument is that *keitai* participate in the construction of social order as much as they participate in its destabilization. After first presenting the methodological and theoretical framework for our study, this chapter presents three technosocial situations enabled by *keitai* e-mail: *keitai* text chat, ambient virtual co-presence, and the augmented "flesh meet."[1]

Method and Conceptual Framework

Our Research

This chapter draws from ongoing ethnographic research on mobile phone use and location, centered at Keio Shonan Fujisawa Campus near Tokyo. Our description is grounded in two different sets of data. One is a set of ethnographic interviews conducted by Ito in the winter of 2000 with twenty-four high school and college students about their use of media, including mobile phones. The central body of data is a set of communication diaries and interviews collected between July and December 2002 by Okabe and Ito. We aimed to collect detailed information on where and when particular forms of mobile communications were used by a diverse set of people. We sought direct observational records in addition to interview data because it is difficult to capture the fleeting particularities of mobile communication after the fact. We adapted data collection methods pioneered by Rebecca Grinter and Margery Eldridge (2001), where they asked ten teenagers to record the time, content, length, location, and recipient (or sender) of all text messages for seven days. As with interviews, this data collection method still relies on second-hand accounting but has the advantage of providing much more detail on use than can be recalled in a stand-alone interview.

We expanded the communication log to include voice calls and the *keitai* Internet, and details about the location and context of use. Participants were asked to keep records of every instance of *keitai* use, including voice, short messages, e-mail, and Web use, for a period of two days. They noted the time of use, whom they were in contact with, whether they received or initiated the contact, where they were, what kind of communication type was used, why they chose that form of communication, who was in the vicinity at the time, if there were any problems associated with the use, and the content of the communication. After completion of the diaries, we conducted in-depth interviews that covered general attitudes and background information relevant to mobile phone use and detailed explications of key instances of use recorded in the diaries. Our study involved seven high school students aged 16–18, six college students aged 18–21, two housewives in their forties with teenaged children, and nine professionals aged 21–51. The gender split was roughly equal, with eleven males and thirteen females. A total of 594 instances of communication were collected for the high school and college students, and 229 for the adults. The majority of users were in the Tokyo Kanto region. Seven were recruited in the Osaka area in southern Japan to provide some geographic variation.

Technosocial Situations and Settings

Much work on mobile phone use and place has examined how mobile phones operate within particular social settings, particularly in public spaces where their use is often perceived as rude or disruptive. Most work in this vein has relied on observations in particular locations, analyzing how use is keyed to physical location and situation (Ito forthcoming; Ling 2002; Murtagh 2002; Plant 2002; Weilenmann and Larsson 2001; see also chapter 10). Somewhat less common have been efforts that focus on the settings being constructed by the mobile phone communications themselves. Some exceptions are studies that document the particularities of short text message communications (Grinter and Eldridge 2001; Kasesniemi 2003; Taylor and Harper 2003), studies of communication conventions in voice calls (Schegloff 2002; Weilenmann 2003), and studies of how workers construct mobile workspaces (Brown and O'Hara 2003; Laurier 2002; Schwarz 2001; see also chapter 11).

All of these approaches to the problem of mobile communication and place address key aspects of the equation: the local setting and the networked setting assembled (Laurier 2002) through mobile practice. The two approaches differ, however, in whether they take an existing, physically localized setting (restaurant, public place) as the social situation to be analyzed or the networked technosocial setting as primary. The latter type of study hinges on unique methodological and conceptual challenges. First, the researcher must track particular mobile communications across different physical locations. Second, the analysis must conceptualize these communications as located within a hybrid infrastructure of place and technology. As yet insufficiently theorized are the new kinds of social settings enabled through mobile communication that differ fundamentally from prior settings such as workplace, restaurant, face-to-face interaction, or fixed-line telephony. We believe it is necessary to examine the integration of technology, social practice, and place in an integrated technosocial framework.

In his review of Erving Goffman's theories of social situation, Joshua Meyrowitz (1985) suggests that the presumed isomorphism between physical space and social situation needs to be questioned when we take into account the influence of electronic media. He sees the work of Goffman and other situationists[2] as presenting the essential insight that social identity and practice are embedded in and contingent on particular social situations. He suggests, however, that these theories fail to take into account how electronic media cross boundaries between situations previously held to be distinct. In particular, he proposes that the information flows enabled by television in the 1960s related to the erosion of the boundaries between social identities and the positions of man and woman, adult and child, the powerful and not so powerful. This chapter builds on Meyrowitz's key insight that social situations are structured by influences that are outside the boundaries of physically co-present and interpersonal encounters. We retain, however, the analytic focus of situationists like Goffman, examining interpersonal communication, interaction, and situation rather than mass communication, information, and identity, on which Meyrowitz focuses. Our focus on interpersonal media leads to a substantially different set of conclusions than those that Meyrowitz suggests.

We propose the term "technosocial situations" as a way of incorporating the insights of theories of practice and social interaction[3] into a framework that takes into account technology-mediated social orders. As Meyrowitz proposes, more and more social orders are built through the hybrid relation between physically co-located and electronically mediated information systems. From a somewhat different perspective, anthropologists have analyzed this dynamic in terms of global or translocal flows

(Appadurai 1996; Clifford 1997; Gupta and Ferguson 1992). We believe that it is crucial to remain attentive to the local particulars of setting, context, and situation in the face of these translocal flows if we are to avoid a technical determinist argument that these technologies necessarily lead to a blurring of spatial and social boundaries. Electronic media have effects that break down certain prior social boundaries, as Myerowitz proposes, but they also have effects of constructing and reifying other social boundaries. We draw broadly from approaches in social and cultural studies of technology that see the technical and social as inseparable outcomes of ongoing and historically contextualized practice (e.g., Bijker, Hughes, and Pinch 1993; Callon 1986; Clarke and Fujimura 1992; Haraway 1991; Latour 1987).

This chapter is an effort to present examples of concrete technosocial situations that span a range of physical locations but still retain a coherent sense of location, social expectation, and role definition exhibited in Goffman's analyses and other practice-based studies. Our general conclusion is that mobile phones do undermine prior definitions of social situations, but they also define new technosocial situations and new boundaries of identity and place. To say only that mobile phones cross boundaries, heighten accessibility, and fragment social life is to see only one side of the dynamic social reconfigurations heralded by mobile communications. Mobile phones create new kinds of bounded places that merge the infrastructures of geography and technology, as well as technosocial practices that merge technical standards and social norms. The remainder of this chapter introduces mobile e-mail as a new communication modality and then presents three different technosocial situations that are built on mobile e-mail: mobile text chat, ambient virtual co-presence, and the augmented "flesh meet."

Keitai E-mail and Youth

Mobile messaging has been prevalent among the youth population since the early 1990s, when pagers were adopted as a way of sending text messages among high school and college students (see chapters 1 and 2). Although technology platforms have changed considerably from the days of numeric pagers, current messaging practices and related norms for social behavior are built on this decade-long history. Japanese keitai e-mail use thus represents a degree of social stabilization that may not be as prevalent in areas where text messaging is a more recent development. We use the term keitai e-mail to refer to all types of textual and pictoral transmission via mobile phones. This includes short message services as well as the wider variety of e-mail communications enabled by the keitai Internet.

Until recently, youth led the Japanese population in terms of adoption rates, but as of 2002 the gap has largely closed. Even as keitai use has become common in all age groups, however, younger people have a higher volume and unique patterns of use that differentiate them from older users. In contrast to the ¥5,613 average monthly payments of the general population, students pay an average of ¥7,186 for their monthly bills (IPSe 2003). Particularly distinctive is use of *keitai* e-mail. Among student *keitai* users, 95.4 percent describe themselves as *keitai* e-mail users, in contrast to 75.2 percent in the general population (Video Research 2002). Teens send twice as many e-mails as people in their twenties—approximately 70 per month versus 30 per month (Mobile Communication Research Group 2002). In contrast to the general population (68.1 percent), almost all students (91.7 percent) report that they send over five messages a day. They also tend to be more responsive to the e-mail that they receive. Almost all students (92.3 percent) report that they view a message as soon as they receive it, whereas a slimmer majority of the general population (68.1 percent) is as responsive. Many older users say that they view a message when convenient to them or at the end of the day (Video Research 2002). Elsewhere we have described some of the unique features of youth populations that make them more avid *keitai* e-mail users than the older generation. This includes their historical relationship to mobile communications as well as their lack of communication channels and legitimate spaces for assembly with peers and lovers (Ito forthcoming; see also chapter 4). We now turn to technosocial situations that are beginning to be structured through the exchange of *keitai* e-mail. These new social conventions are concentrated among, though not restricted to, the younger population.

Mobile Text Chat

A type of technosocial situation that has close analogues in other communication forms is the text-based chat. Here we introduce two examples of mobile text chat conducted between couples, both on private transportation, to highlight the unique qualities of *keitai* e-mail. The first example is from the communication diaries of one female college student who carried on a text message conversation while moving between different forms of public transportation. She has just finished work and made contact with her boyfriend after boarding the bus. Each time-coded entry is a separate text message.[4]

22:20 [Boards bus]
22:24 (*Send*) Ugh. I just finished. (>_<) I'm wasted! It was so busy.
22:28 (*Receive*) Whew. Good job. (>_<)
22:30 (*Send*) I was running around the whole time. Are you okay?
22:30 [Only other passenger leaves. Student makes voice call. Hangs up after two minutes when other passengers board]
22:37 (*Send*) Gee, I wish I could go see fireworks. (;_;)
22:39 (*Receive*) So let's go together! I asked you!
22:40 [Gets off bus and moves to train platform]

22:42   (*Send*) sniff sniff sniff (;_;) Can't if I have a meeting! I have to stay late!

22:43   (*Receive*) You can't come if you have to stay late?

22:46   (*Send*) Um, no . . . I really want to go. (;_;)

22:47   (*Receive*) Can't you work it out so you can make it?

22:48   [Boards train]

22:52   (*Send*) Oh, . . . I don't know. If I can finish preparing for my presentation the next day. I really want to see you. (>_<) I am starting to feel bad again. My neck hurts and I feel like I am going to be sick. (;_;) Urg.

22:57   (*Receive*) I get to see you tomorrow so I guess I just have to hang in there! (^o^)

23:04   [Gets off train]

23:05   (*Send*) Right, right. I still have a lot of work tonight. I can't sleep!

In our interview, the student described how her messaging embeds subtle clues that indicate her status and availability for communication keyed to her physical location.

You talk about the fireworks for about ten minutes. Is this the kind of thing you usually communicate about over e-mail?

Part of the way through, it becomes just something to keep the conversation going just for the sake of continuing. Around when this fireworks topic comes up. All I wanted to really say was about my workday, but since I still have some time to kill. I didn't really care about the fireworks. Oh, I shouldn't be saying this.

So you didn't really want to go.

It isn't exactly a lie, but since I couldn't go, I wrote "Oh, I don't know," kind of grinning.

So the important thing was to keep the conversation going.

Yes, that's right. And after I started feeling that this was going on too long, I suddenly changed topic to my physical condition.

    This last change of topic that she describes happens just as she is getting ready to get off the train, and it is an indicator that the conversation should be coming to an end. She has enlisted a companion on her solitary bus ride, successfully filling dead time with small talk, ending it at precisely the moment when she arrived at her destination. Just as with people who are physically co-present, the exchange with her boyfriend is exquisitely coordinated with her crossing the boundaries between different social situations, beginning when she had a free moment to sit and punch keys on her *keitai* and ending when she would begin the walk to her home, which would make it harder for her to attend to an active conversation. Like two people who might start a conversation as they run into each other in an elevator, ending it smoothly as one is stepping out, her conversation also has a clear, physically demarcated beginning and end that is socially recognized through implicit conversational cues. Nuanced texting conventions signal the end of a conversation without her ever having to explicitly state, "I need to go now."

Another one of our research subjects, a young businessman, engages in extensive *keitai* e-mail exchanges with his girlfriend while he is traveling in public transportation, regularly sending long messages (100 characters or more). He described the social situation created by the convergence of dead time in public transportation and the affordances of the *keitai* Internet. In this case, we did not have access to the full text of the communications, but he described them in his communication diary as "mutterings" and his girlfriend's responses as "mutterings in reply."

When you are on the train, are you generally using your keitai*?*
Yes, or just looking at things on my *keitai*. I like to be doing something. I don't like to think of it as wasted time.
In terms of the content of your messages, are they pretty lightweight? In your diary, you wrote "mutterings" and "mutterings in reply."
Yes. It really is mostly lightweight, content-free communication, but it can be pretty extensive. Here is one that is 110 characters. I can get pretty deep into things in my messages.

The rhythm of his exchanges with his girlfriend fluctuates between quick status updates and comments and longer monologues that transform situations that would have otherwise been occupied by solitary musings into an intimate interpersonal space. His time on public transportation is occupied with these conversations with his girlfriend and shorter bursts of exchanges with friends and work colleagues coordinating meetings and other details. These are two among many examples of text chat that we gathered. Approximately half of the students in our study engaged in this sort of chat sequence while in transit.

What is unique about mobile text chat is the way it is keyed to presence in different physical spaces. We observed mobile text chat in diverse settings: homes, classrooms, and public transportation. Like Internet chat and voice calls, mobile text chat can be used whenever two parties decide to engage in a focused "conversation." What is unique to mobile text chat, however, is that it is particularly amenable to filling even small communication voids, gaps in the day where one is not making interpersonal contact with others, particularly in settings such as public transportation, where there are prohibitions on voice calls (see chapter 10). Mobile text chat can accommodate changes in physical location as well as pauses in conversation as one gets on and off a train, scans a bus schedule, changes channels on the TV, jots down some notes, or answers a question from the teacher. During a focused set of chat exchanges, small pauses are permissible, although longer pauses of over 10 or 15 minutes are generally ended with an apology or some allusion to an interruption. The closest analogue in prior social situations would be the conversations that people have as they share a train ride or pause at an open office door. These are spontaneous conversations with sudden beginnings and ends, sometimes expanding into a longer and deeper conversation, but with an underlying expectation that the interaction will be brief, a colonization of an in-between space rather than a planned and extended meeting. The technosocial situation of the text chat is flexible but has clear social expectations and rhythms.

Ambient Virtual Co-Presence

Unlike voice calls, which are generally point-to-point and engrossing, messaging can be a way of maintaining ongoing background awareness of others, and of keeping multiple channels of communication open. This is like keeping instant message channels open in the background while going about one's work, but the difference is that the *keitai* is carried around just about everywhere by heavy users. The rhythms of *keitai* messaging fluctuate between what we have characterized as chat and a more lightweight awareness of connection with others through the online space. In our interviews with heavy users of *keitai*, all reported that they were only in regular contact with approximately two to five, at most ten, close friends despite having large numbers of entries in their mobile address books. This is what Nakajima, Himeno, and Yoshii (1999) describe as a "full-time intimate community" (see chapters 1 and 6) and Habuchi (chapter 8) calls a telecocoon. For people who are heavy *keitai* e-mail users, there is often a social expectation that these intimates should be available for communication unless they are sleeping or working. Text messages can be returned discreetly during class, on public transportation, or in restaurants, all contexts where voice communication would be inappropriate. Many of the messages that we saw exchanged between this close peer group or between couples included messages that informants described as insignificant or not urgent. Some examples of messages in this category are "I'm walking up the hill now," "I'm tired," "I guess I'll take a bath now," "Just bought a pair of shoes," "Groan, I just woke up with a hangover," "The episode today sucked today, didn't it?"

These messages define a social setting that is substantially different from direct interpersonal interaction characteristic of a voice call, text chat, or face-to-face one-on-one interaction. These messages are predicated on the sense of ambient accessibility, a shared virtual space that is generally available between a few friends or with a loved one. They do not require a deliberate opening of a channel of communication but are based on the expectation that someone is in "earshot." From a technology perspective, this is not a persistent space like an online world that exists independent of specific people's logging in (Mynatt et al. 1997). As a technosocial system, however, people experience a sense of a persistent social space constituted through the periodic exchange of text messages. These messages also define a space of peripheral background awareness that is midway between direct interaction and noninteraction. The analogue is sharing a physical space with others whom one is not in direct communication with but whom one is peripherally aware of. Many of the e-mails exchanged present information about one's general status that is similar to the kind of awareness of another that one would have when physically co-located—a sigh or smile or glance that calls attention to the communicator, a way of entering somebody's virtual peripheral vision. This kind of virtual tap on the shoulder may result in changing the situation into a more direct form of interaction such as a chat sequence via texting or a voice call, but it might also be ignored if the recipient is not available for focused interaction.

Of particular interest are the logs of one teenage couple in our study that is a somewhat more intense version of couples communication that we saw in other instances. Their typical pattern is to begin sending a steady stream of e-mail messages to each other after parting at school. These messages continue through homework, dinner, television shows, and bath, and culminate in voice contact in the late evening, lasting for an hour or more. A trail of messages might follow the voice call, ending in a good night exchange and revived again upon waking. On days that they were primarily at home in the evening, they sent 34 and 56 messages to each other. On days they were out, the numbers dwindled to 6 and 9. The content of the messages ranged from in-depth chat about relational issues, to coordination of when to make voice contact, to lightweight notification of their current activities and thoughts. In this case, and to a smaller degree for other couples living apart, messaging becomes a means for experiencing a sense of private contact and co-presence with a loved one even in the face of parental regulatory efforts and the inability to share any private physical space.

While *keitai* have become a vehicle for youths to challenge the power geometries (Massey 1994) of places such as the home, the classroom, and the street, it has also created new disciplines and power geometries, the need to be continuously available to friends and lovers, and the need to always carry a functioning mobile device. These disciplines are accompanied by new sets of social expectations and manners. One teenage interviewee explains,

I am constantly checking my mail with the hopeful expectation that somebody has sent me a message. I always reply right away. With short text messages I reply quickly so that the conversation doesn't stall.

When users are unable to return a message right away, there is a sense that a social expectation has been violated. When one girl did not notice a message sent in the evening until the next morning, she says, she felt terrible. Three of the students in our diary study reported that they did not feel similar pressure to reply right away. Yet, even in these cases, they acknowledged that there was a social expectation that a message should be responded to within about 30 minutes unless one had a legitimate reason, such as being asleep. One describes how he knows he should respond right away but doesn't really care. Another, who had an atypical pattern of responding with longer, more deliberate messages hours later, said that her friends often chided her for being so slow. In one instance, a student did not receive a reply for a few hours, and his correspondent excused himself by saying he didn't notice the message. The recipient perceived this as a permissible white lie that got around an onerous social expectation. All students who were asked about responses delayed an hour or more said that they would generally make a quick apology or excuse upon sending the tardy response. These exceptions to the norm of immediate response trace the contours of the technosocial situation as much as do conforming practices.

With couples living apart, there is an even greater sense of importance attached to the ongoing availability via messaging. The underside to the unobtrusive and ubiquitous nature of *keitai* e-mail is that there are few legitimate excuses for not responding, particularly in the evening hours when one is at home. Five of the ten student couples in our study were in ongoing contact during the times when they were not at school, and all these couples had established practices for indicating their absence from the shared online space. They invariably sent a good night e-mail to signal unavailability and often sent status checks during the day, such as "Are you awake?" or "Are you done with work?" We saw a few cases where they would announce their intention to take a bath, a kind of virtual locking of the door. He: "Just got home. Think I'll take a bath." She: "Ya. Me too." Just as mobile workers struggle to maintain boundaries between their work and personal lives, youths struggle to limit their availability to peers and intimates. The need to construct and mark these boundaries attests to the status of this ambient virtual peer space as an increasingly structuring and pervasive type of technosocial setting.

The Augmented "Flesh Meet"

Those who are not heavy *keitai* users are often baffled by the sight of Japanese teens congregated at a fast food restaurant, staring at their *keitai* rather than talking to one another. The assumption is that the virtual connection is detracting from the experience of the face-to-face encounter. What nonusers often don't realize is that *keitai* can augment the experience and properties of physically co-located encounters rather than simply detracting from them. Teens use *keitai* to bring in the presence of other friends who were not able to make it to the physical gathering, or to access information that is relevant to that particular time and place. The boundaries of a particular physical gathering, or "flesh meet," are becoming extended through the use of mobile technologies, before, during, and after the actual encounter.

Before the Meeting

Keitai have transformed the experience of arranging meetings in urban space. In the past, landmarks and times were the points that coordinated action and convergence in urban space. People would decide on a particular place and time to meet, and converge at that time and place. We recall hours spent at landmarks such as Hachiko

Square in Shibuya or Roppongi crossing, making occasional forays to a pay phone to check for messages at home or at a friend's home. Now people in their teens and twenties often do not set a fixed time and place for a meeting. Rather, they agree on a general time and place (Shibuya, Saturday late afternoon) and exchange approximately five to fifteen messages that progressively narrow in on a precise time and place, two or more points eventually converging in a coordinated dance through the urban jungle. As the meeting time nears, contact via messaging and voice becomes more concentrated, eventually culminating in face-to-face contact. A case will make this dynamic more evident. It involves a college student, Erin, and her meeting with three other friends in Shibuya to go to a live performance. In an interview, Erin describes the meeting:

At the last event we went to together, friend A and I had decided that we would meet up early for a cup of tea. So I contacted friend A first. Friend S lives kind of on the outskirts of the city, so she isn't always able to come to the shows. I hadn't communicated with her about when we were going. At the last event the three of us did have the conversation that it would be nice if we could all meet up for tea. Friend C isn't a regular at the events, so I wasn't sure if she was coming.

We reconstructed the following sequence of events from the communication log and interview:

E-mail sent from home to friend A:
11:30   Around what time are you going to Shibuya today?
E-mail received at home from friend A:
11:56   About 3.
E-mail sent from home to friend A:
12:00   Okay, I'll get in touch when I'm there.
E-mail received at station from friend A:
13:56   I'll be at Shibuya in about 10 minutes.
E-mail sent from train to friend A:
15:00   Me, too, around then.
E-mail sent from Shibuya station to friend A:
15:06   I'll wait at the back entrance of Quattro.
The two friends meet up at Quattro, the performance hall, then go to a café.
E-mail sent from café to friend S:
6:32   About what time are you coming?
There is no reply from friend S, and the two go on to Tower Records.
A call received at Tower Records from friend S:
17:02   (Two-minute voice conversation about where the two parties are)
Friend S meets them at Tower Records.
A call received at bookstore from friend C:
17:50   (Three-minute voice conversation about where the two parties are and what time the show starts)
The four meet at Quattro.

This kind of coordination is the opening sequence of a face-to-face meeting, extending the parameters of what it means to be together beyond the boundaries of physical contact. There is generally a point where participants converge in virtual space prior to converging in physical space as they begin to microcoordinate (Ling and Yttri 2002) about precisely where and when they are and how they will meet up as they are in transit to the gathering spot. This could be just minutes before physical contact, as in the case of friends S and C, or it could be as a person boards a train to head toward a meeting spot. A person is not generally considered late to a particular gathering if she is present in the virtual space at roughly the appointed time. Informants have reported that they do not apologize for being 30 or more minutes late in arriving at a particular place as long as they have been in e-mail contact in the interim. By contrast, if they had left their phones at home or had let the battery die, the meeting would have been considered a failure. In other words, presence in the virtual communication space is considered an acceptable form of initial showing up for an appointed gathering time.

Although the older generation often describes these practices as "loose" in terms of commitments to time and place, a slackening of manners, we can actually see some consistency in the integrity of the social norms and expectations attached to such gatherings. As in the case of meetings with appointed time and place, with these more flexibly arranged gatherings the consistent rule is that you should not keep somebody waiting in a particular place. If the partner has already shown up in virtual space by announcing where they are, mobile phone users can go off to a bookstore or take care of an errand rather than wait at an arbitrary spot. Here is an example of a meeting between two women in their twenties, again at Shibuya. As is typical in cases like this, lateness is a matter to be announced but not apologized for. The one being "kept waiting" has been attending to other matters about town rather than waiting at an appointed spot. (The following messages are extrapolated from brief descriptions of content rather than being exact transcripts.)

On this day, I had decided to go see a performance with my friend. We had decided to meet in Shibuya but had not decided where. We were supposed to meet at 4:00 [16:00], but by 3:00 I knew I was going to be late, so I sent a message that I would be 30 minutes late as I was leaving the house. She replied right away, "Okay."

[Bus stop]	15:00	(*Send*) I'll be about 30 minutes late.
[Bus stop]	15:01	(*Receive*) Okay.
[Shibuya station]	16:32	(*Send*) I've arrived at Shibuya.
[Shibuya station]	16:33	(*Receive*) Where in Shibuya are you?
[Shibuya station]	16:34	(*Send*) Hachiko Square.
[Shibuya station]	16:35	(*Receive*) Wait there. I'll be right over.
[Shibuya station]	16:36	(*Send*) Okay. Will wait.
[Shibuya station]	16:40	(*Voice call*) "Where are you? Oh, there, okay, I see you."

The first message announcing lateness signals initial presence in a shared techno-social situation of getting together. Subsequent exchanges between the two are highly responsive, with a one-minute turn-around time, and indicate that they are keeping close track of incoming messages. Though this is a case of lateness becoming socially permissible, gatherings arranged via mobile technologies do not necessarily have the effect of shortening the length of time for physical co-presence. This same research subject described how on a prior day she had made plans with some friends to meet after work for drinks. They had decided on 7:30 p.m. as a meeting time, but she got off work early that day and was able to contact a friend to meet earlier because "it would be boring to be by myself until then." In both these cases, social voids and waiting times have been filled with contact and coordination via mobile messaging.

During the Meeting

After participants have converged in physical space, *keitai* communications do not necessarily end, particularly for social gatherings. In contrast to work meetings, where *keitai* communications are largely excluded, among gatherings of young people *keitai* is a common accessory. In the case of Erin's meeting, we saw how the phone was used in the context of a gathering of friends to check on the status of others who might or might not be joining them. When an e-mail message comes into a friend's *keitai*, it is quite common to ask who it is from and then converse about that person. Young people generally reported that they had no reservations about making contact with others via *keitai* when they were with a group of friends, although they might make a brief apology if a one-on-one gathering was interrupted with a voice call.

In other cases, *keitai* messages are used as a way of making contact when the recipient is just out of visual range or is unavailable for voice contact. Messaging during class or lectures is one common way that people in a shared space make contact that would otherwise not be possible in a given setting. We saw e-mail being sent during class in only two of our communication diary cases, but almost all students reported in their interviews that they would receive and send messages in class, hiding their phones under their desks. Here is an interview with one of the high school students who we saw using her phone during class:

In what sorts of places and situations do you use your phone a lot?
At school, during class. I leave my phone on my desk and it vibes.
Your teacher doesn't care?
Well, the teacher pretty much knows. He doesn't do anything about it.
Really? You can leave it out?
Everyone has them out. Some kids even let their *keitai* ring, and the teacher is like, "Hey, it's ringing...." I think this is just our school.
Do you take voice calls during class?

No. That would be going too far.

Oh, so you wouldn't answer. What kinds of exchanges do you have over e-mail during class? Do you send e-mail to people sitting in the same classroom?

Yes, I do that too.

What do you say?

"This is boring."

And you get a reply?

Yes.

When you write your e-mail, do you hide it?

Yes. When the teacher is facing the blackboard, I quickly type it in.

Like this student, three other students described conversations with students in the same classroom, making comments like "This sucks," "This is boring," or "Check it out, the teacher buttoned his shirt wrong." More commonly, students reported that they conducted "necessary" communications during class, such as arranging a meeting or responding to an e-mail from somebody with a specific query.

Another example of augmented co-presence was when one of our informants was standing in a long line for a bus and saw her friend near the front of the line. She sent her friend a message to look back so that she could see her and wave. In other cases, students have described how they will message their friends upon entering a large lecture hall to ask where they are sitting. In all of these cases, *keitai* e-mail augments the properties of a particular place, enabling contacts and communications that would not otherwise be available. One observation documented by Ito of a group of high school students on a bus illustrates this dynamic:

I am sitting near the front of a bus that is not very crowded. Most seats are occupied, but there is nobody standing. A group of five high school boys in black uniforms are congregated just behind me, speaking rather loudly across the middle aisle. They are discussing some kind of gathering they are arranging. I am not able to determine exactly the nature of the gathering, but they are involved in a heated dispute over who is coming and why some members are not coming, and suspicions that somebody is sabotaging the arrangements.

"Ask Ken," one boy suggests. "Ask him if he is coming." One of the boys carrying a *keitai* punches in a message. As they await the reply, they continue to debate about what is going on. "If the girls are coming, the guys will come, too." "Send a message to Ken then, or do you want me to send it?" "No, you send it." In the meantime, Ken has responded that he thinks he will come, a fact that the recipient announces to the group. "Okay, then he's not the problem," they all agree. The exchange continues in this manner, with the boys making selective contact with their friends and collectively developing a theory of the fate of their planned gathering.

After the Meeting

Just as *keitai* e-mail extends the prior and present parameters of social contact, it also extends the possibilities for contact after a gathering. In all the gatherings we saw between heavy *keitai* e-mail users, a trail of messages was scattered after a physical gathering as people continued the conversation, mentioned a forgotten bit of information, or thanked the person who had organized the gathering. "I forgot to give you back your CD." "Thanks for the lift." "Thanks for coming out with me today." In the past, the common practice was to say, "Thanks for last time," on the next occasion of a phone call or a meeting. A newly emergent norm is that these exchanges happen as people scatter to return home on foot or public transportation. The dead time in transit on the way home is now occupied by the fading embers of conversation and contact.

While all planned encounters have always had some element of pre-contact (making a phone call to arrange a meeting, confirming by e-mail, etc.) and post-contact (saying thanks the next time one calls or meets), the mobile phone makes these situations contiguous rather than disjunctive, stitching them together into a technosocial gathering that extends beyond the time and space of physical co-presence. This stretching of prior boundaries of what it means to be together does not mean, however, that the social order attached to these gatherings has eroded. In all the cases presented, we can see the construction of new technosocial settings and situations that have clear expectations for interaction and role performance.

From Technosocial Situations to Technosocial Orders

The cases we have presented of technosocial situations and settings enabled by mobile e-mail hint of broader shifts in what Meyrowitz (1985) has called the information systems that structure the availability of information and social contact in particular situations. We consider these situations as part of changing technosocial orders that structure urban life in Japan. While it is impossible to extrapolate from these cases what the overall contours of these technosocial orders are, we can hazard a few guesses as to certain trends and changes in experience that these cases suggest.

One shift we are seeing is a change in our sense of what it means to be co-present, to share a social and physical space. *Keitai* e-mail constructs a space of connectivity that relies on a pulsating movement between background and foreground awareness and interaction as people shift from lightweight messaging to chat to "flesh meets." Occasionally, contact is severed entirely, as when someone enters a business meeting or the bathtub, but for many heavy *keitai* e-mail users, it is becoming more the norm than the exception that a certain level of ongoing connectivity is expected. Particularly in the case of couples who have a sense of intimate connection but who do not share a home, *keitai* e-mail inscribes a concrete new social setting for private and pervasive togetherness.

Another shifting balance is between serendipitous and intentional contact. The experience of contact with anonymous others in dense urban spaces like Tokyo has changed with the widespread adoption of the *keitai* Internet. Public transportation and meeting places were previously sites for people-watching and occasional light-weight contact with strangers. While these long-standing forms of urban interaction persist, there is a broadened palette of social possibilities in these settings. Through *keitai* contact, these locations can also become settings for intimate and private contact with physically absent others. This is related to what Matsuda describes as "selective sociality" (chapter 6). Even when traveling through urban space solo, mobile messengers make lightweight contact with others, updating friends about their whereabouts, beaming some news about a hot sale, transmitting a photo of a celebrity-sighting. Urban space has become a socially networked space criss-crossed with the flow of messages.

This chapter has used the case of *keitai* e-mail to extend practice-based theories of place and setting to take into account the role of a new media technology. We have argued that new networking technologies structure emergent technosocial orders as much as they erode the boundaries of prior ones. Research must be attentive to emergent technosocial places as well as prior senses of place to avoid a technically determinist assumption that electronic media necessarily erode social boundaries and the integrity of place. While it is crucial to attend to the structuring influences of physical places and architectures, networked and online infrastructures are also becoming increasingly salient. The concept of technosocial situation relies on a formulation of place and setting that accounts for flows of electronic media but is still grounded in material architectures and structuring social orders. Unlike Internet ethnography, which has often been productively pursued with a focus on a bounded online space, *keitai* ethnography insists on accountability to both electronic and traditionally located architectures. It is thus a productive object for examining hybrid technosocial situations that navigate settings defined by long-standing senses of place, such as the train, restaurant, or classroom, and those enabled by the exchange of electronic media. We have described mobile text chat, ambient virtual co-presence, and the augmented "flesh meet" as technosocial situations located in concrete social, electronic, and material architectures that span geographically bounded definitions of place.

Notes

This research was supported by the DoCoMo House design cottage at Keio University Shonan Fujisawa Campus. It has benefited from discussions with DoCoMo House lab leader Kenji Kohiyama and students associated with the lab. In particular, we would like to thank the graduate students who helped with the recruiting, data collection, and transcribing for this project: Kunikazu Amagasa, Joko Taniguchi, Hiroshi Chihara, and Eri Aoki.

1. When examining the practices of those who engage substantially in electronic interpersonal communication, one sees that physical gatherings are often the marked category, as in the description "flesh meet" for a physical gathering of an online community, often called *offukai* (off-

line meeting) in Japan. We borrow the term "flesh meet" to refer to a physically co-located gathering.

2. By "situationist" Meyrowitz is referring to sociological work that focuses on the details of situated action, not to the political and artistic movement known as Situationist International.

3. Although we depart from the methodological details of conversation and interaction analysis, our work is broadly inspired by the related commitment to studying emergent and dialogic practice (e.g., Duranti and Goodwin 1992; Tedlock and Mannheim 1995). Studies of how people incorporate technologies and artifacts into everyday practice are particularly relevant to our perspective (e.g., Lave 1988; Suchman 1987).

4. The messages have been translated into standard English, although the Japanese text includes abbreviated forms and idioms similar to those found in texting in other languages. The Japanese-style emoticons have been left as is. The notes in brackets are actions reconstructed through the communication log and the subsequent interview.

V | Emergent Developments

14 | *Keitai* Use among Japanese Elementary and Junior High School Students

Yukiko Miyaki

In contemporary Japan, children are growing up with *keitai* close to them from a young age. Moreover, *keitai* use, first prevalent among high school students and adults in their twenties, has recently spread to younger age groups: elementary and junior high school students.[1] Though only a few of these actually own *keitai* in Japan, many elementary and junior high school students have ample opportunity to use *keitai* of family members and friends. Since the market among people in their twenties and thirties is reaching saturation, the communications industry and various terminal manufacturers have begun to actively direct their marketing toward children and elders. Various *keitai* services and terminals especially made for children have become available in the past few years.

Amidst the diversification of payment methods, the emergence of prepaid phones marked a major shift. Here was a billing system that charged only for the amount of actual use instead of for possession and maintenance of the phone. With prepaid *keitai*, users purchase a prepaid card and input a code printed on the back of the card into the phone. They can then use the phone up to the amount purchased on the card. This style of prepaid service was first introduced in 1999; at first, these *keitai* were remodeled versions of phones that had become outdated because of the rapid pace of handset obsolescence. Prepaids are an easy purchase because there is no base monthly charge and comparatively little financial impact in the case of theft or loss (the handset is cheap, and there is a prepaid limit to the amount a thief could use). With this insurance against overuse, prepaid phones helped establish the *keitai* market for children, and the purchase of prepaid phones by parents for their children increased rapidly. For standard *keitai*, generally the month's charges are automatically debited from a bank account the following month, but with a prepaid, since the phone can't be used beyond the amount already paid, it is easier for parents to manage charges. These types of *keitai* are sold not only at telecommunications retailers but also at toy stores and convenience stores.

Standard carriers also established subscription plans in order to prevent overuse, where subscribers could set the upper limit for monthly totals on voice calls, packet charges, and Internet use. On the day after reaching this limit, communications are stopped on that phone. These schemes for payment and control have established a mobile environment suitable for children. It is becoming a tradition to present *keitai* as a graduation present to elementary and middle school students.

Children and *Keitai*

Children's lifestyles typically involve spending time alone at home in the daytime while parents are working. Kids today have fewer siblings because of the growth of nuclear families and the increase in the number of two-income families. Moreover, the children of today often go out at night to attend cram classes or other private lessons. According to the General Affairs Bureau (2000a; 2000b), 74.7 percent of elementary school students and 65.2 percent of junior high school students attend lessons of some kind outside of school. Though the average frequency of lessons is two or three times a week, some children are kept busy attending lessons four or five times a week. Since these lessons primarily take place after school hours, we can assume that modern Japanese children are often moving around alone outside of the home in the evening or at night.

There has been a conspicuous increase in crimes involving children, in which not only the victims but also the assailants are elementary or junior high school students. As a matter of fact, according to the research data collected by Kou Nakamura (2000) on elementary schools, approximately 40 percent of girls and boys encounter some kind of situation of physical endangerment by the time they reach the upper grades of elementary school. For young children, the risks associated with traveling outside the home are high regardless of gender.

This is the backdrop for the use of *keitai* by children; they serve as tools for self-defense on the children's part and assurance on the parents' part as they keep track of their children's activities and confirm their safety.

Convenience, confirmation of safety and whereabouts, and help in emergencies —these are all positive uses of *keitai* for children. However, problems for children with mobile devices have begun to emerge, including expenses for overuse, lying about one's whereabouts, and involvement with drugs and sex industries. The use of *keitai* for organized crime, drug dealing, and prostitution is widely reported in the media. As the adoption of *keitai* increases, children are bound to face a higher risk of being involved with such criminal activity, which understandably aggravates parents' concerns for their safety. Though the *keitai* serves as a tool for confirming children's safety, it also raises the risk of children's becoming involved with unsavory communications.

Today's children hold different views toward phones than their parents' generation did. Moreover, the conventional telephone use rules are gradually disintegrating even among parents. In the recent past there were certain rules for telephone use for fixed-line subscription phones in the Japanese home, such as "long phone conversations are not good," "it is not appropriate to make calls late at night," or "telephones should not be used for idle chatter." There were appropriate social considerations behind each of these rules. For example, in the case of long phone conversations, "it is a problem if someone else tries to make a call and cannot get through." In the case of calls late at night, "the phone ringing could be a nuisance in the other house-hold," or "they might become alarmed that there was a disaster" (the phone can be used late at night if it is an emergency). Since phone tolls were expensive, "using the phone for idle chatter" or "spending money on idle chatter" was unacceptable. These rules were accepted in nearly all households and were developed though an implicit consensus.

However, even before *keitai* were adopted by children, call waiting (introduced in 1970) abolished the busy signal. Further, the ability to adjust ringer volumes on different receivers and the trend toward proliferating receivers in each room of the house meant elimination of inappropriate phone ringing and the gradual crumbling of these phone rules. The prior rules for telephone use became outdated norms with no real reason behind them, and children increasingly neglected to follow them. Children came to feel that it was not inappropriate to chat at night with friends they had seen during the day even if this entailed phone costs, so conflicts arose with parents who tried to intervene. In other words, even before the adoption of *keitai* and the social changes associated with them, there was an existing gap in values between adults and children with regard to use of the telephone. Further, there is increasing diversity in perception about telephone use not only between children and adults but among adults themselves. In this way, stable rules for use of the telephone are disappearing in Japan.

In such environment, how are *keitai* viewed by parents and children? What is the value basis for the decision making about the pros and cons of *keitai*? What kinds of problems exist?

The *keitai* has become an indispensable tool for building and maintaining networks of friends for Japanese young people. Moreover, the age of *keitai* users has fallen along with falling mobile service prices, leading to an increase in the adoption of mobile devices among elementary and junior high school students. Observing children who are already familiar with *keitai*, researchers have found these primary uses: self-defense on the streets at night, parents' confirmation of the child's whereabouts, checkup by parents, calling home before going home from private crammer classes, contact with friends, and emergency calls between parents and children.

A child or a parents can initiate *keitai* ownership by a child. Either way, children who have no income inevitably depend on their parents to pay the maintenance and calling charges. Hence, *keitai* ownership by children requires parents' agreement and support. Furthermore, the feeling of security engendered by owning *keitai* fulfills both the child's and the parents' needs.

Because both the parents' and the children's opinions must be taken into account to understand *keitai* use among children, I conducted a questionnaire survey with both elementary and junior high school students and their parents.[1]

The Survey

An outline of the survey can be found in table 14.1. Those surveyed included 497 children aged 10–14 (fifth and sixth graders and junior high school students) and their parents, and 140 parents of children aged 8–9 (third and fourth graders), chosen from the nationwide pool of monitors of the Daiichi Seimei Corporation (formerly Life Design Institute). (Children are generally in elementary school between the ages of 6 and 11, and in junior high school between the ages of 12 and 14 years.) Among the respondents, children aged 10–14 and their parents were surveyed in pairs.

Rates of *Keitai* Use among Children

Of all respondents, 42.7 percent answered yes when asked, "Do you ever use *keitai*?" (figure 14.1). This total includes those who responded, "I have my own *keitai*," "I occasionally borrow *keitai* of family members," and "I occasionally borrow *keitai* of friends." According to this research, more girls than boys, and more junior high school students than elementary school students, have experience with *keitai* use. However, still only a few children possess their own *keitai*; the majority borrows *keitai* of family members. A total of 9.8 percent of all children possess their own *keitai*—13.1 percent of girls and 6.6 percent of boys.

How Children Acquire Their Own *Keitai*

Children in possession of *keitai* were asked how they became owners of *keitai* (figure 14.2). The most common answer (64.3 percent) was, "I asked Mother or Father to let me own one because I wanted to," which indicates that the main cause for ownership is the children's active desire.

On the other hand, 21.4 percent claimed, "I didn't necessarily want one but Mother or Father insisted," which indicates that ownership was decided based upon the parents' wishes. Other reasons include "I was given one as a gift by somebody other than my parents" and "I bought one myself."

Although the comparison may not be rigorous because only a few children actually own their own *keitai*, one can say that in most cases ownership by elementary school students was insisted on by parents, whereas ownership by junior high school students came about because the children actively desired it. According to the survey results, more girls than boys actively desire *keitai* ownership.

Table 14.1
Survey Outline

Area surveyed:	Nationwide
Respondents:	Group A—Children aged 10–14 and their mothers[a] Group B—Mothers of children aged 8–9[a]
No. of samples:	637 (A—497 pairs; B—140 people)
Sampling population:	Monitors of the Life Design Institute and their families
Methodology:	Postal mail—questionnaires
Time period:	September 2000
Total no. of responses:	618 (A—482 pairs; B—136 people)
Response rate:	97.0%
No. of valid responses:	614 (A—480 pairs; B—134 people)
Valid-response rate:	96.4%

Age of Parent	Percent	Gender of Child, Group A	Percent
20s	0.8	Boys	42.1
30s	36.7	Girls	56.3
40s	55.6	NA	1.7
50s	4.0		
60s	2.6	Gender of Child, Group B	
NA	0.3	Boys	44.0
		Girls	56.0
Age of Child, Group A			
10	13.3	Occupation of Parent	
11	16.7	Homemaker	44.5
12	22.9	Employee, public official, corporate worker	10.8
13	21.0	Self-employed	3.4
14	22.7	Part-time	37.3
15	0.4	Other	1.6
NA	2.9	NA	2.4
Age of Child, Group B			
8	44.0		
9	56.0		

a. Mothers were asked to answer the parents' questionnaire except in exceptional cases. Hence, 92.4 percent of parent respondents are mothers.

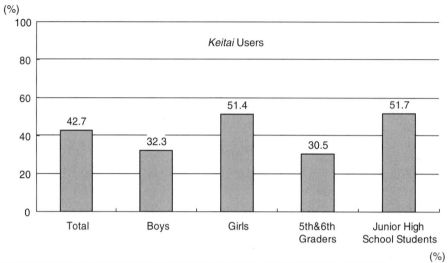

	Total	Boys	Girls	5th&6th Graders	Junior High School Students
n =	480	199	265	165	266
Own *keitai*	9.8	6.6	13.1	3	15.7
Occasionally borrow *keitai* of family members	29.8	23.2	34.4	26.2	31.8
Occasionally borrow *keitai* of friends for usage	3.1	2.5	3.9	1.2	4.2
Do not use *keitai* at all	55.6	67.7	48.6	69.5	48.3
NA	1.7	0	0	0	0

Figure 14.1
Children's rates of *keitai* use, by gender and grade level (children's responses).

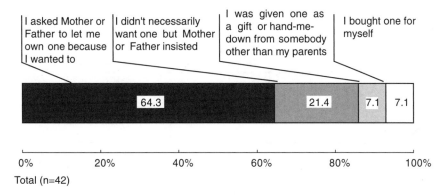

I asked Mother or Father to let me own one because I wanted to

I didn't necessarily want one but Mother or Father insisted

I was given one as a gift or hand-me-down from somebody other than my parents

I bought one for myself

Figure 14.2
Acquiring *keitai* (children's responses).

Children's Lifestyles and *Keitai* Use

What kinds of lifestyles do these these *keitai*-using children have? Children who commute to school by bus or train have a greater tendency to use *keitai* than those who use other means for commuting (figure 14.3). Private school students are more familiar with the use of *keitai* than public school students.

Children who attend cram classes for reviewing school work or preparing for school entrance exams or who take piano, choir, drawing, or any sort of artistic lessons have a greater tendency to use *keitai* than those who are not involved in activities outside of school. (The *keitai* use rate among children who attend cram classes is 46.7 percent versus 40.8 percent for children who do not. The *keitai* use rate for children who take artistic lessons is 47.4 percent versus 41.5 percent for those who do not.)

Children's Friendship Patterns and *Keitai* Use

Rates of *keitai* use among children vary depending on children's interest in socializing. As a whole, children who take the initiative to build relationships seem keener to use *keitai* than those who don't (figure 14.4). When asked whether they "want to make a lot of new friends," "belong to a particular clique of friends," "can make friends with anybody," or "occasionally argue with friends," a higher number of *keitai* users ascribed those characteristics to themselves compared to nonusers.

Nonowners' Desire to Possess *Keitai*

Among children who don't own *keitai*, 60 percent claimed desire for ownership of a *keitai* exclusively their own (figure 14.5). The need for *keitai* is felt more urgently by girls than boys, and by junior high school students than elementary school students. The need is exceptionally high among girls; over 70 percent claim the need for *keitai*

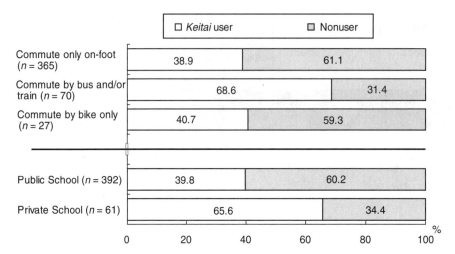

Figure 14.3
Children's rates of *keitai* use, by private versus public schools and means of transportation (children's responses).

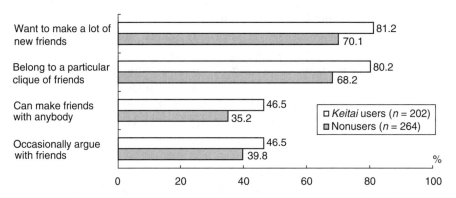

Figure 14.4
Relation between children's friendship patterns and *keitai* use (children's responses).

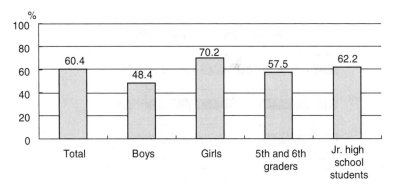

		Total	Boys	Girls	Elementary school students	Junior high school students
	n =	429	186	228	160	222
I always wish I had my own *keitai*		24	16.7	29.4	20	26.6
I sometimes wish I had my own *keitai*		36.4	31.7	40.8	37.5	35.6
I don't care whether I have my own *keitai* or not		24.7	30.6	19.7	26.3	22.1
I don't need my own *keitai*		14.9	21	10.1	16.3	15.8

Figure 14.5
Nonowners' desire for *keitai*, by gender and grade level (children's responses).

ownership. Although 14.9 percent of all respondents claimed "I don't need *keitai*," a significant difference was seen between the genders; merely 10.1 percent of girls contributed to this number compared with 21.0 percent of boys. Among nonowners, more children with siblings than without siblings expressed a desire for *keitai* (figure 14.6).

Children's Reasons for Desiring *Keitai* Ownership
Why do children desire *keitai* ownership? We asked children who claimed they needed one. The most common reason, shared by 41.5 percent of these respondents, was "it would be convenient for being able to make phone calls from anywhere" (figure 14.7). This was followed by 31.0 percent who claimed the desire for an "exclusive

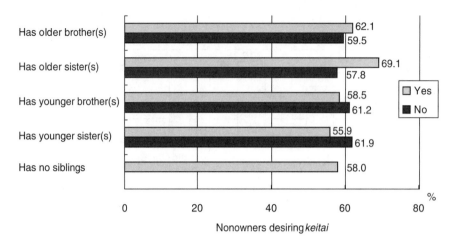

Figure 14.6
Nonowners' desire for *keitai*, by whether they have siblings (children's responses).

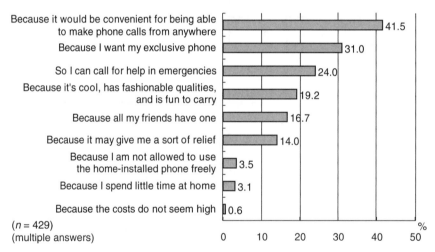

Figure 14.7
Children's reasons for desiring *keitai* (children's responses).

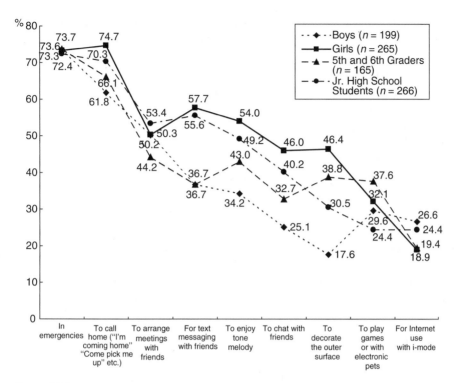

Figure 14.8
Children's intended use of *keitai*, by gender and grade level (children's responses).

phone," which surpassed "call for help in emergencies" (24.0 percent). For children, it may well be assumed that *keitai* has a higher value as an exclusive phone than as a mobile phone. It is also interesting that *keitai* is more highly evaluated as being "cool" or "fun" (19.2 percent) than as giving "a sort of relief" (14.0 percent), which indicates their low expectations of fulfilling security needs with *keitai*.

How Children Intend to Use *Keitai*

Figure 14.8 shows children's intended use of *keitai*, by gender and grade level. The top answer was "in emergencies," but girls' intended use for the purpose of calling home (74.7 percent) surpassed "in emergencies," the top answer of total respondents. In addition, girls have a strong urge to send text messages (57.7 percent versus 36.7 percent for boys) and to customize ring tone melodies (54.0 percent versus 43.0 percent for boys). Girls show more personalized interest toward *keitai* compared to boys, in the sense that they are more particular about chatting with friends and decorating the outer surface of the *keitai* handset.

Girls in general have a stronger urge for communication than boys do, and take especially strong initiative in text messaging. According to past research (Life Design Institute 1996) with fourth through sixth graders, 64.6 percent of girls compared to 1.7 percent of boys claimed that they "intend to keep good relationships with friends by exchanging letters or diaries." Girls have a tendency to keep in close touch with friends using various methods including text messaging, although they speak to each other every day. The widespread capacity for text messaging on *keitai* today developed from the fad for pagers among female high school students in the late 1980s and early 1990s (see chapter 2). As the versatility of *keitai* progressed, not only voice communication but also text messaging became highly valued. It has been observed that text messaging is already attracting the attention of girls in elementary school.

Children's Views on *Keitai* Use by Children
What kind of image does *keitai* have from the viewpoint of children? Survey results convey the impression that children are not lacking common sense about the rules of *keitai* use and are rather realistic in the way they view it (figure 14.9).

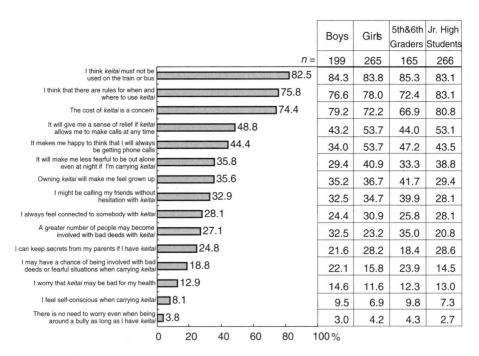

	Boys	Girls	5th&6th Graders	Jr. High Students
n =	199	265	165	266
I think *keitai* must not be used on the train or bus — 82.5	84.3	83.8	85.3	83.1
I think that there are rules for when and where to use *keitai* — 75.8	76.6	78.0	72.4	83.1
The cost of *keitai* is a concern — 74.4	79.2	72.2	66.9	80.8
It will give me a sense of relief if *keitai* allows me to make calls at any time — 48.8	43.2	53.7	44.0	53.1
It makes me happy to think that I will always be getting phone calls — 44.4	34.0	53.7	47.2	43.5
It will make me less fearful to be out alone even at night if I'm carrying *keitai* — 35.8	29.4	40.9	33.3	38.8
Owning *keitai* will make me feel grown up — 35.6	35.2	36.7	41.7	29.4
I might be calling my friends without hesitation with *keitai* — 32.9	32.5	34.7	39.9	28.1
I always feel connected to somebody with *keitai* — 28.1	24.4	30.9	25.8	28.1
A greater number of people may become involved with bad deeds with *keitai* — 27.1	32.5	23.2	35.0	20.8
I can keep secrets from my parents if I have *keitai* — 24.8	21.6	28.2	18.4	28.6
I may have a chance of being involved with bad deeds or fearful situations when carrying *keitai* — 18.8	22.1	15.8	23.9	14.5
I worry that *keitai* may be bad for my health — 12.9	14.6	11.6	12.3	13.0
I feel self-conscious when carrying *keitai* — 8.1	9.5	6.9	9.8	7.3
There is no need to worry even when being around a bully as long as I have *keitai* — 3.8	3.0	4.2	4.3	2.7

Figure 14.9
Children's views on *keitai*, by gender and grade level (children's responses).

We can see a difference between the two genders concerning *keitai* use. Boys have negative yet realistic concerns about the costs of *keitai* use or its leading to involvement with "bad deeds or fearful situations," whereas girls hold positive and feeling-based images, responding, "Being able to make phone calls whenever I want to gives me a sense of relief," "It makes me feel secure when going out alone at night," or "It makes me feel connected with somebody at all times." According to research I have conducted since 1996, gender differences in views toward *keitai* use by children are characteristic of young people just under and over the age of twenty as well.

Keitai Use among Mothers

In the course of the research, parents were also questioned about their own *keitai* use. A total of 40.7 percent of mothers claimed ownership of an exclusive *keitai* (figure 14.10). Moreover, 18.3 percent of them "use *keitai* of family members." This means that 59.0 percent of mothers are familiar with *keitai* use in one way or another. When observing the relation between the use of *keitai* among mothers and children, I found that children whose mothers used *keitai* tended to become *keitai* users themselves (figure 14.11). I would suggest that mothers using *keitai* will, in many cases, share use with a child and will have less resistance to the child's using *keitai*.

When observing the relation between what the parents considered a dangerous situation for children and the rate of use of *keitai* among children, I found that 52.2 percent of parents whose children are *keitai* users claimed concern for the child's safety on the way and from school compared with 35.6 percent of parents whose children are nonusers. Furthermore, parents of *keitai*-using children saw danger in more areas than did parents of nonusing children (table 14.2).

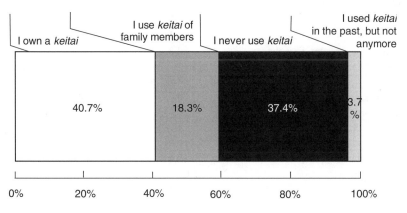

Figure 14.10
Mothers' rates of *keitai* use (parents' responses).

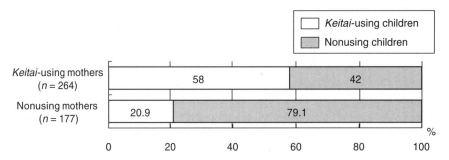

Figure 14.11
Relation between mothers' and their children's *keitai* use.

Table 14.2
Parents' Concerns for Their Children's Safety (Parents' Responses)

	Percent of Parents Citing Occasion/Location of Concern	
	Child Uses *Keitai* N = 205	Child Doesn't Use *Keitai* N = 267
Bad weather (rain, snow, typhoon)	59.5	53.2
On route to and from crammer class	53.7	55.4
On route to and from school	52.2	35.6
When out at night	32.2	35.6
Playtime after school	18.0	19.1
Weekends/holidays	10.2	5.2
While at school	2.4	3.4
Streets	70.7	65.9
Busy uptown streets	55.1	45.7
Amusement facilities (e.g., arcades, karaoke booths)	52.2	50.9
Parks	32.7	26.6
On transportation (e.g., trains, buses)	30.7	15.0
Elevators and stairs of apartment buildings	26.3	23.6
Shopping streets	21	15.7
Train stations	20.5	12.4
Large recreational facilities (e.g., amusement parks, zoos)	10.2	6.4
Friends' houses	4.4	6.4
School playground	3.9	1.1
School classrooms	3.4	1.9
Classrooms of crammer class and other activities outside school	1.5	0.4

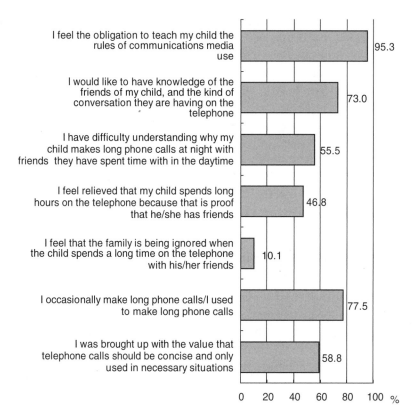

Figure 14.12
Parents' views on children's communications media use (parents' responses).

Parents' Views on Communications Media Use by Children

How do parents view children's use of communications devices? A total of 95.3 percent of parents "feel the obligation to teach my child the rules of communications media use" (figure 14.12).[2] And 73.0 percent of parents declared, "I would like to have knowledge of the friends of my child, and the kind of conversation they are having on the telephone."

On the other hand, 55.5 percent of parents claimed they "have difficulty in understanding why my children make long phone calls at night with friends whom they have spent time with in the daytime." Among parents with children aged 16–24, 70.3 percent of those over 50 felt that they "have difficulty" compared with 60.8 percent of those under 50. It may be surmised that younger parents have a better understanding of children's attitudes towards telephone use. However, younger parents were found to feel more obligated to educate the child about communications media use. In addition, these parents were more conscious of the kind of friends their children interact with over the telephone compared with the older parents. While understanding their children's views of *keitai* use, current parents of elementary and junior high school students often feel that telephone use among children should be under parental supervision.

When probing further into the views of telephony among parents, I found that 58.8 percent affirmed that they were "brought up with the value that telephone calls should be concise and only used in necessary situations," and 77.5 percent claimed that they "occasionally make/used to make long phone calls." In an early study I conducted in 1997, 52.9 percent of parents under the age of 50 with children 16–24 years old answered yes to the question, "Did you yearn to or enjoy making long phone calls when you were young?" compared with 32.3 percent of parents over the age of 50. Also, 65.8 percent of parents aged under 50 and 82.3 percent of parents aged over 50 claimed that they were brought up holding the value that telephone calls need be concise and used only in necessary situations." Thus we see that views and values toward telephone use vary among different generations.

Parents' Views on *Keitai* Use by Children

How do parents view *keitai* use by children? Including parents of children who do and do not use *keitai*, respondents expressed three top answers, related to rules, responsibility, and costs of service (figure 14.13). A positive reason advanced for children's *keitai* use is "the sense of relief that *keitai* brings. However, a majority of parents also claimed that *"keitai* use seems to accelerate children's conducting activities behind parents' backs" and expressed concerns about "electromagnetic waves." Although there has been some research on the effects of electromagnetic waves radiated from *keitai*, the need for further examination grows as *keitai* use proliferates among children in the midst of physical growth and development. On the other hand, fewer than half of the respondents answered that *keitai* causes failure of telephone use rules in the home or that *"keitai* eases control of children," demonstrating that most parents' views do not conform to what public discourse generally sees as the effects of *keitai* use among children.

Between parents whose children use or do not use *keitai*, the largest difference was in the numbers of those who claimed, "I am not worried even when my child is on his/her way home at night as long as he/she is carrying a *keitai*, since it enables contact at all times." This view was shared by 87.7 percent of parents whose children are *keitai* users and 69.8 percent of those whose children do not use *keitai*. Moreover, 45.3 percent of parents of *keitai* users, and 63.1 percent of parents of nonusers agreed with the statement that *"keitai* use seems to accelerate children's conducting activities behind parents' backs." Also, 44.3 percent of parents of *keitai* users compared to 30.4 percent of parents of nonusers agreed that *"Keitai* use eases control of children"; 37.6 percent of parents of *keitai* users and 29.4 percent of parents of nonusers said that "mo-

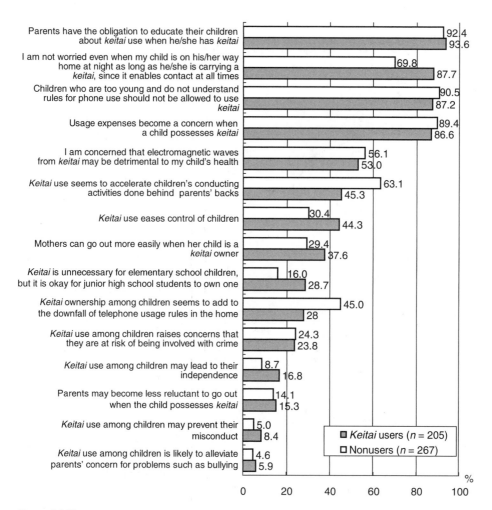

Figure 14.13
Parents' views on *keitai* use by children (parents' responses).

thers can go out more easily when the child owns a *keitai*"; and 28.0 percent of parents of *keitai* users and 45.0 percent of parents of nonusers claimed that "*keitai* ownership among children seems to add to the downfall of telephone use rules in the home."

Overall, parents of *keitai*-using children tend to have positive feelings toward the convenience of *keitai* and are less concerned about the ill effects it may have upon their children. On the other hand, parents whose children are nonusers of *keitai* have a negative image of the device, fearing that *keitai* aggravates in parent-child relationships and accelerates the downfall of traditional family rules.

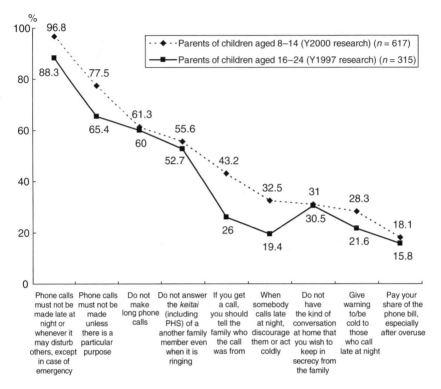

Figure 14.14
Family telephone use rules (parents' responses).

Family Telephone Use Rules

The current diversification of the telecommunications environment is contributing to the gradual breakdown of the conventional rules for telephone use in the home. This is raising questions such as, "Why aren't long phone calls allowed when the call-waiting service is available?" "Why aren't long phone calls permitted when the person on the other side of the line lives alone or uses *keitai*?" However, many parents still hold values in accordance with the conventional rules of telephone use, which conflict with the views held by their children. I polled families with children in elementary and junior high schools to examine this situation (figure 14.14). A total of 96.8 percent of the respondents agreed that "phones calls must not be made late at night or whenever it may disturb others except in case of emergency." Moreover, 77.5 percent of respondents agreed that "phone calls must not be made unless there is a particular purpose." Four fifths of children today say they "make long phone calls," but three fifths of their parents still observe conventional telephone rules in the home, stating that "long phone calls must not be made." This demonstrates that such rules still exist even among the generation of parents who currently have young children. According to research I conducted in 1997, parents of children in elementary and junior high school set rules similar to, or even stricter than, parents of children aged 16–24 for use of the telephone in the home. More than a few parents make active efforts to be aware of relationships their children have with friends, stating, "I make them report to the family of the child to whom they were talking on the phone."

Parents' Concerns about *Keitai* Use by Children

I collected significant information in the free-answer format. Here I present answers that stood out from the rest.

Keitai ownership does bring relief for parents because it enables them to be aware of their children's whereabouts and confirm their safety. On the other hand, it also brings about anxiety. For example, there is no way to confirm the truth when the parent asks, "Where are you right now?" and the child answers, "Cram school." There are cases in which the parent starts to doubt the child. Some parents even insist on having the child carry a PHS, which offers a location detection service in order to keep track of the child's whereabouts.

Many parents mention that their understanding of their children's surroundings had been easily grasped through calls on fixed-line home telephones. That has become more difficult with the use of *keitai*. There is also some fear that children may become more dependent upon parents' instructions via *keitai* and that as a result children could become incapable of coping with problems on their own. Some complain that when children interact directly with their friends via *keitai*, they miss the opportunity to speak with adults other than their own parents, such as their friends' parents. There was one case in which a child's *keitai* bill grew enormously, since the surrounding children became dependent upon it as if it were a public phone. In short, the adoption of *keitai* among children brings both relief and anxiety to parents.

The free answers portrayed the ways in which parents judge the relevance of *keitai* ownership for their children; these contradict each other, demonstrating different levels of mutual trust in the parent-child relationship:

I make my child own *keitai* because I trust him/her. It is okay to allow the child to own a *keitai* because he/she is able to use it with proper manners and common sense.

I will not allow my child to own a *keitai* because I trust him/her. The *keitai* is unnecessary because he/she tells me whenever he/she is going out and frequently makes contact with home of his/her whereabouts.

I make my child own a *keitai* because I can't trust him/her. The *keitai* is necessary because I need to check up on the activities of the child and his/her whereabouts.

I will not allow my child to own a *keitai* because I can't trust him/her. I cannot let my child own a *keitai* because he/she does not act according to the rules, and therefore I have no idea of how he/she may use it.

Though the views of *keitai* differ greatly between parents and children, the positive and negative effects caused by its adoption by children may be seen as a reflection of the state of mutual trust and the level of communication in the parent-child relationship, family rules, and discipline taught at home.

Schools' Views on *Keitai*

So far, our observations have been about the effects of *keitai* on parent-child relationships. However, children's use of *keitai* also raises the problem of its control at school. As mentioned, there are many cases in which children possess *keitai* for use in emergencies and to reassure their parents of their safety while going to and from school. However, while confirming its purpose as a tool for safety, elementary and junior high school administrators find *keitai* a cause for concern because it has caused various problems at school.

A written questionnaire was sent via postal mail (some interviews took place on the phone) to public and private elementary school and junior high school administrators in the Setagaya ward of Tokyo in January 2001.[3] The majority of the respondents were head teachers or counselors at the schools.

Adoption of *Keitai* among Students

Over half of the schools with valid answers reported that "*keitai* adoption among children is highly progressed" (over half of the students in a class are *keitai* owners) or "*keitai* adoption among children is somewhat progressed" (around a quarter of the students in a class are *keitai* owners). While most elementary school administrators said only that "*keitai* adoption among children is somewhat progressed," a large number of junior high school administrators said that "*keitai* adoption is highly progressed" among students in their school.

Rules and Punishment

The rules pertaining to *keitai* use by children vary among schools. Many schools have rules banning the carrying of *keitai*; when children are found in possession of *keitai* at these institutions, a report is sent to parents, and the handset is confiscated and then returned to the student after school hours. Other schools establish a system of applying for possession, basically giving tacit approval. Although not enough samples were examined to compare the difference between public schools and private schools, it may be surmised that the handling of *keitai* use varies among schools. For example, many parents whose children attend private schools insist that their children carry *keitai* as a tool for self-protection. These children often travel longer distances on public transportation to attend school.

School Administrators' Views on Children's *Keitai* Use

On the subject of *keitai* possession among children at school, many administrators responded, "Though it is not preferable, nothing can be done about it" or "It is not preferable." None of the administrators answered that possession among their students is "preferable." Though many schools do accept possession of the *keitai* among children, their concern about use among children remains.

Responses at the PTA

When a poll was taken among parent-teacher associations as to whether *keitai* possession among children caused problems, 60 percent of respondents said yes. Their responses were either "many parents disapprove of *keitai* possession among children" or "groups of parents who approve of and disapprove of *keitai* possession among children coexist," which indicates that the majority of parents do not demonstrate approval for their children's owning *keitai*.

Problems at Elementary and Junior High Schools

Approximately 40 percent of school administrators said yes when asked whether saw problems with students' possession of *keitai*. Particular problems included use, ringing, text messaging during class; and misplacement and robbery. Also mentioned in the free-answer section was that *keitai* use is related to harassment and that it aggravates troubles with neighboring schools because it enables students to organize "gangs" more easily. It also appears that the schools are having difficulty establishing rules about *keitai* use because parents in the PTA hold split opinions of whether possession is necessary. In the research conducted among parents and children, some parents declared, "Although the school bans its possession, I have my child carry *keitai* because I am worried." Furthermore, there was a case in which the parent filed a claim after confiscation of a child's *keitai*. There was even a case in which a parent's *keitai* rang during a graduation ceremony. While many schools maintain the view that parents should educate their children on manners for *keitai* use, others claimed that parents themselves often lack manners in its use.

Conclusion

The results of my research on children's, parents', and schools' views on *keitai* are summarized here.

There are two motives for the children's use of *keitai*: an active motive, in which the child desires possession for *keitai*'s convenience, use as a tool for self-defense, exclusivity, and fashionable quality; and a passive motive, where the child carries a *keitai* device because of parents' orders. Girls are taking the leading role in setting the trend, strongly desiring *keitai* ownership for leisure purposes. A distinctive feature of children's desire for *keitai* ownership is that their main purpose is privacy, not the convenience of mobility.

On the other hand, parents give permission for *keitai* use by children primarily to secure the child's safety. Safety concerns, costs of use, possible bad influences, and the relationship of mutual trust between parent and child are factors weighed by parents when making the decision of whether their child should own *keitai*. However, there is a large difference between groups of parents who take the initiative for the child's *keitai* ownership and those who merely permit ownership because of the child's desires. Parents experience both relief and anxiety about their children's possession of *keitai*. Parents' own *keitai* use has consequences; children whose parents own *keitai* have many opportunities for use, and the effect has been that children's use correlates highly with their parents'. Ultimately, parents' attitudes toward *keitai* play a large part in determining the child's level of *keitai* use.

School administrators have the strong tendency to judge *keitai* as unnecessary; however, they are having difficulty establishing or enforcing rules banning its use in school. Use during class, robbery, and bullying are among the problems raised by the schools. In addition, school administrators are having difficulty coping with the nascent social rules and manners of *keitai* use, leaving no solution for dealing with the split opinions among parents toward *keitai*. Though many schools have instructed children not to bring "unnecessary items" to school, there is a controversy over whether *keitai* are "unnecessary."

In general, strenuous effort is required to maintain the balance between children, parents, and schools when it comes to *keitai* communication. It is not rare for a child to be caught in a complex situation in which the parents and the school each hold different views on the possession of *keitai*. Though both parents and schools feel the need to establish rules or control of *keitai* use, there are often no grounds for doing so. In society today, where rules for *keitai* use are not fully established, no single being or organization can establish a single standard for *keitai* use.

Keitai adoption among children is still in the developing stage. However, after observing the strong need and its expected convenience among children, it may well be assumed that the prevalence of phones for kids will continue to grow in the future. The basis for establishing rules about *keitai* for children in the home and in society is still vague today. As long as *keitai* serves its purpose as a means of communication, it will definitely be closely involved with building and maintaining relationships among children. Even when parents ban its possession, children are apt to have opportunities to use the *keitai* of friends. It is not uncommon for parents to give permission for possession when children insist, "Everybody owns one."

Establishing relationships between family members or parents, children, and the children's friends through the use *keitai* is difficult. Balancing these concerns with the rules of schools is critical as well. There is a risk of lingering on the trivial debate of whether *keitai* use among children brings a sense of relief or involves danger. Instead, we must direct our attention toward studying *keitai* use among children inside the home and examining the setting of rules between parents and children and in the schools. This will be key to obtaining "a positive effect" from the use of *keitai* by children.

Notes

1. "Elementary school and junior high school students" as used in this research refers strictly to students of the upper grades of elementary school (fifth and sixth graders) and junior high school students because *keitai* use is growing among such age groups and for the convenience of the research. However, the research conducted on the parents surveyed in this research included some with children in third and fourth grades as well.

2. The percentages in figure 13.12 indicate the total of "yes" and "I rather agree" answers of parents with children aged 16–24 (research conducted in 1997), $N = 617$. The present questionnaire is basically in correspondence with the 1997 research. However, the statement, "I occasionally make long phone calls/I used to make long phone calls" is a revision of the statement, "I enjoyed making long phone calls/I wished to make long phone calls when I was young" of the 1997 research.

3. The ward was chosen because it has the highest population and number of schools among all regions of the Tokyo prefecture. *List of Registered Schools of Tokyo* (2000 edition) was used for reference.

Fumitoshi Kato, Daisuke Okabe, Mizuko Ito, and Ryuhei Uemoto

Now that *keitai* are well established in our social landscape, demand for new handsets is driven by users seeking new features and added functions. As the size of a basic *keitai* terminal shrinks, built-in camera and GPS (global positioning system) functions are becoming popular and accepted add-ons. There is always a *keitai* within hand's reach, and information necessary in everyday life is now being stored in *keitai* memory or is accessible over the *keitai* Internet. As Okada has described chapter 2, *keitai* are increasingly incorporating nonvoice functions and are now used most frequently in Japan for e-mail rather than voice calls. These e-mail exchanges are coming to include photos and moving images. In particular, the introduction of a terminal with the *Sha-Mail* (photo mail) function by J-Phone (now Vodafone) in November 2000 changed *keitai* communication, making it possible to send photographs with *keitai*.

This chapter describes how *keitai* camera services started in Japan and how they are being used in everyday life. As such, it is a report on emergent developments and research in progress rather than a completed study. We introduce the present status of the *keitai* camera business and technology and describe use patterns based on some preliminary studies. Then we discuss emergent social and cultural issues surrounding the use of *keitai* cameras: a new photography practice, efforts at regulation, and a social dynamic of peer-to-peer visibility.

Keitai Cameras

Adoption of *Keitai* Cameras

Chapters 1 and 2 described how *keitai* have been widely adopted by users of all ages in Japan. While the number of new *keitai* subscribers is leveling out, sales of *keitai* with cameras are steadily growing and expanding the source of income for each carrier. J-Phone was the first to introduce *keitai* cameras, followed by KDDI and NTT DoCoMo. On January 14, 2003, NTT DoCoMo reached the five million mark in number of i-shot terminals sold, and the next day J-Phone announced it had broken the eight million mark in number of *Sha-Mail* terminals sold. As of the end of December 2002, KDDI had sold 2.47 million *keitai* camera terminals. Today it is estimated that over a quarter of the population of Japan has a *keitai* camera (see chapter 2). Now when people buy *keitai*, the characteristics of the built-in camera often provide the criteria for product selection.

A report from the Multimedia Research Institute (2003) indicates that the *keitai* market is reaching saturation, with the number of subscriptions topping 75 million. Consequently, there is little expectation of attracting new subscribers, and the *keitai* terminal market is beginning to show dependence on the demand created by users' replacing terminals. The *keitai* camera is acknowledged to be one attractive feature to help expand the market. In fact, the percentage of *keitai* with cameras as a portion of all domestic *keitai* shipments has increased dramatically, from approximately 7 percent in 2001 to 60 percent in 2002. According to estimates by the Telecommunication Carriers Association, as of May 2004 *keitai* cameras constitute 60.7 percent of all *keitai* (the breakdown by carrier is 54.7 percent for NTT DoCoMo, 72.6 percent for KDDI's au, and 79.6 percent for Vodafone). The trend is toward *keitai* cameras' becoming a standard feature of mobile phones in Japan.

The phrase, "Is a camera really necessary in a *keitai*?" appeared in an October 2002 advertisement for a new TU-KA terminal. Its intent was to emphasize the thinness and lightness of the model, which did not have a built-in camera. But its implicit message was that *keitai* with no camera is noteworthy—an indication that *keitai* with a camera now represents the default category.

Keitai as a Digital Camera

In the past few years, *keitai* camera development has advanced dramatically in a number of areas, particularly in the resolution of images. High-end terminals can reach the million-pixel level, and they might include features such as autofocus, optical zoom, and removable memory cards; indeed, they are coming to rival stand-alone digital cameras. Now having reached a certain level of functionality, development efforts are turning to optimizing and improving design, usability, and integration with other functions.

At the same time, most users see cameras in *keitai* as "nice to have, but not necessary." Targeting this base of users, manufacturers are producing numerous low-cost entry-level terminals with a resolution of 100,000–300,000 pixels. The market is likely to continue to polarize into segments for high-end models and for entry-level models (table 15.1).

The average resolution of standard *keitai* cameras is equivalent to that of early digital cameras (approximately 300,000 pixels), but the essential difference is the built-in transmission function. In other words, the *keitai* camera can be perceived as a digital camera with a transmission function as a standard feature. It is easy to take photographs with a *keitai* camera, and it is also easy to send the photographic image data to another *keitai* or to a PC. In research by Info-Plant (2002), on reasons why people want

Table 15.1

Keitai Camera Features, June 2004

	High-end Model	Entry-level Model
Camera module	1 million–2 million pixels (3.2 million pixels maximum)	100,000–300,000 pixels
	Two camera modules (interior, exterior)	One camera module
Camera functionality	Autofocus	Digital zoom
	Digital zoom (20X maximum)	Light
	Optical zoom (currently only one model, 2X)	
	Light or flash	
	Movie	
Terminal functionality	Larger body	Smaller body
	Landscape orientation	Portrait orientation
	Memory card slot	Built-in memory only
	Additional extras: rotating lens, lens cover, rotating and reversible main display	
	USB PC connectivity	
	Infrared data communication (bluetooth)	
Transmission standards	3G or 2.5G	2G

to buy a *keitai* camera, 70 percent said "because it is convenient to take photos with it." Approximately 50 percent gave reasons such as "it seems fun" and "want to attach photos to e-mail," which illustrates the growth of an awareness of *keitai* as a camera.

Although media appliances have become increasingly lighter and smaller in recent years, it is still more convenient to carry a *keitai* camera than to carry both a *keitai* and a digital camera. We can expect to see increasingly better picture quality in these mobile phone digital cameras, and we can also expect to see the *keitai* camera make visual communication using photographs commonplace. New functions like moving-image recording and GPS are being added to *keitai*, but taking still photographs with *keitai* seems to have already captured the imagination of many users.

Current Use Patterns

Taking Photos

Social and cultural studies of camera phone use in Japan (and elsewhere) are still relatively sparse, but work is beginning to emerge that describes practices of picture taking and sharing that differ both from the uses of the stand-alone camera and the kinds of

Figure 15.1
Taking a photograph with a *keitai* camera.

Figure 15.2
Keitai camera photos of scenes from everyday life.

social sharing that happened via *keitai* voice and text communication. We draw from existing survey research as well as ethnographic research in progress by Okabe and Ito.[1]

Info-Plant conducted a survey on what kind of photographs people take with their *keitai* cameras (2002). Those surveyed were allowed to give multiple answers; almost half of the men and the women said "my own face" or "friends' faces." Other popular answers included "scenery," "my child(ren)," and "pets." In their free answers, women noted a greater variety of photographed subjects (e.g., "food"). In a survey of three hundred Internet users (men and women between the ages of 10 and 50), japan.internet.com (June 2, 2003), found that 65 percent reported using a *keitai* camera. In response to the question, "In what kinds of settings and for what purposes to you actually use the camera function?" the most common answer (75 percent) was "recording and commemorating moments with family, friends, acquaintances," followed by "recording and commemorating interesting or unusual things in everyday life" (69 percent) and "travel photos, such as of scenery" (39 percent).

Although many of the categories cited, such as travel photos and photos of friends, family, and pets, are shared with the traditional camera, the high number of responses in the category of "interesting or unusual things in everyday life" points to a new mode of more pervasive photo taking when a camera is always at hand. Okabe and Ito (2003) found that *keitai* camera users are taking photos of serendipitous sightings and moments in everyday life (rather than of special planned events that have traditionally been documented by amateur photography): a scene from an escalator in a familiar train station, a delicious-looking cake at a café, or a friend who just fell into a puddle. With the *keitai* camera, the mundane is elevated to a photographic object.

One important capability of the *keitai* camera is being able to view the photographed image on the spot. In that sense, the photo can be experienced immediately, an advantage digital cameras share with *puri-kura* (instant sticker photo booths) and instant cameras. Users can check the images, and if they didn't turn out well, they can retake the photo right away. In interviews with users, Okabe and Ito (2003) found that camera phone users often described the act of taking and viewing photos when gathered with friends as itself a focus of social activity.

Sharing and Transmitting Photos

The *keitai* camera represents not only an ever-present image-capture device but also an ever-present image-sharing and transmission device. Just as the pervasive presence of a camera invites users to take photos more integrated with the stream of everyday life, the ability to share photos from an ever-present *keitai* screen, or to send photos via *keitai*, means that image sharing is also much more commonplace. While processes of sharing photos that rely on *keitai* cameras are still in their early development among users, current research documents that the devices can support a higher rate of photo sharing than was previously the norm with analog print photography.

In a multiple-choice survey by japan.internet.com (July 16, 2003), when asked how they used photos taken by *keitai* camera, almost 90 percent of respondents answered that they "view them on their *keitai* handset," followed by 60 percent who "use them as *keitai* wallpaper," over 50 percent who "e-mail them to friends and family," and 35 percent who "upload them to a PC." Okabe and Ito's (2003) ethnographic research shows a similar pattern, with most users reporting that they rarely send photos to another person's *keitai* or to a PC but often show people photos on the *keitai* handset screen. Photos are generally only transmitted when the content is genuinely newsworthy or timely (as in a celebrity sighting), or if the photo is being sent as an intimate gesture (as in a new haircut shot to a boyfriend or when a father sends a photo of a child to a mother away on business). Our pilot studies of photo sharing among young couples and peer groups show that couples and close groups of female friends have the highest volume of image transmission; among male friends and less intimate relations, images are rarely transmitted, although they may be shared from the handset when people are physically co-present. In other words, visual sharing is most prevalent among "full-time intimate communities" (chapter 6; Nakajima, Himeno, and Yoshii 1999).

For photos that are e-mailed, users typically send an image taken with the *keitai* camera to another *keitai* by attaching it to an e-mail. As the name implies, J-Phone's *Sha-Mail* was designed specifically for photo transmission between *keitai*. At the time of *Sha-Mail*'s introduction, there were only a few models of *keitai* cameras, and it was not possible to send photos between different operators. Even so, people began accepting the *keitai* camera as an easy-to-carry digital camera. Taking photos with a camera was already a trend among the youth (with small and fashionable instant cameras like Xiao and Cheki), which could have been a factor behind the smooth acceptance of the *keitai* camera (see chapters 2 and 4).

Transmission of photos from *keitai* to PC or to a Web site is a less common but emergent practice. Since most current *keitai* models can exchange e-mail over the Internet, it is possible to send images taken with a *keitai* camera to a PC by attaching the images to regular e-mail. At least initially, operators designed *keitai* camera handsets for the exchange of photos between *keitai*, but there is growing demand for linkages with PCs, particularly as image quality improves. When one archives images taken with *keitai*, the data must be copied onto a separate medium because there is a limit to the capacity in the *keitai* handset itself. It is also troublesome to send the photos one by one. In light of such needs, some models have slots for a memory card to store image data, and there are also models that connect directly to a PC with a cable, or more recently, via infrared or Bluetooth. With the growth of photo journal, moblog, and image archiving sites, applications, and services, users are increasingly encountering a variety of options for saving, sharing, and sending photos.

Emergent Social and Cultural Dynamics

Small-Scale Photography

Susan Sontag (1977) has described how photographs have produced a new practice of mediated vision, or "looking with photographs." As cameras became a popular commodity, "looking with photographs" became more common in our everyday lives. The experience of seeing people and objects framed with a small camera and viewed on a *keitai* monitor may result in a new way of seeing things. In our everyday lives the convenience of taking, storing, and viewing photos on a *keitai* handset is part of the recording of the social and cultural context in which we find ourselves or the recalling of the situation in which the recording took place.

In many cases, the size of a photograph taken with a *keitai* camera is small. This size (approximately 120×120 pixels) itself can be perceived as a new photographic trait. Small photographs are commonly known as index prints (as in contact sheets) and thumbnails. Thumbnails, which are useful in getting an overview of a group of photographs, are based on the assumption that there are original photographs in a larger size. In other words, photographs in the form of thumbnails are the previews of the originals (including the size and quality).

By contrast, photographs taken with and stored in a *keitai* camera are not previews; typically they are compact to begin with. These small and light photographs are transmitted quickly and easily from the handset. When we take a photograph with a *keitai* camera, we reacknowledge the importance of focus in terms of having to decide what to fit inside the small rectangular frame. We also make an immediate judgment as to whether to save, send, or delete the photograph. Moreover, when we look with keitai, our vision captures the scene inside the rectangular frame, but at the same time our attention is drawn to the surroundings (outside the frame) more than it is with the conventional photograph. These photos stimulate the imagination by conjuring absence cited by the image in the frame. Further, they are reflections of individual viewpoints and subjectivity that provide insight on the attention and interests of the person who took them. These inherent characteristics of photographs are accentuated by the lightweight, low-profile, tiny photographs taken by *keitai*.

This new type of photography enabled by *keitai* cameras was exemplified by a *keitai* photo that won an award in the "News" category of the 24th Yomiuri Photograph Grand Prix (2002). It was a photo taken by a businessman of an exploding ATM in a business district in Osaka. Yoshihiro Tachiki, a well-known professional photographer, commented that the "resolution is low, but the tension of the incident is clearly conveyed. The penetration of the *keitai* camera has made a photojournalist of everyone in the country. News photographs will break the traditional barriers and will become more interesting than ever." Although the performance of the *keitai* camera is still limited, its dynamism, being able to take and immediately transmit an image, as well as its ubiquity are key features that challenge prior photographic frameworks. The photographs taken by the *keitai* camera may come to record a frame (or a fragment) in our daily lives while obscuring the boundary between amateur and professional.

Among early exhibitions devoted to camera phone photography were SENT: America's First Phonecam Art Show, sponsored by Motorola in Los Angeles in July 2004, "the first to place camera phones in the hands of professional artists, to investigate the creative potential of mobile imaging devices,"[2] and the Mobile Phone Photo Show at the Rx Gallery in San Francisco in May/June 2004, in which "anyone anywhere in the world with a camera phone is invited to participate."[3]

Regulation

Matsuda (chapter 1) describes the evolving public discourses, concerns, and moral panics that have characterized both the early and the more mature years of *keitai* adoption. The newly ubiquitous image-capture functions of *keitai* cause concern today similar to the concerns surrounding *keitai* manners common in the mid-nineties. Then, *keitai* was perceived as violating the boundaries of public and private space, particularly in public transportation. Just as voice calls were gradually regulated out of existence on Japanese public transportation (see chapter 10), we are seeing efforts to regulate and discipline uses of *keitai* cameras. As *keitai* cameras have become ubiquitous, the media have widely taken up the issue of using them to spy and take unauthorized stealth photographs. While *keitai* cameras have been celebrated as a device for grassroots and peer-to-peer journalism (Hall 2002; Rheingold 2003), they are also viewed with suspicion as devices for "up-skirt" and locker room photography or tools for industrial espionage.

One of the first efforts at regulating *keitai* camera use in Japan came from bookstore owners. In the summer of 2003 booksellers popularized the term *digitaru manbiki* (digital shoplifting) to describe the practice of taking photos of pages from books and magazines in stores. For example, *Asahi Shinbun* (2003) wrote,

Publishers and bookstores are facing problems with customers browsing their offerings who take snaps of a portion of a magazine with their *keitai* cameras. This practice, called "digital shoplifting," has emerged with the improved resolution of these cameras. The Japan Magazine Publishers Association has started a campaign to appeal to customers' manners. According to the association, informational magazines are the most commonly photographed, particularly movie times, restaurant information, and recipes. Currently, there are no laws to regulate this behavior, and it is unclear whether personal use of these photographs would be considered copyright infringement. However, since these customers don't buy the magazines they have photographed, bookstores and publishers have continued to raise voices of concern that "this goes beyond the bounds of appropriate browsing."

In terms of functionality, the cameras in *keitai* are no different from stand-alone digital cameras, but because the photography is integrated with a communications technology that is now accepted as pervasive, it raises a new set of issues. While cameras have routinely been prohibited at certain performances and exhibits, and it would be suspicious to pull out a camera in a locker room, *keitai* have not traditionally been prohibited in these settings. Although *keitai* cameras are all designed with a shutter sound that cannot be turned off, users can mute the sound with a finger or a piece of clothing. Just as efforts to regulate *keitai* voice calls intensified as the technology became more prevalent in the late 1990s, we are currently seeing intensifying efforts to regulate *keitai* cameras in settings such as corporate facilities, locker rooms, bookstores, and special events.

Toward a Shared Visibility Society?

Even as the media report on sensational transgressions like "up-skirt" photography, such *keitai* camera crimes themselves might be photographed in turn by somebody else's *keitai* camera. The distinction between the photographer and the photographic subject has become increasingly ambiguous given a social environment where a large proportion of the population is equipped with image-capture devices ready at hand and ready to transmit. In an environment constructed by digital media and network technology, monitoring is simple. We are no longer surprised to see small surveillance/security cameras set up in convenience stores, banks, and elevators. The camera's eye is dispersed throughout society, and we are starting to develop self-identity as its subjects (Kato 1998).

This monitoring function is often referred to in discussions of the "surveillance society," where control and power relationships are depicted elements of our present society (e.g., Saito 2002; Gandy 1993; Lyon 1994). This is a scenario in which our behavior patterns, buying behavior, and even interpersonal relationships are inevitably compiled as data, where our personal information is associated with our actions as constantly monitored by cameras dispersed throughout society. The N System of the highway system and the surveillance camera network in Kabuki-cho of Shinjuku are some examples in Japan. Media reporting of incidents of voyeurism and stealth photography are often placed in this frame, where our privacy and personal integrity is violated. The camera's eye, which has previously been set up inside buildings or in public places (physically fixed to a particular location), is now built into *keitai* and functions at a more intimate range. *Keitai* with a camera's eye are mobile and becoming ubiquitous in our daily lives.

The *keitai* camera differs from institutionalized surveillance systems in that it is a peer-to-peer rather than a top-down form of monitoring. Even without the *keitai* camera, there are strangers' eyes everywhere in society, and communication behavior is developed by our awareness of mutually seeing and being seen. The mutual observation of one another in itself is nothing new. The problem lies in the possibility that we may be watched without our knowledge (being seen without seeing) and the possibility that these images can now be easily distributed over worldwide networks. The acceptance and the growing ubiquity of the *keitai* camera may provide an opportunity to reflect upon these scenes in everyday life and to redefine the separation of public and private affairs, the different kinds of authority and responsibility possible and desirable in a society where peer-to-peer visibility may be assumed.

Notes

1. Their camera phone study is an extension of the diary-based study of *keitai* use described in chapter 13. In addition to the diary recording of voice and text exchanges, users are asked to keep records of photos taken, received, and shared with *keitai*.

2. ⟨http://sentonline.com/pr.html⟩.

3. ⟨http://rxgallery.com/mpps/index.html⟩.

References

Agar, Jon. 2003. *Constant Touch: A Global History of the Mobile Phone*. Cambridge: Icon Books.

Aizawa, Masao. 2000. "Keitai" ni Hitokoto (One Word for "*Keitai*"). *Nihongo-Gaku* 19 (October): 79–80.

Akiyama, Toyoko. 1998. Seikatsu Mokuhyo to Ikikata (Goals in Life and Lifestyle). In *Gendai Nihonjin no Ishiki Kozo* (*Structures of Values Orientation of the Modern Japanese*), 4th ed., 13–34. Tokyo: Nihon Hoso Shuppan Kyokai.

Akiyoshi, Mito. 2004. Unmediating Community: The Non-Diffusion of the Internet in Japan. Doctoral diss., Department of Sociology, University of Chicago.

Amano, Hiroko, Setsu Ito, and Masumi Mori. 1994. *Seikatsu Jikan to Seikatsu Bunka* (*Time Use and Culture of Daily Life*). Tokyo: Kouseikan.

Appadurai, Arjun. 1996. *Modernity at Large: Cultural Dimensions of Globalization*. Minneapolis: University of Minnesota Press.

Aronson, Sidney H. 1971. The Sociology of the Telephone. *International Journal of Comparative Sociology* 12: 153–167.

Asahi Shinbun. 1995. Keitai Denwa, Benri na Hanmen Hata Meiwaku ... Tsuukin Denshanai deno Shiyou ni Kujou Zou (*Keitai*, Useful but Bothersome to Others ... Complaints about Use in Commuter Trains on the Rise). November 29.

———. 2000. i-mode Has Made Its Appearance in the Workplace—Easy to Use and Carry Around. September 16, evening ed.

———. 2003. *Keitai* Camera Page Capture, Industry Protests, Restaurant Information and Recipes from Magazines. July 5.

Barnes, Stuart J., and Sid L. Huff. 2003. Rising Sun: iMode and the Wireless Internet. *Communications of the ACM* 46 (11): 79–84.

Baudrillard, Jean. 1986. *Amérique. America*, trans. Chris Turner. New York: Norton, 1989.

Beck, Ulrich, Anthony Giddens, and Scott Lash. 1994. *Reflexive Modernization: Politics, Tradition and Aesthetics in the Modern Social Order.* Cambridge: Polity Press.

Berger, Peter, and Thomas Luckmann. 1966. *The Social Construction of Reality: A Treatise in the Sociology of Knowledge.* New York: Bantam Dell.

Bernard, H. Russell, Peter D. Killworth, Eugene C. Johnsen, Gene A. Shelley, Christopher McCarty, and Scott Robinson. 1990. Comparing Four Different Methods for Measuring Personal Networks. *Social Networks* 12: 179–215.

Bessatsu Takarajima. 1995. Kekkon no Okite (A Rule of Marriage). Takarajima-sha.

Bijker, Wiebe E. 1992. The Social Construction of Fluorescent Lighting, or How an Artifact Was Invented in Its Diffusion Stage. In *Shaping Technology/Building Society: Studies in Sociotechnical Change,* ed. W. Bijker and J. Law, 75–104. Cambridge, Mass.: MIT Press.

Bijker, Wiebe E., Thomas P. Hughes, and Trevor F. Pinch, eds. 1993. *The Social Construction of Technological Systems: New Directions in the Sociology and History of Technology.* Cambridge, Mass.: MIT Press.

Bijker, Wiebe E., and John Law, eds. 1992. *Shaping Technology, Building Society: Studies in Sociotechnical Change.* Cambridge, Mass.: MIT Press.

Black, Jeremy. 1998. *Maps and Politics.* Chicago: University of Chicago Press.

Bødker, Susanne, and Kaj Grønbæk. 1995. Users and Designers in Mutual Activity: An Analysis of Cooperative Activities in Systems Design. In *Cognition and Communication at Work,* ed. Y. Engeström and D. Middleton, 130–158. Cambridge: Cambridge University Press.

Brown, Barry, Nicola Green, and Richard Harper, eds. 2002. *Wireless World: Social and Interactional Aspects of the Mobile Age.* London: Springer-Verlag.

Brown, Barry, and Kenton O'Hara. 2003. Place as a Practical Concern of Mobile Workers. *Environment and Planning A.* 35 (9): 1565–1587.

Button, Graham, and Richard Harper. 1993. Taking the Organization into Account. In *Technology in Working Order: Studies of Work, Interaction, and Technology,* ed. G. Button, 98–107. New York: Routledge.

Cabinet Office. Japan. 2002a. *White Paper on the National Lifestyle 2001.* Gyosei.

———. 2002b. *Jouhouka Shakai to Seishounen: Dai 4-Kai Jyohou Shakai to Seishounen ni Kansuru Chousa Hokokusho (Youth and the Information Society: Report on a Survey of Fourth-Stage Information Society and Youth).* Director-General for Policy Planning.

———. 2003. *Gekkan Yoron Chosa* (Monthly Public Opinion Survey). Minister's Secretariat of Public Relations. February.

Callon, Michel. 1986. Some Elements of a Sociology of Translation. In *Power, Action, and Belief: A New Sociology of Knowledge,* ed. J. Law, 196–233. New York: Routledge.

———. 1987. Society in the Making: The Study of Technology as a Tool for Sociological Analysis. In *The Social Construction of Technological System*, ed. W. Bijker, T. Hughes, and T. Pinch, 83–103. Cambridge, Mass.: MIT Press.

Callon, Michel, and John Law. 1997. After the Individual in Society: Lessons on Collectivity from Science, Technology, and Society. *Canadian Journal of Society* 22: 165–182.

Castells, Manuel. 2000. *The Rise of the Network Society*, 4th ed. Oxford: Blackwell.

Castells, Manuel, Imma Tubella, Teresa Sancho, Maria Isabel Diaz de Isla, and Barry Wellman. 2003. *The Network Society in Catalonia: An Empirical Analysis*. Barcelona: Universitat Oberta de Catalunya.

Chae, Minhee, and Jinwoo Kim. 2003. What's So Different about the Mobile Internet? *Communications of the ACM* 46 (12): 240–247.

Chen, Wenhong, and Barry Wellman. 2004. *Charting and Bridging Digital Divides: Comparing Socioeconomic, Gender, Life Stage, Ethnic, and Rural-Urban Internet Access in Eight Countries*. Report to the AMD Global Consumer Advisory Board. October.

———. 2004. Charting Digital Divides within and between Countries. In *Transforming Enterprise*, ed. W. Dutton, B. Kahin, R. O'Callaghan, and A. Wyckoff. Cambridge, Mass.: MIT Press.

Chen, Wenhong, Jeffrey Boase, and Barry Wellman. 2002. The Global Villagers: Comparing Internet Users and Uses Around the World. In *The Internet in Everyday Life*, ed. B. Wellman and C. Haythornthwaite, 74–113. Oxford: Blackwell.

Cherny, Lynn, and Elizabeth Weise, eds. 1996. *Wired Women*. Seattle: Seal Press.

Chomsky, Noam. 1975. *Reflections on Language: The Whidden Lectures*. New York: Random House.

———. 1986. *Knowledge of Language: Its Nature, Origin, and Use*. Westport, Conn.: Praeger.

Churchill, Elizabeth F., and Nina Wakeford. 2002. Framing Mobile Collaborations and Mobile Technologies. In *Wireless World: Social and Interactional Aspects of the Mobile Age*, ed. B. Brown, N. Green, and R. Harper, 154–179. London: Springer-Verlag.

Clarke, Adele, and Joan Fujimura. 1992. *The Right Tools for the Job: At Work in Twentieth Century Life Sciences*. Princeton, N.J.: Princeton University Press.

Clifford, James. 1997. *Routes: Travel and Translation in the Late Twentieth Century*. Cambridge, Mass.: Harvard University Press.

Cockburn, Cynthia, and Ruza Furst-Dilic. 1994. *Bringing Technology Home: Gender and Technology in a Changing Europe*. Berkshire, U.K.: Open University Press.

Cohen, Stanley. 1972. *Folk Devils and Moral Panics*. London: MacGibbon and Kee.

———. 1972. *A Law Enforcement Guide to United States Supreme Court Decisions*. Springfield, Ill.: Charles C. Thomas.

Crook, Stephen. 1998. Minotaurs and Other Monsters: "Everyday Life" in Recent Social Theory. *Sociology: The Journal of the British Sociological Association* 32 (3): 523–540.

de Gournay, Chantal. 2002. Pretense of Intimacy in France. In *Perpetual Contact: Mobile Communication, Public Performance*, ed. J. E. Katz and M. Aakhus, 193–205. Cambridge: Cambridge University Press.

Deal, Terrence, and Allan Kennedy. 1982. *Corporate Cultures: The Rites and Rituals of Corporate Life.* Reading, Mass.: Addison-Wesley.

Dentsu Soken. 2000. *Keitai de Mietekita Nihon Gata Jyoho Kakumei (A Japanese Style Information Revolution Visible through* Keitai). Dentsu Soken Report 5.

Dixon, Nancy. 2000. *Common Knowledge: How Companies Thrive by Sharing What They Know.* Boston: Harvard Business School Press.

Dourish, Paul. 2001. *Where the Action Is: The Foundations of Embodied Interaction.* Cambridge, Mass.: MIT Press.

du Gay, Paul, Stuart Hall, Linda Janes, Hugh Mackay, and Keith Negus. 1997. *Doing Cultural Studies: The Story of the Sony Walkman.* Berkshire, U.K.: Open University Press.

Duranti, Alessandro, and Charles Goodwin. 1992. *Rethinking Context: Language as Interactive Phenomenon.* Cambridge: Cambridge University Press.

Engeström, Yrjo, and David Middleton. 1996. *Cognition and Communication at Work.* Cambridge: Cambridge University Press.

Erikson, Erik H. 1968. *Identity: Youth and Crisis.* New York: Norton.

Feather, Frank. 2000. *Future Consumer.com: The Webolution of Shopping to 2010.* Toronto: Warwick Publishing.

Feld, Scott L. 1982. Social Structural Determinants of Similarity Among Associates. *American Sociological Review* 47: 797–801.

Feyerabend, Paul. 1975. *Against Method.* New York: Norton.

Fischer, Claude S. 1976. *The Urban Experience.* New York: Harcourt Brace.

———. 1982. *To Dwell Among Friends: Personal Networks in Town and City.* Chicago: University of Chicago Press.

———. 1992. *America Calling: A Social History of the Telephone to 1940.* Berkeley: University of California Press.

Fortunati, Leopoldina. 2002. Italy: Stereotypes, True and False. In *Perpetual Contact: Mobile Communication, Private Talk, Public Performance*, ed. J. E. Katz and M. Aakhus, 42–62. Cambridge: Cambridge University Press.

Foucault, Michel. 1975. *Surveiller et Punir: Naissance de la Prison.* Paris: Gallimard.

Fujimoto, Kenichi. 1985. Seisei Gengo Riron no Tetsugaku-teki Kiso (Philosophical Implications of Transformational Generative Linguistic Theories: Focused on Chomsky and Quine). Master's thesis, Osaka University.

———. 1997. *Poke-beru Shojo Kakumei* (*Girls' Pager Revolution: Introduction to Media Folklore*). Osaka: Etre.

———. 1999a. Mobile no Bunka-Shakaigaku (Cultural Sociology of Mobiles: Media Metamorphosis of Home, Car, Object, and Human Body over 300 Years). *Fashion-Kankyo* 8 (3).

———. 1999b. Tokumei, Zatsuon, Kisei (Anonym, Noise and Parasite: X and Y Icono-Logics on Pocket-Bell Friends). In *Gendai-Huzoku 1998–99* (Modernology Annual 1998–99). Kyoto: The Society of Modernology.

———. 2000. Syntony, Distony, Virtual Sisterhood, and Multiplyng Anonymous Personalities: Invisible Pseudo-Kinship Structure through Mobile Media Terminal's Literacy. In *Information and Communication*, ed. T. Umesao, W. Kelly, and M. Kubo. Senri Ethnological Studies 52. Osaka: National Museum of Ethnology.

———. 2002a. Will Mobile Media Terminals Replace Traditional Pleasures in the Information Society? In *The Senses, Pleasure, and Health*, ed. D. Warburton, E. Sweeney, and ARISE (Associates for Research into the Science of Enjoyment). Reading: University of Reading.

———. 2002b. Ousei Dakusei: Kyaah to Dami wo meguru Keitai Kukan/Bungaku-Ron (Yellow Voices and Brown Voices: Mobile Space, Mobile Literature Focused on Feminine Shrill Voices and Masculine Raspy Thick Voices). In *Datsu-Bungaku to Cho-Bungaku* (*De-Literature and Super-Literature*), ed. M. Saitoh. Tokyo: Iwanami Shoten.

———. 2002c. Keitai no Ryuko-Gaku (Fashion-Discourse Theory of Mobile Phone). In *Keitai-Gaku Nyumon* (*Understanding Mobile Media*), ed. T. Okada and M. Matsuda. Tokyo: Yuhikaku.

———. 2002d. Media to Kansei Jyouhou (Understanding Media and Shikohin). *Sen-I Gakkai-Shi* 58 (12).

Fujimoto, Kenichi, ed. 2003a. *Territory Machine: Gendai-Huzoku 2003* (*Territory Machine: Modernology Annual 2003*). Kyoto: Society of Modernology.

———. 2003b. Shin-Shitstu ni Uzumaku "Kawaii" Keitai Kukan (My Own "Pretty" Space Whirled in the Bedroom). In *Nemuri-Gi no Bunka-Shi* (Ethnography of Sleep Clothes and Sleep Environments), ed. S. Yoshida and Research Institute of Sleep Study (RISS). Tokyo: Toseisya Publishing.

———. 2004. Keimo no Bensyouho karamita Keitai no Yaban to Bigaku (Violent Barbarism and Esthetics of *Keitai* from Viewpoint of Dialectic of Enlightenment). *Joho Bigaku Kenkyu*.

Fujimura, Joan. 1996. *Crafting Science: A Sociohistory of the Quest for the Genetics of Cancer*. Cambridge, Mass.: Harvard University Press.

Fujita, Hidenori. 1991. Gakkou-ka Jyoho-ka to Ningen Keisei Kukan no Henyo (The Irony and Contradictions of the Highly Schooled and High Information Society as Educational Environment). *Gendai Shakaigaku Kenkyu* 4: 1–33.

Fukutomi, Tadakazu. 2003. *Hitto Shouhin no Butaiura (The Back Stage of Hit Products)*. Tokyo: Ascii.

Funk, Jeffrey L. 2001. *The Mobile Internet*. Pembroke, Bermuda: ISI Publications.

Gadamer, Hans-Georg. 1960. *Wahrheit und Methode. Truth and Method*, ed. J. C. Weinsheimer and D. G. Marshall. New York: Continuum International, 1989.

Gandy, Oscar H. 1993. *The Panoptic Sort: A Political Economy of Personal Information*. Boulder, Colo.: Westview Press.

Geertz, Clifford. 1983. *Local Knowledge: Further Essays in Interpretive Anthropology*. New York: Basic Books.

General Affairs Bureau Youth Relations Headquarters. 2000a. *Report on Values Held by the Youth*.

———. 2000b. *Research Report on Youths and the Mobile Phone*.

General Institute of Information Communications. 2001. *Information Communications Handbook 2001*.

Geser, Hans. 2004. Towards a Sociological Theory of the Mobile Phone. University of Zurich. ⟨http://socio.ch.mobile/t_geserl.htm⟩.

Giddens, Anthony. 1990. *The Consequences of Modernity*. Stanford, Calif.: Stanford University Press.

———. 1991. *Modernity and Self-Identify: Self and Society in the Late Modern Age*. Cambridge: Polity Press.

———. 1992. *The Transformation of Intimacy: Sexuality, Love and Eroticism in Modern Societies*. Cambridge: Polity Press.

Goffman, Erving. 1959. *The Presentation of Self in Everyday Life*. New York: Doubleday.

———. 1963. *Behavior in Public Places: Notes on the Social Organization of Gatherings*. New York: Free Press.

———. 1967. *Interaction Ritual: Essays in Face-to-Face Behavior*. New York: Doubleday Anchor Books.

Goodwin, Charles, and Naoki Ueno. 2000. Vision and Inscription in Practice. *Mind, Culture and Activity* 7 (1–2): 1–3.

Gottlieb, Manette, and Mark McLelland, eds. 2003. *Japanese Cybercultures*. New York: Routledge.

Greer, Scott. 1962. *The Emerging City: Myth and Reality*. New York: Free Press.

Grinter, Rebecca E., and Marge A. Eldridge. 2001. y do tngrs luv 2 txt msg? Paper presented at the Seventh European Conference on Computer-Supported Cooperative Work, Bonn, Germany.

Grudin, Jonathan. 1990. The Computer Reaches Out. *Proceedings of the ACM Conference on Human Factors in Computing Systems. CHI '90*, 261–296.

Gumpert, Gary. 1987. *Talking Tombstones and Other Tales of the Media Age*. Oxford: Oxford University Press.

Gupta, Akhil, and James Ferguson. 1992. Space, Identity, and the Politics of Difference. *Cultural Anthropology* 7: 6–23.

Habuchi, Ichiyo. 2002a. Keitai ni Utsuru Watashi (The Me Reflected in the *Keitai*). In *Keitai-Gaku Nyumon* (*Understanding Mobile Media*), ed. T. Okada and M. Matsuda, 101–121. Tokyo: Yuhikaku.

———. 2002b. Media Kankyo no Henyo to Haji (Transformation and Shame of the Media Environment). *Kyoiku to Igaku* 50 (8): 44–50.

———. 2003. Keitai Denwa Riyo to Network no Doshitsusei (Uses of the *Keitai* and Network Homogeneity). *Jinbun Shakai Ronso*: 73–83. Faculty of Humanities, Hirosaki University, Aomori, Japan.

Hall, Edward T. 1966. *The Hidden Dimension*. Garden City, N.Y.: Doubleday.

Hall, Justin. 2002. Mobile Reporting: Peer-to-Peer News. *The Feature.* ⟨http://www.thefeature.com/article?articleid=14274⟩.

Hampton, Keith. 2001. Living the Wired Life in the Wired Suburb: Netville, Globalization and Civic Society. Doctoral diss., Department of Sociology, University of Toronto.

Hampton, Keith, and Barry Wellman. 2002. The Not So Global Village of Netville. In *The Internet in Everyday Life*, ed. B. Wellman and C. Haythornthwaite, 345–371. Cambridge: Blackwell.

———. 2003. Neighboring in Netville: How the Internet Supports Community and Social Capital in a Wired Suburb. *City and Community* 2 (3): 277–311.

Hara, Kiyoharu, and Kazuo Takahashi. 2003. Keitai Denwa wo Riyo Shita Souhoukou Jyugyo no Arikata ni Kansuru Jisshoteki Kenkyu (Empirical Study on Interactive Lesson Using Cellular Phone). *Kansai Kyoiku Gakkai Kiyo* 27: 96–100.

Haraway, Donna. 1991. *Simians, Cyborgs, and Women: The Reinvention of Nature*. New York: Routledge.

Harper, Richard. 2003. Are Mobiles Good or Bad for Society? In *Mobile Democracy: Essays on Society, Self and Politics*, ed. K. Nyiri, 185–214. Vienna: Passagen Verlag.

Hashimoto, Yoshiaki. 1998. Personal Media to Communication Kodo (Personal Media and Communication Behavior). In *Media Communication Ron*, ed. Ikuo Takeuchi, Kazuto Kojima, and Yoshiaki Hashimoto, 117–138. Tokyo: Hokujyu Syuppan.

———. 2002. The Spread of Cellular Phones and Their Influence on Young People in Japan. *Review of Media, Information and Society* 7: 97–110.

Hashimoto, Yoshiaki, Akiko Komatsu, Masaaki Kurihara, Koji Hanme, and Kasharp Anurag. 2001. Shutoken Jyakunen Sono Communication Kodo: Internet, Keitai Mail Riyo wo Chushin ni (Communication Behavior of the Youth in the Metropolitan Area: Focusing on the Use of the Internet, Mobile Phone and Short Message System). *Tokyo Daigaku Shakai Joho Kenkyusho Chosa Kenkyu Kiyo* 16: 94–210.

Hashimoto, Yoshiaki, Ron Korenaga, Kenichi Ishii, Daisuke Tsuji, Isao Nakamura, and Yasutoshi Mori. 2000. Keitai Denwa wo Chushin to suru Tsushin Media Riyo ni Kansuru Chosa Kenkyu (Survey Research on Uses of Cellular Phone and Other Communication Media in 1999). *Tokyo Daigaku Shakai Joho Kenkyusho Chosa Kenkyu Kiyo* 14: 83–192.

Haythornthwaite, Caroline, and Barry Wellman. 1998. Work, Friendship, and Media Use for Information Exchange in a Networked Organization. *Journal of the American Society for Information Science* 49 (12): 1101–1114.

————. 2002. The Internet in Everyday Life: An Introduction. In *The Internet in Everyday Life*, ed. B. Wellman and C. Haythornthwaite, 3–44. Oxford: Blackwell.

Hegel, Georg. 1808. *Phaenomenologie des Geistes* (*The Phenomenology of Mind*), tr. J. B. Bailey. Mineola, N.Y.: Dover, 2003.

Hine, Christine. 2000. *Virtual Ethnography*. London: Sage.

Hirano, Hideaki, and Osamu Nakano. 1975. *Copy Taiken no Bunka: Kodoku na Gunsyu no Kouei* (*A Culture of Copy Experience: Descendents of a Lonely Crowd*). Tokyo: Jiji Tsuushin Sha.

Hjorth, Larissa. 2003. Cute@keitai.com. In *Japanese Cybercultures*, ed. N. Gottlieb and M. McLelland, 50–59. New York: Routledge.

Hogan, Bernie. 2003. Media Multiplexity: An Examination of Differential Communication Usage. Paper presented at the Association of Internet Researchers Conference, Toronto.

Holden, Todd Joseph Miles, and Takako Tsuruki. 2003. *Deai-kei*: Japan's New Culture of Encounter. In *Japanese Cybercultures*, ed. N. Gottlieb and M. McLelland, 34–49. New York: Routledge.

Horkheimer, Max, and Theodor Adorno. 1947. *Dialektik der Aufkraerung. Dialectic of Enlightenment: Philosophical Fragments*, ed. Gunzelin Schmid Noerr, trans. Edmund Jephcott. Stanford, Calif.: Stanford University Press, 2002.

Hoshino, Hiromi. 2001. Keitai wo Kinshi Sureba Sumu Mondai Ka: Kireru Wakamono Yori Kowai Mono (Will Prohibiting Mobile Phones Solve the Problem? Something More Frightening Than Short-Tempered Youth). *Chuo Kouron* 116 (20): 210–219.

Hosokawa, Shuhei. 1981. *Walkman no Shujigaku* (*The Rhetoric of the Walkman*). Tokyo: Asahi Shuppan.

Howard, Philip N., Lee Rainie, and Steve Jones. 2002. Days and Nights on the Internet: The Impact of a Diffusing Technology. In *The Internet in Everyday Life*, ed. B. Wellman and C. Haythornthwaite, 45–73. Oxford: Blackwell.

Hughes, Thomas P. 1979. The Electrification of America: The System Builders. *Technology and Culture* 20 (1): 124–162.

Hutchins, Edwin. 1988. The Technology of Team Navigation. In *Intellectual Teamwork: Social and Technical Bases for Cooperative Work*, ed. J. Galegher, R. E. Kraut, C. Egido, 191–220. Hillsdale, N.J.: Erlbaum.

Ikeda, Ken'ichi. 2002. Patterns of Mobile Phone Use: A Social Psychological Viewpoint. In *The Information Society and Youth*, 287–301. Report of the Fourth National Survey on the Information Society and Youth. Cabinet Office, Japan.

Ikeda, Ken'ichi, Tetsuro Kobayashi, and Kakuko Miyata. 2003. The Social Implications of Internet Use in Japan: Collective Use of the Internet Can Be a Lubricant of Democracy. Paper presented at the Association of Internet Researchers Conference, Toronto.

Ilahiane, Hsain. 2004. Mobile Phones, Globalization, and Productivity in Morocco. Paper presented at the Intel Workshop on Information Technology, Globalization, and the Future. Hillsborough, Oregon.

Illich, Ivan. 1981. *Shadow Work*. London: Boyars.

Info-Plant. ⟨http://www.info-plant.com/⟩.

Inoue, Shoichi. 1989. *Nostalgic Idol: Ninomiya Kinjiro*. Tokyo: Shinjuku Publishing.

IPSe. 2003. *Third Annual Consumer Report: Survey Results from Research on Mobile Phone Use*. Tokyo: IPSe Communications.

Ishiguro, Itaru. 2002. Snowball Sampling—ho ni yoru Daikibo Chosa to Sono Yukosei ni tsuite-02: Hirosaki Chosa Data wo Mochiita Ippanteki Shinrai-gainen no Kento (On Large-Scale Research by Snowball Sampling Method and Its Effectiveness; Examination of the Common Concept of Trust by Using Hirosaki Research Data). *Jinbun Shakai Ronso*: 85–98. Faculty of Humanities, Hirosaki University, Aomori, Japan.

Ishii, Hisao. 2003. Keitai Denwa de Musubareta Seishounen no Ningen Kankei no Tokushitsu (The Features of Adolescent Personal Relationships Formed through Cell Phones). *Kodomo Shakai Kenkyu* 9: 42–59.

Ishii, Kenichi. 2004. Internet Use via Mobile Phone in Japan. *Telecommunications Policy* 28 (1): 43–58. ⟨http://www.sciencedirect.com/science/journal/03085961⟩.

Ishikawa, Masato. 2000. Media ga Motarasu Kankyo Henyo ni Kansuru Ishiki Chosa: Densha Nai no Keitai Denwa Shiyo wo Rei nishite (An Attitude Survey on Environmental Changes Caused by the Media: Focusing on the Use of Mobile Phones in the Train). *Jyoho Bunka Gakkai-shi* 7 (1): 11–20.

Ishikawa, Minoru. 1988. Toshi no Tokumeisei to Shiteki Kukan: Toshi Shakai-teki Sogosayo-ron no Ichizuke (Anonymity in Cities and Private Space—Positioning of the Theory of Urban Social Interaction). *Toshimondai Kenkyu* 40 (2).

Ishita, Saeko. 1998. *Yumeisei to iu Bunka Souchi* (*The Cultural Apparatus of Fame*). Tokyo: Keiso Shobo.

Ito, Masayuki. 2002. Net Renai no Spirituality (Spirituality of Internet Romance). In *Spirituality wo Ikiru: Atarashii Kizuna wo Motomete* (*Living in Spirituality*), ed. N. Kashio. Tokyo: Serika Shobo.

Ito, Mizuko. Forthcoming. Mobile Phones, Japanese Youth, and the Replacement of Social Contact. In *Mobile Communication and the Re-Negotiation of the Public Sphere*, ed. R. Ling and P. Pedersen. London: Springer-Verlag.

Ito, Mizuko, Vicki L. O'Day, Annette Adler, Charlotte Linde, and Elizabeth D. Mynatt. 2001. Making a Place for Seniors on the Net: SeniorNet, Senior Identity, and the Digital Divide. *Computers and Society* 31 (3): 15–21.

Ivy, Marilyn. 1995. *Discourses of the Vanishing*. Chicago: University of Chicago Press.

Iwabuchi, Koichi. 2003. *Recentering Globalization: Popular Culture and Japanese Transnationalism*. Durham, N.C.: Duke University Press.

Iwagami, Mami. 2003. *Lifecourse to Gender de Yomu Shakai* (*The Sociology of the Family: Gender and Lifecourse Perspectives*). Tokyo: Yuhikaku.

Iwata, Kou. 2001. Keitai Denwa no Riyou to Yujin Kankei: Keitai Sedai no Communication (Use of Mobile Phone and Friendship of Senior High School Students: The Characteristics of Communications among the Generation Which Came to Use Mobile Telephone). *Mono-Gurafu Koukousei* (*Report of Survey on Senior High School Students*) 63: 12–33. Tokyo: Benesse Educational Research Center.

JASRAC (Japanese Society for Rights of Authors, Composers and Publishers). 2003. Royalty Collection for 2002. ⟨http://www.jasrac.or.jp/release/03/05_2.html⟩.

Johou Tsushin Handbook 2004 (*Information and Communication in Japan 2004*). Tokyo: Tsushin Jyoho Kenkyujo.

Jones, Stephen G., ed. 1995. *Cybersociety: Computer-Mediated Communication and Community*. Thousand Oaks, Calif.: Sage.

———. 1998. *Cybersociety 2.0: Revisiting Computer-Mediated Communication and Community*. Thousand Oaks, Calif.: Sage.

Kadono, Yukihiro, Kenichi Fujimoto, Shinya Hashizume, and Michio Ito, eds. 1994. *Osaka no Hyougen-Ryoku* (*The Vernacular Style of Presentations in Osaka*). Tokyo: Parco Publishing.

Kageyama, Yuri. 2003. NTT Tests Superfast Mobile Phone. *Associated Press*, December 7.

Kamimura, Shuichi, and Mieko Ida. 2002. Will the Internet Take the Place of Television? From a Public Opinion Survey on "The Media in Daily Life." In *Broadcasting Culture & Research* 19. NHK Broadcasting Culture Research Institute. ⟨http://www.nhk.or.jp/bunken/book-en/b4-e.html⟩.

Kant, Immanuel. 1781. *Kritik der Reinen Vernunft. Critique of Pure Reason*. Cambridge: Cambridge University Press, 1999.

Kasesniemi, Eija-Liisa. 2003. *Mobile Messages: Young People and a New Communication Culture*. Tampere, Finland: Tampere University Press.

Kasesniemi, Eija-Liisa, and Pirjo Rautiainen. 2002. Mobile Culture of Children and Teenagers in Finland. In *Perpetual Contact: Mobile Communication, Private Talk, Public Performance*, ed. J. E. Katz and M. Aakhus, 170–192. Cambridge: Cambridge University Press.

Kashima, Shigeru. 2000. *Sailor-Fuku to Eiffel-Toh* (*Sailor Uniforms for Schoolgirls and Eiffel Tower*). Tokyo: Bungei-Shunjyu Publishing.

Kashimura, Aiko. 2002. Daitaiteki Seikatsu Sekaiteki Communication no Tenkai (Development of Alternative Lifeworld Communication). In *Tsunagarino Naka no Iyashi* (*Healing in the Connection*), ed. S. Tanabe and S. Shimazono, 211–249. Tokyo: Senshu Daigaku Shuppan Kyoku.

Kato, Fumitoshi. 1998. Denshi Network no Naka no Shisen (Eyes Inside the Electronic Network). In *Media ga Kawaru, Chi ga Kawaru: Network Kankyo to Chi no Collaboration* (*Media Changes, Knowledge Changes: Collaboration of Network Environment and Knowledge*), ed. T. Inoue and M. Umegaki, 121–141. Tokyo: Yuhikaku.

Kato, Haruhiro. 2001. *Media Bunka no Shakai-gaku* (*Sociology in Media Culture*). Tokyo: Fukumura Shuppan.

Kato, Hidetoshi. 1958. Aru Kazoku no Communication Seikatsu (Communication Life of One Family). *Shiso*, 92–108. Tokyo: Iwanami-shoten.

———. 2002. *Kurashi no Sesoshi* (*Changes in Lifestyle*). Tokyo: Chuko Shinsho.

Katz, James E., ed. 2003. *Machines That Become Us: The Social Context of Personal Communication Technology*. New Brunswick, N.J.: Transaction Publishers.

Katz, James E., and Mark Aakhus, eds. 2002. *Perpetual Contact: Mobile Communication, Private Talk, Public Performance*. Cambridge: Cambridge University Press.

Kawatoko, Yasuko. 2003. Machines as a Social System. *Journal of the Center for Information Studies* 5: 20–24.

Kawaura, Yasuyuki. 1992. Keitai-Jidosya Denwa to Communication Kukan (Mobile-Car Phone and Communication Space). *Yokohama Ichiritsu Daigaku Ronso* 43 (2–3): 307–331.

———. 2002. Koritsu no Fuan, Kodoku no Fuan: Keitai Bunka-ron (Concern about Isolation, Concern about Loneliness: The *Keitai* Culture Theory). *Gendai no Esupuri* 421: 167–178.

Kikasete.net. 2002. Survey on the "Keitai Camera": Trend-Watch No. 62. ⟨http://www.kikasete. net/⟩.

Kim, Shin Dong. 2002. Korea: Personal Meanings. In *Perpetual Contact: Mobile Communication, Private Talk, Public Performance*, ed. J. E. Katz and M. Aakhus, 63–79. Cambridge: Cambridge University Press.

Kimura, Tadamasa. 2001a. *Digital Divide toha Nanika* (*What Is the Digital Divide?*). Tokyo: Iwanami Shoten.

———. 2001b. Internet to i-mode kei Keitai Denwa no Hazama (The Gap between Internet and i-mode Type *Keitai*). *Nihongo-Gaku*, September, 54–71.

Kimura, Yuichi. 1994. *Arashi no Yoruni* (*One Stormy Night*), tr. Lucy North. Tokyo: Kodansha. 2003.

Kinsella, Sharon. 1995. Cuties in Japan. In *Women, Media, and Consumption in Japan*, ed. L. Skov and B. Moeran, 220–254. Honolulu: University of Hawaii Press.

Kioka, Yasumasa. 2003. "Dating Sites and the Japanese Experience." In *Children, Mobile Phones and the Internet: Proceedings of the Experts' Meeting*, Tokyo. ⟨http://www.iajapan.org/hotline/2003mobile-en.html⟩.

Kitada, Akihiro. 2002. *Koukoku Toshi Tokyo: Sono Tanjo to Shi* (*Advertisement Metropolitan Tokyo: Birth and History*). Tokyo: Kosaido Shuppan.

Kobayashi, Kouichi. 1995. Multichannel–jidai kara Multimedia Jidai e: Hoso e no Chosen to Hoho kara no Chosen (From Multichannel to Multimedia, Challenge Toward and from Broadcasting). *Hosogaku Kenkyu* 45: 7–65.

Kobayashi, Senju. 2001. *Hayakawari Jisedai Keitaidenwa IMT-2000* (*Easy Description of IMT-2000*). Tokyo: Kousyobou.

Kohiyama, Kenji. 1996. *Chikyu Shisutemu toshiteno Maluchimedia* (*Multimedia as a Global System*). Tokyo: NTT Shuppan.

Kohiyama, Kenji, and Satoshi Kurihara. 2000. *Shakai Kiban to shiteno Joho Tsushin* (*Information and Telecommunications Systems as Social Foundation*). Tokyo: Kyoritsu Shuppan.

Koku, Emmanuel, Nancy Nazer, and Barry Wellman. 2001. Netting Scholars: Online and Offline. *American Behavioral Scientist* 44 (10): 1750–1772.

Koku, Emmanuel, and Barry Wellman. 2004. Scholarly Networks as Learning Communities: The Case of Technet. In *Designing for Virtual Communities in the Service of Learning*, ed. S. Barab, R. Kling, and J. Gray, 299–337. Cambridge: Cambridge University Press.

Kopomaa, Timo. 2000. *The City in Your Pocket: The Birth of the Mobile Information Society*. Helsinki: Gaudeamus.

Kotamraju, Nalini, and Nina Wakeford. 2002. "It's Just Easier to Text, Really": Young People and the New Communication in the United Kingdon and the United States. Paper presented at the Annual Meeting of the Society for the Social Studies of Science, Milwaukee.

Kotani, Satoshi, ed. 1993. *Wakamono Ron wo Yomu* (*Reading Youth Theory*). Tokyo: Sekai Shiso Sha.

Kotera, Atsushi. 2002. Koukousei Daigakusei no Keitai Denwa Riyo ni kansuru Chosa Kekka to Bunseki (Result and Analysis of a Survey on Mobile Phone Use by High School Students and University Students). *Kokusai Bunka Gaku* 6: 119–142.

Krogh, George von, Kazuo Ichijo, and Ikujiro Nonaka. 2000. *Enabling Knowledge Creation: How to Unlock the Mystery of Tacit Knowledge and Release the Power of Innovation*. Oxford: Oxford University Press.

Kuhn, Thomas. 1962. *The Structure of Scientific Revolutions*. Chicago: University of Chicago Press.

———. 1977. *Essential Tension*. Chicago: University of Chicago Press.

Kurihara, Masaki. 2003. Wakamono no Taijin Kankei ni okeru Keitai Mail no Yakuwari (Role of E-mail through Cellular Phone in Young People's Interpersonal Relationships). *Johotsushin-Gakkaishi* 21 (1): 87–94.

Kurita, Nobuyoshi. 1999. Iconic Communication through "Print-Club" Photo Stickers. *Mass Communication Kenkyu* (55): 131–152.

Larimer, Tim. 2000. Rage for the Machine. *Time Asia*, May 1. ⟨http://www.time.com/time/asia/magazine/2000/0501/cover1.html⟩.

Latour, Bruno. 1987. *Science in Action*. Cambridge, Mass.: Harvard University Press.

———. 1993. *We Have Never Been Modern*. Cambridge, Mass.: Harvard University Press.

Laurier, Eric. 2002. The Region as a Sociotechnical Accomplishment of Mobile Workers. In *Wireless World: Social and Interactional Aspects of the Mobile Age*, ed. B. Brown, N. Green, and R. Harper, 46–60. London: Springer-Verlag.

Lave, Jean. 1988. *Cognition in Practice*. Cambridge: Cambridge University Press.

Licoppe, Christian, and Jean-Philippe Heurtin. 2002. France: Preserving the Image. In *Perpetual Contact: Mobile Communication, Private Talk, Public Performance*, ed. J. E. Katz and M. Aakhus, 94–109. Cambridge: Cambridge University Press.

Lie, Merete, and Knut H. Sørensen. 1996. *Making Technology Our Own? Domesticating Technology into Everyday Life*. Oslo: Scandinavian University Press.

Life Design Institute. 1996. *Research Report of Friendships among Children*. Tokyo: Life Design Institute.

———. 1998. *The Penetration of Personal Communications Media and the Values among Parents and Children*. Tokyo: Life Design Institute.

———. 1999. *Relationship Between the Use of Communications Media and Friendships among Youths*. LDI Report, July. Tokyo: Life Design Institute.

Lin, Nan. 2001. *Social Capital: A Theory of Social Structure and Action*. Cambridge: Cambridge University Press.

Ling, Richard. 1998. "One Can Talk about Common Manners!": The Use of Mobile Phones in Inappropriate Situations. *Telektronik* 94: 65–76.

———. 2001. *Adolescent Girls and Young Adult Men: Two Subcultures of the Mobile Telephone*. R&D Report 34. Telenor, Oslo.

———. 2002. The Social Juxtaposition of Mobile Telephone Conversations and Public Spaces. Paper presented at the Conference on the Social Consequences of Mobile Telephones, Chunchon, Korea.

———. 2004. *The Mobile Connection: The Cell Phone's Impact on Society*. San Mateo, Calif.: Morgan Kaufmann.

Ling, Richard, and Birgitte Yttri. 2002. Hyper-coordination via Mobile Phones in Norway. In *Perpetual Contact: Mobile Communication, Private Talk, Public Performance*, ed. J. E. Katz and M. Aakhus, 139–169. Cambridge: Cambridge University Press.

Livingstone, Sonia. 1992. The Meaning of Domestic Technologies: A Personal Construct Analysis of Familial Gender Relations. In *Consuming Technologies: Media and Information in Domestic Spaces*, ed. R. Silverstone and E. Hirsch, 113–130. New York: Routledge.

Lyon, David. 1994. *The Electronic Eye: The Rise of Surveillance Society*. Cambridge: Polity Press.

Maeda, Hiroo. 2001. Gakkou Seikatsu to Keitai Denwa ha Kyozon Dekirunoka (Can School Life and the *Keitai* Co-Exist?). *Bousei*, 44–49. Tokyo: Tokai Daigaku Shuppan-kai.

Mannheim, Karl. 1929. *Ideologie und Utopie. Ideology and Utopia: An Introduction to the Sociology of Knowledge*. New York: Harcourt, 1985.

Marcus, George. 1995. Introduction. In *Technoscientific Imaginaries: Conversations, Profiles, and Memoirs*, ed. G. Marcus, 1–9. Chicago: University of Chicago Press.

Masataka, Nobuo. 2003. *Keitai wo Motta Saru (Monkey with a Mobile Phone)*. Tokyo: Chukou Shinsho.

Massey, Doreen. 1994. *Space, Place, and Gender*. Minneapolis: University of Minnesota Press.

Matsuba, Hitoshi. 2002. *Keitai no Naka no Yokubou (Desire in the Keitai)*. Tokyo: Bungeishunju.

Matsuda, Misa. 1996a. Keitai Denwa Riyo no Case Study (A Case Study on Mobile Telephone Use). *Tokyo Daigaku Shakai Joho Kenkyusho Chosa Kenkyu Kiyo* 7: 167–189.

———. 1996b. Fukyu Shoki ni okeru Media no Uwasa (The Rumor of the Media on the Early Stage in Diffusion). *Tokyo Daigaku Shakai Joho Kenkyusho Kiyo* 52: 25–46.

———. 1996c. Gender no Kanten kara no Media Kenkyu Saiko: Gender to Media no Shakai-teki Kosei ni Shoten wo Atenagara (Reviewing Media Research from Gender Perspectives: Focusing on the Social Framework of Gender and Media). *Mass Communication Kenkyu* 48: 190–203.

———. 1997. Toshi Densetsu: "Keitai no Denjiha ga Abunai" (Urban Legend: "Microwaves of *Keitai* are Dangerous"). In *Poke-beru Keitai Shugi! (Pager and Keitai Manifesto!)*, H. Tomita, K. Fujimoto, T. Okada, M. Matsuda, and N. Takahiro, 142–164. Tokyo: Just System.

———. 1999a. Personalization. In *Shakai Joho-gaku II: Media (Social Information Studies II: Media)*, 157–175. University of Tokyo Institute of Socio-Information and Communication Studies. Tokyo: University of Tokyo Press.

———. 1999b. Wakamono Keitai Denwa Bunka Ron (Discussion about Young People's Mobile Phone). *Communication* 81: 22–25.

———. 2000a. Wakamono-no Yujin-kankei to Keitai Denwa Riyo: Kankei Kihakuka-ron kara Sentaku-teki Kankei-ron e (Friendship of Young People and Their Use of Mobile Phones: From the View of "Superficial Relation" to "Selective Relation"). *Shakai Johogaku Kenkyu* 4: 111–122.

———. 2000b. Denwa to Gender (Telephone and Gender). In *Jyoho-Tsushin to Shakaishinri* (*Social Psychology of Information Communication*), ed. O. Hiroi and M. Funatsu, 71–93. Tokyo: Hokujyu Syuppan.

———. 2001a. Personal Phone, Mobile Phone, Private Phone. *Gendai no Esupuri: Keitai Denwa to Shakai Seikatsu* 405 (Spring): 126–138.

———. 2001b. Daigakusei no Keitai Denwa Denshi-mail Riyou Jyokyo 2001 (The Use of Mobile Phones and E-mail among University Students in 2001). *Jyoho-Kenkyu* 26: 167–179.

———. 2002. Mobile Shakai no Yukue (Future of Mobile Society). In *Keitai-Gaku Nyumon* (*Understanding Mobile Media*), ed. T. Okada and M. Matsuda, 205–227. Tokyo: Yuhikaku.

———. 2003. Mobile Communication Bunka no Seiritsu (Birth of Mobile Communication Culture). In *Series Shakai Joho Gaku eno Sekkin 2: Denshi Media Bunka no Shinso* (*Approaching Social Information Studies 2: The Depth of Electronic Media Culture*), ed. M. Ito, K. Kobayashi, and T. Masamura, 173–194. Tokyo: Waseda Daigaku Shuppan-kai.

Matsuda, Misa, Hidenori Tomita, Kenichi Fujimoto, Ichiyo Habuchi, and Tomoyuki Okada. 1998. Idotai Media no Fukyo to Henyo (Diffusion and Transformation of Mobile Media). *Tokyo Daigaku Shakai Joho Kenkyusho Kiyo* 56: 89–104.

Matsui, Takeshi. 1998. Shohin no Shakaiteki Teigi no Rekishiteki Tenkai—1 (Historical Development of Social Definitions of a Commodity, Part 1: A Case Study of the Common Images of Pager in Japanese Society). *Shohin Kenkyu* 48 (3–4): 25–37.

———. 1999a. Shohin no Shakaiteki Teigi no Rekishiteki Tenkai—2 (Historical Development of Social Definitions of a Commodity, Part 2: A Case Study of the Common Images of Pager in Japanese Society). *Shohin Kenkyu* 49 (1–2): 27–36.

———. 1999b. Shohin no Shakaiteki Teigi no Tayousei (Multiplicity of Social Definitions for Commodities). *Soshiki Kagaku* 33 (2): 105–115.

Matsumiya, Makoto. 1999. Fuman ga Takamaru Chugaku 2, 3-nensei: Kodomo Mondai Yoron Chosa kara (Rising Dissatisfaction among Second and Third Year Junior High School Students: From the Results of the Public Opinion Poll on Child Issues). In *Hoso-kenkyu to Chosa* (*NHK Monthly Report on Broadcast Research*), 24–37. January.

Matsumoto, Yasushi. 1991. Toshi Bunka (Urban Culture). In *Shakaigaku no Riron de Toku Gendai no shikumi* (*The Mechanism of Modernity*), ed. T. Yoshida, 173–187. Tokyo: Shinyosha.

———. 1992. Toshi wa Nani wo Umidasuka: Urbanism Riron no Kakushin (What the Cities Will Create: Reform in Urbanism Theory). In *Toshi Shakaigaku no Frontier 2 Seikatsu, Kankei, Bunka* (*The Frontiers of Urban Sociology 2: Life, Relationships, Culture*), ed. K. Morioka and Y. Matsumoto, 33–68. Tokyo: Nihon Hyoronsha.

————. 1995. Gendai toshi no Henyo to Community, Network (Changes in the Modern City and Community, Network). In *Zoshoku suru Network* (Multiplying Networks), ed. Y. Matsumoto, 1–90. Tokyo: Keisoshobo.

Matsunaga, Mari. 2000. *i-mode Jiken* (*The Happening of i-mode*). Tokyo: Kadokawa Syoten.

————. 2001. *The Birth of i-mode*. Singapore: Chuang Yi Publishing.

McCullough, Malcolm. 2004. *Digital Ground: Architecture, Pervasive Computing, and Environmental Knowing*. Cambridge, Mass.: MIT Press.

McGray, Douglas. 2002. Japan's Gross National Cool. *Foreign Policy*, May/June.

McLuhan, Marshall. 1962. *The Gutenberg Galaxy: The Making of Typographic Man*. Toronto: University of Toronto Press.

————. 1964. *Understanding Media: The Extensions of Man*. New York: McGraw-Hill.

Meguro, Yoriko. 1987. *Kojinka suru Kazoku* (*Individualization of the Family*). Tokyo: Keisoshobo.

————. 1992. Between the Welfare and Economic Institutions: Japanese Families in Transition. *International Journal of Japanese Sociology* 1: 35–46.

Meyrowitz, Joshua. 1985. *No Sense of Place: The Impact of Electronic Media on Social Behavior*. Oxford: Oxford University Press.

Michael, Mike. 2000. *Reconnecting Culture, Technology and Nature: From Society to Heterogeneity*. New York: Routledge.

Mikami, Shunji. 2001. Keitai Denwa no Manner ni miru Koushi no Yuragi (Changes in Public and Private in *Keitai* Manners). *Gendai no Esupuri: Keitai Denwa to Shakai Seikatsu* 405 (Spring): 96–105.

Mikami, Shunji, Ron Korenaga, Isao Nakamura, Takehide Kenjo, Yasutoshi Mori, Kaga Yanagisawa, Yasuko Mori, and Naoya Sekiya. 2001. Keitai Denwa • PHS no Riyo Jittai 2000 (Survey Research on Uses of Cellular Phone and PHS in 2000). *Tokyo Daigaku Shakai Joho Kenkyusho Chosa Kenkyu Kiyo* 15: 145–253.

Milgram, Stanley. 1970. The Experience of Living in Cities. *Science* 167: 1461–1468.

————. 1977. *The Individual in a Social World: Essays and Experiments*. Boston: Addison-Wesley.

Miller, Daniel, and Don Slater. 2000. *The Internet: An Ethnographic Approach*. Oxford: Berg.

Ministry of Public Management, Home Affairs, Posts and Telecommunications. Japan. 1999. *Sekai no Seinen tono Hikaku kara mita Nihon no Seinen. Dai 6-kai Seikai Seinen Ishiki Chosa Hokokusho* (*Observations on Japanese Youth in Comparison to the Youth of the World. Report on the Sixth World Youth Awareness Survey*). Youth Affairs Administration. Tokyo: National Printing Bureau.

————. 2002. *Communications Use Trend Survey 2002*. Tokyo: National Printing Bureau.

————. 2003. *Communications Use Trend Survey 2003*. Tokyo: National Printing Bureau.

Mitra, Ananda. 2003. Online Communities, Diasporic. In *Encyclopedia of Community*, ed. K. Christensen and D. Levinson, 1019–1020. Thousand Oaks, Calif.: Sage.

Mitsuya, Keiko, Hiroshi Aramaki, and Sachiko Nakano. 2002. Hirogaru Internet, Shikashi Terebii towa Taisa: 'IT Jidai no Seikatsu Jikan' Chosa kara (IT Use Increases but Lags Far Behind TV: From the Survey on "Time Use in the IT Age"). In *Hoso Kenkyu to Chosa* (*NHK Monthly Report on Broadcast Research*), 2–21. April.

Mitsuya, Keiko, and Sachiko Nakano. 2001. Job-holders Working Longer Hours in the Economic Downturn: From the Survey on Japanese Time Use 2000. *Broadcasting Culture & Research.* ⟨http://www.nhk.or.jp/bunken/index-e.html⟩.

Miyadai, Shiniji. 1994. *Seifuku Shojo tachi no Sentaku* (*Choices of Girls in Uniforms*). Tokyo: Kodansha.

———. 1997. *Maboroshi no Kougai: Seijyuku-shakai wo Ikiru Wakamono-tachi no Yukue* (*The Illusionary Suburbs*). Tokyo: Asahi Shinbunsha.

Miyake, Kazuko. 2000. Keitai to Gengo Kodo • Hi Gengo Kodo (*Keitai* and Verbal and Non-Verbal Language). *Nihongo-Gaku*, October, 6–17.

Miyaki, Yukiko Abe. 1997. *The Mobile Communication Life Led by the Youth of Today.* Tokyo: Life Design Institute.

———. 1999. *Seinenso no Tsushin Media Riyo to Yujin Kankei* (*Young People's Communication Media Use and Friendship Relations*). Life Design Reseach Center Report (July), 27–49.

Miyata, Kakuko. 2001. Keitai Denwa Riyou to Taijin Kankei: Nenri to Seibetsu no Shiten Kara (The Interpersonal Relationship by *Keitai*: The Perspective of Age and Gender). *Meiji Gakuin Daigaku Shakai Gakubu Fuzoku Kenkyusho Nenpo* 31: 65–80.

Miyoshi, Masao, and Harry D. Harootunian, eds. 1989. *Postmodernism and Japan.* Durham, N.C.: Duke University Press.

Mobile Communication Research Group. 2002. *Keitai Denwa Riyo no Shinka to Sono Eikyo* (*The Evolution of the Uses of* Keitai *and Its Influence*).

———. 2003. *Net Heavy Shakai Kankoku no Jitsuzo* (*The Reality of Heavy Use Network in Korea*).

Mori, Kumiko, and Yasuhiko Ishida. 2001. Meiwaku no Seisei to Jyuyo ni kansuru Kisoteki Kenkyu (Why Do People Feel Cellular Phones Are a Menace?: A Content Analysis of Discussions on the USENET Newsgroups and Japanese Newspapers). *Aichi Shukutoku Daigaku Ronshu—Communication Gakubu Hen* 1: 77–92.

Morioka, Masahiro. 1993. *Ishiki Tsushin* (*Awareness Communication*). Tokyo: Chikuma Shobo.

Morita, Akio, with Edwin M. Reingold and Mitsuko Shimomura. 1986. *Made in Japan: Akio Morita and SONY.* New York: E. P. Dutton.

Morley, David. 1986. *Family Television: Cultural Power and Domestic Leisure.* New York: Routledge.

———. 2000. *Home Territories: Media, Mobility and Identity*. New York: Routledge.

Morley, David, and Kevin Robins. 1995. *Spaces of Identities: Global Media, Electronic Landscapes and Cultural Boundaries*. New York: Routledge.

Multimedia Research Institute. 2003. *Nendo Kokunai Keitai Denwa Tanmatsu Shukka Gaikyo 2002* (*Overview of Domestic* Keitai *Terminal shipments in FY 2002*). ⟨http://www.m2ri.co.jp/newsreleases/030410.htm⟩.

Murtagh, Ged M. 2002. Seeing the "Rules": Preliminary Observations of Action, Interaction, and Mobile Phone Use. In *Wireless World: Social and Interactional Aspects of the Mobile Age*, ed. B. Brown, N. Green, and R. Harper, 81–91. New York: Springer-Verlag.

Myerson, George. 2001. *Heidegger, Habermas and the Mobile Phone*. Lanham, Md.: National Book Network.

Mynatt, Elizabeth, Annette Adler, Mizuko Ito, and Vicki O'Day. 1997. Network Communities: Something Old, Something New, Something Borrowed . . . *Computer Supported Cooperative Work* 6: 1–35.

Nagai, Susumu, ed. 1994. *Gendai Terecomu Sangyo no Keizai Bunseki* (*Economic Analysis of Today's Telecommunication Industry*). Tokyo: Housei Daigaku Shuppan Kyoku.

Nagai, Yoshikazu. 1986. Toshi no Tokumeisei to Itsudatsu Koi: Inpei to Hakken no Kanosei (Anonymity in Cities and Deviant Behavior: Possibility of Concealment and Discovery). *Sociology* 30 (3): 77–96.

Nakajima, Ichiro, Keiichi Himeno, and Hiroaki Yoshii. 1999. Ido-denwa Riyo no Fukyu to sono Shakaiteki-imi (Diffusion of Cellular Phones and PHS and Its Social Meanings). *Joho Tsuushin Gakkai-shi* 16 (3): 79–92.

Nakamura, Isao. 1996a. Keitai Denwa no "Riyo to Manzoku": Sono Kozo to Jyokyo Izonsei (The Uses and Gratifications of the Cellular Telephone). *Mass Communication Kenkyu* 48: 146–159.

———. 1996b. Wakamono no Ningen Kankei to Poketto Beru Riyo (Young People's Interpersonal Relationships and Pager Use). In *Nihon Shakai Shinri Gakkai Dai 37 Kai Taikai Happyou Ronbunshu* (*Proceedings of the 37th Meeting of the Japan Social Psychology Society*).

———. 1996c. Denshi Media no Personal-Ka: Sono Katei to Riyou Henka no Tokushitsu (Personalization of Electronic Media: Its Process and Features of Change of Use). In *Jouhou Koudou to Chiiki Jouhou System* (*Information Behavior and Local Information System*), 168–194. Institute of Socio-Information and Communication Studies. Tokyo: University of Tokyo Press.

———. 1997. Ido-tai Tsuushin Media ga Wakamono no Ningen-kankei oyobi Seikatu-koudou ni Ataeru Eikyou "Pocket-bell • PHS Riyo ni Kansuru Paneru Chosa no Kokoromi" (Effects of Mobile Telecommunication Media on Personal Relationships and Behavior in Daily Life of the Young: Panel Survey on the Use of Pager and PHS Telephone). In *Heisei 8 Nendo Jyouhou Tsusuhin-gakkai Nenpou* (*JSICR Annual Report 1996*), 27–40.

———. 2000. Denwa to Ningen Kankei (Telephone and Interpersonal Relationships). In *Joho Tsushin to Shakai Shinri* (*Information Communication and Social Psychology*), ed. O. Hiroi and M. Funatsu, 45–70. Tokyo: Hokuju Shuppan.

———. 2001a. Keitai Mail no Ningen Kankei (Human Relationships and *Keitai* E-mail). In *Nihonjin no Joho Kodo 2000* (*Japanese Information Behavior 2000*), 285–303. Tokyo: University of Tokyo Press.

———. 2001b. Keitai Denwa no Fukyu Katei to Shakai-teki imi (*Keitai* Adoption Process and Its Social Meaning). *Gendai no Esupuri* 405 (Spring): 46–57.

———. 2003. Keitai Mail to Kodoku (Loneliness and the Uses of the Short Message Service). *Matsuyama Daigaku Ronsyu* 14 (6): 85–99.

Nakamura, Isao, and Osamu Hiroi. 1997. Keitai Denwa to 119 Ban Tsuho (Cellular Telephone and 119 Call). *Tokyo Daigaku Shakai Joho Kenkyusho Chosa Kenkyu Kiyo* 9: 87–103.

Nakamura, Osamu. 2000. *Kodomotachi wa Dokode Hanzai ni Atte Truka* (*Where Children Face Danger*). Tokyo: Shobunsha.

Nakanishi, Shintaro. 2003. Deaikei Saito to Atarashii Deai no Bunka (Dating Sites and the New Culture of Encounter). *Gekkan Seito Shidou* 33 (10): 14–17.

Nakano, Sachiko. 2002. Internet Riyo to Terebi Shicho no Kongo (Internet Use and the Future of TV Viewing). In *Hoso-kenkyu to Chosa* (*NHK Monthly Report on Broadcast Research*), 126–137. August.

National Institute of Population and Social Security Research. 2003. *Dai 12 kai Shussei Doko Kihon Chosa* (*The Twelfth Japanese National Fertility Survey*).

National Police Agency. 1996. *Police White Paper 1996*.

———. 2000. *Police White Paper 2000*.

National Police Department Public Archive. 2002. *Heisei 14-nen no Iwayuru Deai-kei Site ni Kankei shita Jiken Kenkyo Joukyo ni Tsuite* (*Regarding the 2002 Situation of Arrests Related to Incidents Tied to So-Called Deai-kei Sites*).

Natsuno, Takeshi. 2003a. *i-mode Strategy*. Hoboken, N.J.: Wiley.

———. 2003b. *The i-mode Wireless Ecosystem*. Hoboken, N.J.: Wiley.

Newman, Susan. 1998. Here, There, and Nowhere at All: Distribution, Negotiation, and Virtuality in Postmodern Ethnography and Engineering. *Knowledge and Society* 11: 235–267.

NHK Broadcasting Culture Research Institute. 2002. *NHK Data Book 2000: National Time Use Survey*. Tokyo: Japan Broadcast Publishing Co.

Nihongo-gaku. 2001 (September). Tokyo: Meiji Shoin.

Nishioka, Ikuo. 2003. Keitai Denwa no Manner ha Nihon ga Sekaiichi? (Is Japan the Best in the World for *Keitai* Manners)? ⟨http://bizplus.nikkei.co.jp/colm/colCh.cfm?i=t_nishioka33⟩.

Nolan, Jason. 2004. The Technology of Difference: ASCII, Hegemony, and the Internet. In *Communities of Difference: Language, Culture, and the Media*, ed. Peter Trifonas. New York: Palgrave Macmillan.

Nomura Research Institute. 2003. *Cyber Life Observations.*

Nomura Securities Company Ltd. 2001. *Dai 7-kai Kakei to Kosodate Hiyo* (*Seventh Research on Family Budget and Cost of Raising Children*). ⟨http://www.nomura.co.jp/introduc/faundation/angel.html⟩.

Nozawa, S. 1996. Aspects spatiaux de liens personnels dans le Japon moderne. *Bulletin de la Societe Neuchateloise de Geographie* 40: 83–97.

NTT. 1990. Nijuuisseiki no Saabisu Bijon: Shinkoudo Jouhou Tsuushin Saabisu VI&P Genjitsu (A Vision for Twenty-First Century Services: The Current Reality of Information Communication VI&P Services). *NTT Gijyutsu Journal* (May).

NTT DoCoMo. 1999. *DoCoMo Report* No. 8.

NTT DoCoMo. 2002. *NTT DoCoMo Jyuu-nen Shi: Mobairu Furonteia he no Chousen* (*A Ten-Year History of NTT DoCoMo: Challenging the Mobile Frontier*). Tokyo: NTT DoCoMo.

Oda, Kumiko. 2000. Keitai ga Tabete Shimau Wakamono Shouhi (The Youth Consumption Mobile Phone Gobble). *Economist* 78 (27): 48.

Ohira, Ken. 1995. *Yasashisa no Seishin Byouri* (*Psychopathology of Kindness*). Tokyo: Iwanami Shoten.

Ohta, Hiroshi. 2001. *Mail no J-PHONE Tanjo Hiwa* (*The Secret of the Birth of J-PHONE Mail*). Bessatsu Takarajima Real No. 014. Takarajimasha.

Okabe, Daisuke, and Mizuko Ito. 2003. Camera Phones Changing the Definition of Picture-Worthy. *Japan Media Review*, August 29. ⟨http://www.ojr.org/japan/wireless/1062208524.php⟩.

Okada, Tomoyuki. 1993. Dengon Dial to iu Giji-kukan (*Dengon Dial* as Cyberspace). *Gendai no Esupuri* 306 (January): 93–101.

———. 1997. Poke-beru Keitai no "Yasashisa" (The "Kindness" of Pagers and *Keitai*). In *Poke-beru Keitai Shugi!* (*Pager and* Keitai *Manifesto!*), H. Tomita, K. Fujimoto, T. Okada, M. Matsuda, and N. Takahiro, 76–96. Tokyo: Just System.

———. 2002. Keitai kara Manabu to Iukoto (What It Means to Learn from *Keitai*). In *Keitai-Gaku Nyumon* (*Understanding Mobile Media*), ed. T. Okada and M. Matsuda, 3–19. Tokyo: Yuhikaku.

Okada, Tomoyuki, and Ichiyo Habuchi. 1999. *Idotai Media ni Kansuru Gaito Chosa no Kiroku* (*Records from a Public Survey on Mobile Media*), 9. Mukogawa Women's University. Institute of Esthetics in Everyday Life.

Okada, Tomoyuki, and Misa Matsuda, eds. 2002. *Keitai-gaku Nyumon* (*Understanding Mobile Media*). Tokyo: Yuhikaku.

Okada, Tomoyuki, Misa Matsuda, and Ichiyo Habuchi. 2000. Ido-denwa Riyo ni okeru Media Tokusei to Taijin Kankei: Daigakusei wo Taisho toshita Chosa-jirei yori (Specifics of Medium and Interpersonal Relationships in Using Mobile Telephone: The Results of a Survey on University Students). *Joho Tsushin Gakkai Nenpo*, 43–60.

Okada, Tomoyuki, and Hidenori Tomita. 1999. Shiryo: Ido-denwa ni Kansuru Gaito Chosa no Kiroku (Excerpt from the Records of the Street Survey Regarding Mobile Phones). *Kenkyu Sosho: Institute of Economics and Political Studies in Kansai University* 112: 208–233.

Okonogi, Keigo. 1978. *Moratorium Ningen no Jidai* (*The Age of Moratorium People*). Tokyo: Chuo Koronsha.

———. 2000. *Keitai • Net Ningen no Seishin Bunseki: Shonen mo Otona mo Hikikomori no Jidai* (*A Psychological Analysis of Mobile/Net People: Spreading of* Hikikomori *in both Youth and Adults*). Tokyo: Asuka-Shinsya.

Okuno, Masato. 2003. Denshi Media to Mondai Koudou: Deai-kei Saito ni yoru Enjo Kousai Yakubutu Ranyou no Jittai (Electronic Media and Problematic Behavior: The Reality of Youth Prostitution and Drug Abuse Triggered by *Deai-kei* Sites). *Gekkan Seito Shidou* 33 (10): 22–27.

Okuno, Takuji. 2000. *Daisan no Shakai* (*The Third Society*). Tokyo: Iwanami Shoten.

Ono, Hiroshi, and Madeline Zavodny. 2004. Gender Differences in Information Technology Use: A U.S.-Japan Comparison. Working Paper 2004-2. Federal Reserve Bank of Atlanta, Georgia.

Orr, Julian. 1996. *Talking about Machines: An Ethnography of a Modern Job.* Ithaca, N.Y.: Cornell University Press.

Otani, Shinsuke. 1995. *Gendai Toshi-jumin no Personal Network* (*The Personal Network of the Modern City-Dweller*). Tokyo: Minerva Shobo.

———. 1999. Personal Community Networks in Contemporary Japan. In *Networks in the Global Village*, ed. B. Wellman, 279–297. Boulder, Colo.: Westview Press.

Palen, Leysa, Marilyn Salzman, and Ed Youngs. 2001. Discovery and Integration of Mobile Communications in Everyday Life. *Personal and Ubiquitous Computing Journal* 5: 109–122.

Pinch, Trevor, and Wiebe Bijker. 1993. The Social Construction of Facts and Artifacts, or, How the Sociology of Science and the Sociology of Technology Might Benefit Each Other. In *The Social Construction of Technological Systems*, ed. W. Bijker, T. Hughes, and T. Pinch, 17–51. Cambridge, Mass.: MIT Press.

Plant, Sadie. 2002. *On the Mobile: The Effects of Mobile Telephones on Social and Individual Life.* Report for Motorola.

Popper, Karl. 1994. *The Myth of the Framework.* New York: Routledge.

Puro, Jukka-Pekka. 2002. Finland: a Mobile Culture. In *Perpetual Contact: Mobile Communication, Private Talk, Public Performance*, ed. J. E. Katz and M. Aakhus, 19–29. Cambridge: Cambridge University Press.

Putnam, Robert D. 2000. *Bowling Alone*. New York: Simon and Schuster.

Quan-Haase, Anabel, Barry Wellman, James Witte, and Keith Hampton. 2002. Capitalizing on the Internet: Network Capital, Participatory Capital, and Sense of Community. In *The Internet in Everyday Life*, ed. B. Wellman and C. Haythornthwaite, 291–324. Oxford: Blackwell.

Quine, Willard V. M. 1981. *Theories and Things*. Cambridge, Mass.: Harvard University Press.

Rakow, Lana F. 1988. Women and the Telephone: The Gendering of a Communication Technology. In *Technology and Women's Voices*, ed. C. Kramarae, 207–228. New York: Routledge.

Rakow, Lana F., and Vija Navarro. 1993. Remote Mothering and the Parallel Shift: Women Meet the Cellular Telephone. *Critical Studies in Mass Communication* 10: 114–157.

Rheingold, Howard. 1993. *The Virtual Community: Homesteading on the Electronic Frontier*. Reading, Mass.: Addison-Wesley.

———. 2002. *Smart Mobs: The Next Social Revolution*. Cambridge, Mass.: Perseus.

———. 2003. Moblogs Seen as a Crystal Ball for a New Era in Online Journalism. *Online Journalism Review USC Annenberg*. ⟨http://www.ojr.org/ojr/technology/1057780670.php⟩.

Rivière, Carole A., and Christian Licoppe. 2003. From Voice to Text: Continuity and Change in the Use of Mobile Phones in France and Japan. Paper presented at the International Sunbelt Social Network Conference, Cancún, Mexico.

Rogers, Everett M. 1986. *Communication Technology: The New Media in Society*. New York: Free Press.

Rutledge, Bruce. 2003. Conference Panelists See Bright Future for Mobile Publishing. *Online Journalism Review USC Annenberg*. ⟨http://www.ojr.org/ojr/technology/1058998393.php⟩.

Said, Edward. 1978. *Orientalism*. New York: Random House.

Saito, Takao. 2002. *Koizumi Kaikaku to Kanshi Shakai* (*The Koizumi Reform and Surveillance Society*). Booklet 573. Tokyo: Iwanami Shoten.

Sakai, Junko. 1995. Keitai-girai no Dokuhaku (Monologue of a *Keitai* Hater). *Koukoku* 308: 17.

Sankei Shinbun. 1999. Tanoshiku chaku-mero, Watashi-rashisa no Enshutsu (Enjoy *chaku-mero*, Present Your Individuality). Osaka evening ed., February 6, 8.

Schafer, R. Murray. 1977. *The Tuning of the World*. Toronto: McClelland and Stewart.

Schegloff, Emanuel. 2002. Beginnings in the Telephone. In *Perpetual Contact: Mobile Communication, Private Talk, Public Performance*, ed. J. E. Katz and M. Aakhus, 284–300. Cambridge: Cambridge University Press.

Schwarz, Heinrich. 2001. Techno-Locales: Social Spaces and Places at Work. Paper presented at the Annual Meeting of the Society for the Social Studies of Science, Cambridge, Mass.

Schwarz, Heinrich, Bonnie Nardi, and Steve Whittaker. 1999. The Hidden Work in Virtual Work. In *Proceedings of the International Conference on Critical Management*.

Serres, Michel. 1980. *Le parasite. The Parasite*. Columbia, Md.: Daedalus Books, 1982.

Sechiyama, Kaku. 1996. Shufu no Hikaku Shakai Gaku (Comparative Sociology of Housewives). In *Iwanami Kouza Gendai Shakai Gaku 19 'Kazoku' no Shakai Gaku (Iwanami Sociology Lectures 19: Sociology of Families)*, ed. S. Inoue, C. Ueno, M. Osawa, M. Mita, and S. Yoshimi, 217–235. Tokyo: Iwanami Shoten.

Sherry, John, and Tony Salvador. 2001. Running and Grimacing: The Struggle for Balance in Mobile Work. In *Wireless World: Social and Interactional Aspects of the Mobile Age*, ed. B. Brown, N. Green, and R. Harper, 108–120. London: Springer-Verlag.

Shibui, Tetsuya. 2003. *Deai-kei Site to Wakamono-tachi* (Deai-kei *Site and Youth*). Tokyo: Yosensha.

Shimizu, Shinji. 2001. Shijika no Paradox (The Paradox of Family Privatization). *Kazoku Shakai-gaku Kenkyu* 13 (1): 97–104.

Silverstone, Roger, Eric Hirsch, and David Morley. 1992. Information and Communication Technologies and the Moral Economy of the Household. In *Consuming Technologies: Media and Information in Domestic Spaces*, ed. R. Silverstone and E. Hirsch, 15–31. London: Routledge.

Simmel, Georg. 1902. The Bridge and the Door. In *Simmel Shousakushu 12 (Collection of Writings by Simmel)*. Tokyo: Hakusuisha.

———. 1903. Die Grosstadt und das Geistesleben. In *Simmel Shousakushu 12 (Collection of Writings by Simmel)*. Tokyo: Hakusuisha.

Skog, Berit. 2002. Mobiles and the Norwegian Teen: Identity, Gender, and Class. In *Perpetual Contact: Mobile Communication, Private Talk, Public Performance*, ed. J. Katz and M. Aakhus, 255–273. Cambridge: Cambridge University Press.

Smith, Marc. 2000. Some Social Implications of Ubiquitous Wireless Networks. *ACM Mobile Computing and Communications Review* 4 (2): 25–36.

Smith, Marc, and Peter Kollock, eds. 1998. *Communities in Cyberspace*. New York: Routledge.

Sontag, Susan. 1977. *On Photography*. New York: Farrar, Straus, Giroux.

Spigel, Lynn. 1992. *Make Room for TV: Television and the Family Ideal in Postwar America*. Chicago: University of Chicago Press.

Suchman, Lucy. 1987. *Plans and Situated Actions: The Problem of Hunan/Machine Communication*. Cambridge: Cambridge University Press.

Suematsu, Toru, and Hitoshi Joh. 2000. Keitai Denwa ga Personal Space ni oyobosu Eikyo (How Mobile Phones Affect Personal Space). *Ningen Kagaku Kenkyu* 8 (1): 67–77.

Tachikawa, Keiji, Kenji Kohiyama, and Sachio Tokunaga. 1995. *Personal Tsuushin no Subete (All About Personal Communications)*. Japan: NTT.

Takada, Masatoshi. 2003. *Naze Tadano Mizuga Ureru noka? Shikouhin no Bunka-ron (Why Is Free Water Sold High? Cultural Study of Refreshing Favorite Goods)*. Tokyo: PHP Institute.

Takahiro, Norihiko. 1997a. Bokutachi no Multimedia, Poke-beru: Sokubaku no Media kara Kaiho no Media e (Our Multimedia, the Pager: From a Binding to a Liberating Medium). In *Poke-beru Keitai Shugi!* (*Pager and* Keitai *Manifesto!*), H. Tomita, K. Fujimoto, T. Okada, M. Matsuda, N. Takahiro, 32–58. Tokyo: Just System.

———. 1997b. 38 Nenme no Media (The 38th Year Media). *Telecom Shakai Kagaku Gakusei Sho Nyusho Ronbunshu* 6: 1–91.

Takeda, Toru. 2002. *Wakamono ha Naze Tsunagari-tagaru no ka: Keitai Sedai no Yukue* (*Why Do Youngsters Want to Be Connected? The Fate of the Mobile Phone Generation*). Tokyo: PHP.

Takeyama, Masanao, and Kensuke Inomata. 2002. Keitai Denwa wo Mochiita Jyugyo Live Ankeito (Online Survey of a Class Using *Keitai*). *Musashi Kogyo Daigaku Kankyo Jyoho Gakubu Jyoho Media Center Journal* 3: 70–77.

Tamaru, Eriko. 2002. Workplace no Design (Design of Workplace). In *Media Psychology Nyumon* (*Introduction to Media Psychology*), ed. T. Sakamoto, H. Takahashi, H. Yamamoto, 71. Tokyo: Gakubunsya.

Tamaru, Eriko, and Naoki Ueno. 2002. Shakai-Douguteki Network no kochiku tositeno design (Design as Construction of a Sociotechnical Network). *Japanese Society for the Science of Design Journal* 9 (3): 14–21.

Tanaka, Megumi. 2000. Jieigyo-so to Koyosha-so no Yujin Network (Network of Friends of the Self-Employed and Employers). In *Toshi-shakai no Personal Network* (Personal Networks in Urban Society), ed. K. Morioka, 107–124. Tokyo: The University of Tokyo Press.

Taylor, Alex, and Richard Harper. 2003. The Gift of Gab? A Design Oriented Sociology of Young People's Use of Mobiles. *Computer Support Cooperative Work* 12: 267–296.

Taylor, Richard. 2003. Japan Signals Mobile Future. *BBC News World Edition*. September 6.

Tedlock, Dennis, and Bruce Mannheim, eds. 1995. *The Dialogic Emergence of Culture*. Urbana: University of Illinois Press.

Telecommunications Carriers Association. (Web site compiling statistics about the Japanese telecommunications industry) ⟨http://www.tca.or.jp/index-e.html⟩.

Tkach-Kawasaki, Leslie. 2003. Internet in East Asia. In *Encyclopedia of Community*, ed. K. Christensen and D. Levinson, 794–798. Thousand Oaks, Calif.: Sage.

Tobin, Joseph, ed. 2004. *Pikachu's Global Adventure: The Rise and Fall of Pokemon*. Durham, N.C.: Duke University Press.

Tokyo Metropolitan Government. 1997. *Basic Survey on the Healthy Upbringing of the Youth*. Bureau of Citizens and Cultural Affairs. ⟨http://www.seikatubunka.metro.tokyo.jp/index9files/inv6/inv6.htm⟩.

Tokyo Prefecture Board of Education. 2000. *List of Registered Schools of Tokyo*. Tokyo: Hara Shobo.

Tomita, Hidenori. 1994. *Koe no Odyssey: Dial Q² no Sekai: Denwa Bunka no Shakai-gaku* (*The Odyssey of Voice: The World of Dial Q²: Sociology of the Telephone Culture*). Tokyo: Koseisha-koseikaku.

———. 1997. "Jiyu to Kodoku" to Keitai ("Freedom and Loneliness" and *Keitai*). In *Poke-beru Keitai Shugi!* (*Pager and* Keitai *Manifesto!*), H. Tomita, K. Fujimoto, T. Okada, M. Matsuda, N. Takahiro, 59–75 . Tokyo: Just System.

———. 1999. Network Shakai no Naka no Shinmitsu to Soen: Idotai Media Hihan no Shakai-teki Haikei (Intimacy and Remoteness in the Network Society: Social Background of Criticism on Mobile Media). *Kenkyu Sosho: Institute of Economic and Political Studies in Kansai University* 112: 171–191.

Tomita, Hidenori, and Masayuki Fujimura, eds. 1999. *Minna Bocchi no Sekai* (*A World of Being All Together*). Tokyo: Koseisya-Koseikaku.

Tomita, Hidenori, Kenichi Fujimoto, Tomoyuki Okada, Misa Matsuda, and Norihiko Takahiro. 1997. *Poke-beru Keitai Shugi!* (*Pager and* Keitai *Manifesto!*). Tokyo: Just System.

Tönnies, Ferdinand. 1957. *Gemeinschaft und Gesellschaft. Community and Society*, ed. and trans. C. Loomis. East Lansing, Mich.: Michigan State University Press.

Tsuji, Daisuke. 1999. Wakamono no Communication Henyo to Atarashii Media (New Media and Changes in Youth Communication). In *Kodomo • Seishonen to Communication* (*Children and Youth Communication*), ed. Y. Hashimoto and F. Mamoru, 11–27. Tokyo: Hokujyu Shuppan.

———. 2003a. Wakamono ni okeru Idoutai Tsuushin Media no Riyo to Kazoku Kankei no Henyo (Youth Use of Mobile Media and Changes in Family Relationships). *Kenkyu Sosho: Institute of Economic and Political Studies in Kansai University* 133: 73–92.

———. 2003b. Wakamono no Yujin • Okako Kankei to Communication ni kansuru Chosa Kenkyu Gaiyo Hokokusho (Summary Report on a Questionnaire Survey of Young People's Communications and Relationships with Friends and Parents). *Kansai Daigaku Shakai Gakubu Kiyo* 34 (3): 373–389.

Tsuji, Daisuke, and Shunji Mikami. 2001. *Daigakusei ni okeru Keitai Mail Riyo to Yujin Kankei* (*A Preliminary Student Survey on E-mail Use by Mobile Phones*). ⟨http://www2.ipcku.kansai-u.ac.jp/~tsujidai/paper/r02/index.htm⟩.

Tsuji, Izumi. 2003. Keitai Denwa wo Moto nishita Kakudai Personal Network Chosa no Kokoromi: Wakamono no Yujin Kankei wo Chushin ni (Research on Growing Personal Networks on the Memory Bank of a Mobile Phone: From the Viewpoint of Friendship Relations of Contemporary Youth). *Shakai Jyohogaku Kenkyu* 7: 97–111.

Tsukamoto, Kiyoshi. 2000. *Keitai ga Nihon wo Sukuu* (*The* Keitai *Will Save Japan*). Tokyo: Takarajima Shinsho.

Tsunashima, Ritomo. 1992. Heisei Niten Kansoku 8: Keitai Denwa (Heisei Two-Point Observation 8: Mobile Phone). *Shukan Asahi*, February 28, 62–63.

Tsunoyama, Sakae. 1980. *Cha no Sekai-shi* (*World History of Tea*). Tokyo: Chuoh-Koron.

———. 1984. *Tokei no Shakai-shi* (*Social History of Watches*). Tokyo: Chuoh-Koron.

Tsuyama, Keiko. 2000. *NTT & KDDI: Dou naru Tsushin-Gyokai* (*NTT and KDDI: The Future of Tele-communication Industry*). Tokyo: Nihon Jitsugyo Shuppansha.

Tsuzurahara, Ai. 2004. Wakamono no Communication to Denshi Media: Keitai-mail ni miru Gen-jyo Bunseki karano Kousatsu (Young People's Communication and Electronic Media). Unpublished M.A. Thesis, Otsuma Women's University.

Turkle, Sherry. 1988. Computational Reticence: Why Women Fear the Intimate Machine. In *Technology and Women's Voices*, ed. C. Kramarae, 41–61. New York: Routledge.

———. 1995. *Life on the Screen*. New York: Simon and Schuster.

———. 1999. Cyberspace and Identity: Symposium. *Contemporary Sociology Journal of Reviews* 28 (6).

UCLA Center for Communication Policy. 2003. *UCLA Internet Report: Surveying the Digital Future, Year Three.* ⟨http://www.ccp.ucla.edu/⟩.

Ueda, Atsushi, ed. 1994. *The Electric Geisha*. Tokyo: Kodansya.

Ueno, Chizuko. 1994. *Kindai Kazoku no Seiritsu to Shuen* (*The Establishment and the Disintegration of the Modern Family*). Tokyo: Iwanami Shoten.

Ueno, Naoki. 2000. Ecologies of Inscription: Technologies of Making the Social Organization of Work and the Mass Production of Machine Parts Visible in Collaborative Activity. *Mind, Culture and Activity* 7 (1–2): 59–80.

Ueno Naoki, and Tamaru Eriko. 2002. *Joho Ecology ni Motozuita System no Design* (A System Design According to an Information Ecology). *Musashi Institute of Technology Media Center Journal* 3: 2–9.

Ueno, Naoki, and Yasuko Kawatoko. 2003. Technologies Making Space Visible. *Environment and Planning A* 35: 1529–1545.

Ukai, Masaki, Yoshikazu Nagai, and Kenichi Fujimoto, eds. 2000. *Sengo Nihon no Taishuu Bunka* (*The Japanese Popular Culture After World War II*). Kyoto: Syowado.

Video Research. 2002. Mobile Phone Use Situation. Tokyo: Video Research.

Vogel, Ezra. 1979. *Japan as Number One: Lessons for America*. Cambridge, Mass.: Harvard University Press.

Wakabayashi, Mikio. 1992. Denwa no aru Shakai (Society with the Telephone). In *Media to shiteno Denwa* (*Telephone as Media*), S. Yoshimi, M. Wakabayashi, and S. Mizukoshi, 23–58. Tokyo: Kobundou.

Wang Chao Ying. 1996. *Tompa Character*. Maru Sha.

Watanabe, Asuka. 2000. Keitai Denwa • PHS wo Tsukai Konasu Hito ha Oshare Shouhi mo Dai (People Who Are Fluent Users of *Keitai* and PHS Are Also Heavy Fashion Consumers). *Kesho Bunka* 40: 99–111.

Watanabe, Jun. 1989. *Media no Micro Shakai-gaku* (*Micro-Sociology of Media*). Tokyo: Chikuma Shobo.

Watts, Duncan J. 2002. *Six Degrees: The Science of a Connected Age*. New York: Norton.

Weilenmann, Alexandra. 2003. "I Can't Talk Now, I'm in a Fitting Room": Formulating Availability and Location in Mobile Phone Conversations. *Environment and Planning A* 35 (9): 1589–1605.

Weilenmann, Alexandra, and Catrine Larsson. 2001. Local Use and Sharing of Mobile Phones. In *Wireless World: Social and Interactional Aspects of the Mobile Age*, ed. B. Brown, N. Green, and R. Harper, 99–115. London: Springer-Verlag.

Weiser, Mark. 1991. Future Computers. *Scientific American* 265: 94–104.

Weiser, Mark, and John Seely Brown. 1996. The Coming Age of Calm Technology. In *Beyond Calculation: The Next Fifty Years of Computing*, ed. P. Denning and R. M. Metcalfe, 75–85. London: Springer-Verlag.

Wellman, Barry. 1979. The Community Question. *American Journal of Sociology* 84: 1201–1231.

———. 1988. "Structural Analysis: From Method and Metaphor to Theory and Substance." In *Social Structures: A Network Approach*, ed. B. Wellman and S. D. Berkowitch, 19–61. Cambridge: Cambridge University Press.

———. 1992. Men in Networks: Private Communities, Domestic Friendships. In *Men's Friendships*, ed. P. Nardi, 74–114. Thousand Oaks, Calif.: Sage.

———. 1997. An Electronic Group Is Virtually a Social Network. In *Culture of the Internet*, ed. S. Kiesler, 179–205. Mahwah, N.J.: Erlbaum.

———. 2001. Physical Place and Cyberspace: The Rise of Personalized Networks. *International Urban and Regional Research* 25 (2): 227–252.

———. 2002. Little Boxes, Glocalization, and Networked Individualism. In *Digital Cities II: Computational and Sociological Approaches*, ed. M. Tanabe, P. van den Besselaar, and T. Ishida, 10–25. Berlin: Springer-Verlag.

———. 2003. Glocalization. In *Encyclopedia of Community*, ed. K. Christensen and D. Levinson, 559–562. Thousand Oaks, Calif.: Sage.

Wellman, Barry, ed. 1999. *Networks in the Global Village*. Boulder, Colo.: Westview.

Wellman, Barry, Jeffrey Boase, and Wenhong Chen. 2002. The Networked Nature of Community on and off the Internet. *IT and Society* 1 (1): 151–165.

Wellman, Barry, and Caroline Haythornthwaite, eds. 2002. *The Internet in Everyday Life*. Oxford: Blackwell.

Wellman, Barry, and B. Hogan. 2004. The Immanent Internet. In *Netting Citizens: Exploring Citizenship in a Digital Age*, ed. J. MacKay. Edinburgh: St. Andrew Press.

Wellman, Barry, and Barry Leighton. 1979. Networks, Neighborhoods and Communities. *Urban Affairs Quarterly* 14: 363–390.

Wellman, Barry, Anabel Quan-Haase, Jeffrey Boase, Wenhong Chen, Keith Hampton, Isabel Diaz de Isla, and Kakuko Miyata. 2003. The Social Affordances of the Internet for Networked Individualism. *Journal of Computer Mediated Communication* 8 (3). ⟨http://www.ascusc.org/jcmc/vol8/issue3/wellman.html⟩.

Wellman, Barry, and Scot Wortley. 1990. Different Strokes from Different Folks: Community Ties and Social Support. *American Journal of Sociology* 96 (5): 558–588.

White, Merry I. 2002. *Perfectly Japanese: Making Families in an Era of Upheaval.* Berkeley: University of California Press.

White Paper on Communications in Japan. 1986. Ministry of Posts and Telecommunications. Tokyo: National Printing Bureau.

White Paper on Communications in Japan. 1994. Ministry of Posts and Telecommunications. Tokyo: National Printing Bureau.

White Paper on Communications in Japan. 1997. Ministry of Posts and Telecommunications. Tokyo: National Printing Bureau.

White Paper on Communications in Japan. 2000. Ministry of Posts and Telecommunications. Tokyo: National Printing Bureau.

White Paper: Information and Communications in Japan. 2002. Ministry of Public Management, Home Affairs, Posts and Telecommunications. Tokyo: National Printing Bureau.

White Paper: Information and Communications in Japan. 2003. Ministry of Public Management, Home Affairs, Posts and Telecommunications. Tokyo: National Printing Bureau.

Wittgenstein, Ludwig. 1953. *Philosophical Investigations.* Oxford: Blackwell.

Woolgar, Steve, ed. 2002. *Virtual Society? Technology, Cyberbole, Reality.* Oxford: Oxford University Press.

World Internet Project Japan. 2002. *Internet Usage in Japan.* Survey Report. ⟨http://media.asaka.toyo.ac.jp/wip/survey2002e/⟩.

Yamagishi, Toshio. 1998. *Shinrai no Kozo: Kokoro to Shakai no Shinka Game (Structure of Trust: Evolution Game of the Mind and Society).* Tokyo: Tokyo University Press.

Yan, Xu. 2003. Mobile Data Communications in China. *Communications of the ACM* 46 (12): 81–85.

Yanagita, Kunio. [1931] 1990. *Seken-banashi no Kenkyu* (A Study on Small Talk). In *Yanagita Kunio Zenshu 9*, Kunio Yanagita, 511–530. Tokyo: Chikuma Bunko.

Yano, Masakazu. 1995. *Seikatsu Jikan no Shakai Gaku* (*Sociology of Time Use—Social Time and Personal Time*). Tokyo: Tokyo University Press.

Yano Research Institute. 2002. *IT-kaden/IT-house no Shorai Tenbo* (*The Future of IT-appliance/IT-house*). Tokyo: Yano Research Institute.

Yomiuri Photograph Grand Prix. 2002. The Twenty-Fourth Yomiuri Photograph Grand Prix Award Winners. ⟨http://www.yomiuri.co.jp/photogp/grandprix/24th/index-houdo.htm⟩.

Yomiuri Shinbun. 1991. Kochira Shakaibu: Keitai Denwa Shiyou wa Basho wo Wakimaete (Be Selective of Time and Place for *Keitai* Use). September 11.

———. 1995. Letter to the Editor. May 9.

———. 1997. Letter to the Editor. December 26.

———. 1998. Sedai Ishiki Usui Wakamono "Jibun Katte" ga Hannsuu Chikai 49% (Young People with Lack of Consciousness of Other Generations: Selfishness at Close to Half: 49%). June 27.

———. 2002. Letter to the Editor. December 13.

Yoneyama, Toshinao. 1981. *Doujidai no Jinruigaku* (*Contemporary Anthropology*). Tokyo: Nihon Hoso Shuppan Kyokai.

Yoshii, Hiroaki. 2001. Wakamono no Keitai Denwa Kodo (Youth *Keitai* Activity). *Gendai no Esupuri: Keitai Denwa to Shakai Seikatsu* 405 (Spring): 85–95.

Yoshimi, Shunya. 1998. "Made in Japan": Sengo Nihon ni okeru Denshi Rikkoku Shinwa no Kigen ("Made in Japan": The Period of the Myth of Postwar Electronic Nation-Building). In *Joho Shakai no Bunka 3* (*Design, Technology, Market*), ed. A. Shimada, H. Kashiwagi, and S. Yoshimi, 133–174. Tokyo: University of Tokyo Press.

Yoshimi, Shunya, Mikio Wakabayashi, and Shin Mizukoshi. 1992. *Media to shite no Denwa* (*The Telephone as a Medium*). Tokyo: Kobundo.

Yoshino, Kosaku. 1999. Rethinking Theories of Nationalism: Japan's Nationalism in a Marketplace Perspective. In *Consuming Ethnicity and Nationalism: Asian Experiences*, ed. K. Yoshino, 8–28. Honolulu: University of Hawaii Press.

Yuzawa, Takehiko. 1995. *Zusetsu Kazoku Mondai no Genzai* (*Modern Family Issues*). Tokyo: Nihon Hoso Shuppan Kyokai.

Contributors

Jeffrey Boase
University of Toronto
Centre for Urban and Community
Studies
455 Spadina Avenue, Toronto M5S2G8
Canada
Tel: +1-416-978-0250
Fax: +1-416-978-7162
jeff.boase@utoronto.ca

Shingo Dobashi
Musashi Institute of Technology
Faculty of Environmental and
Information Studies
3-3-1 Ushikubonishi Tsuzuki-ku
Yokohama Kanagawa 224-0015
Japan
Tel: +81-45-910-2922
Fax: +81-45-910-2605
dobashi@yc.musashi-tech.ac.jp

Kenichi Fujimoto
Mukogawa Women's University
Institute of Esthetics in Everyday Life
1-13 Tozaki-cho Nishinomiya Hyogo
663-8121
Japan
Tel: +81-798-67-1291
Fax: +81-798-67-1503
sonntag@mwu.mukogawa-u.ac.jp

Ichiyo Habuchi
Hirosaki University
Faculty of Humanities
1 Bunkyo-cho Hirosaki Aomori 036-8560
Japan
Tel: +81-172-39-3205
Fax: +81-172-39-3205
ichiyo@cc.hirosaki-u.ac.jp

Ken'ichi Ikeda
The University of Tokyo
Department of Social Psychology
7-3-1 Hongo
Bunkyo-ku Tokyo 113-8656
Tel: +81-3-5841-3868
Fax: +81-3-3815-6673
ikeken@l.u-tokyo.ac.jp

Mizuko Ito
University of Southern California
Annenberg Center for Communication
734 West Adams Boulevard, Los Angeles
CA 90007
USA
Tel: +1-213-743-1919
Fax: +1-213-743-2962
mito@annenberg.edu

Fumitoshi Kato
Keio University
Faculty of Environmental Information
5322 Endo Fujiswa Kanagawa 252-8520
Japan
Tel: +81-466-49-3619
Fax: +81-466-47-5041
fk@sfc.keio.ac.jp

Haruhiro Kato
Chukyo University
Department of Sociology
101 Tokodachi Kaizu-cho Toyota Aichi
470-0393
Japan
Tel: +81-565-46-6514
Fax: +81-565-46-1298
hkato@sass.chukyo-u.ac.jp

Kenji Kohiyama
Keio University
Graduate School of Media and
Governance
5322 Endo Fujiswa Kanagawa 252-8520
Japan
Tel: +81-466-49-3440
Fax: +81-466-47-5151
kohiyama@sfc.keio.ac.jp

Misa Matsuda
Chuo University
Faculty of Literature
742-1 Higashinakano Hachioji Tokyo
192-0393
Japan
Tel: +81-426-74-3733
Fax: +81-426-74-3738
mmatsuda@tamacc.chuo-u.ac.jp

Yukiko Miyaki
Dai-ichi Life Research Institute, Inc.
Research and Development Center
Life Design Research Unit
1-13-1 Yuraku-cho Chiyoda-ku Tokyo
100-0006
Japan
Tel: +81-3-5521-4767
Fax: +81-3-3212-4470
yukiko.miyaki@nifty.com

Kakuko Miyata
Meiji Gakuin University
Department of Sociology
1-2-37 Shiroganedai Minato-ku Tokyo
108-0071
Japan
Tel: +81-3-5421-5570
Fax: +81-3-5421-5697
miyata@soc.meijigakuin.ac.jp

Daisuke Okabe
Keio University
Graduate School of Media and
Governance
5322 Endo Fujisawa
Kanagawa 252-8520
Japan
Tel: +81-466-47-5000+53664
Fax: +81-466-47-5151
okabe@sfc.keio.ac.jp

Tomoyuki Okada
Kansai University
Faculty of Informatics
2-1-1 Ryozenji-cho Takatuski Osaka 569-
1095
Japan
Tel: +81-72-690-2459
Fax: +81-72-690-2492
okada@res.kutc.kansai-u.ac.jp
Faculty of Informatics

Eriko Tamaru
Fuji Xerox Co., Ltd.
Yokohama Business Park East Tower 13F
134 Goudo-cho Hodogaya-ku Yokohama
Kanagawa 240-0005
Japan
Tel: +81-45-337-6215
Fax: +81-45-336-3803
eriko.tamaru@fujixerox.co.jp

Hidenori Tomita
Bukkyo University
School of Sociology
96 Kitahananobo-cho Murasakino Kita-
ku Kyoto 603-8301
Japan
Tel: +81-75-491-2141
Fax: +81-75-493-9032
h-tomita@bukkyo-u.ac.jp

Ryuhei Uemoto
Tokyo University
Graduate School of Frontier Sciences
Tel: +81-4-7136-4003
Fax: +81-4-7136-4020
uemoto@media.k.u-tokyo.ac.jp

Naoki Ueno
Musashi Institute of Technology
Faculty of Environment
and Information Studies
3-3-1 Ushikubo-nishi Tsuzuki-ku
Yokohama Kanagawa 224-0015
Japan
Tel: +81-45-910-2942
Fax: +81-45-910-2605
nueno@yc.musashi-tech.ac.jp

Barry Wellman
University of Toronto
Centre for Urban and Community
Studies
455 Spadina Avenue, Toronto M5S 2G8
Canada
Tel: +1-416-978-3930
Fax: +1-416-978-7162
wellman@chass.utoronto.ca

Index